Afforestation and Reforestation

Afforestation and Reforestation: Drivers, Dynamics, and Impacts

Special Issue Editors

Gang Dong
Jingfeng Xiao
Ge Sun
Lu Hao
Zhiqiang Zhang

MDPI • Basel • Beijing • Wuhan • Barcelona • Belgrade

Special Issue Editors

Gang Dong
Shanxi University
China

Jingfeng Xiao
University of New Hampshire
USA

Ge Sun
United States Department of Agriculture
Forest Service
USA

Zhiqiang Zhang
Beijing Forestry University
China

Lu Hao
Nanjing University of Information
Science and Technology
China

Editorial Office
MDPI
St. Alban-Anlage 66
4052 Basel, Switzerland

This is a reprint of articles from the Special Issue published online in the open access journal *Forests* (ISSN 1999-4907) from 2018 to 2019 (available at: https://www.mdpi.com/journal/forests/special_issues/afforestation_reforestation)

For citation purposes, cite each article independently as indicated on the article page online and as indicated below:

LastName, A.A.; LastName, B.B.; LastName, C.C. Article Title. *Journal Name* **Year**, *Article Number*, Page Range.

ISBN 978-3-03921-447-1 (Pbk)
ISBN 978-3-03921-448-8 (PDF)

Cover image courtesy of Feng Zhang.

Contents

About the Special Issue Editors

Gang Dong is currently Postdoc at the Chinese Academy of Agricultural Sciences, and Researcher at Shanxi University. He earned his Ph.D. in Ecology from Northeast Normal University in 2011. As Co-Chair of the United States–China Carbon Consortium (USCCC), Dr. Dong's research mainly focuses on synthetic observations of eddy flux towers and remote sensing. He is particularly interested in understanding the ecohydrological processes under global climate change and land use/cover change scenarios, particularly in rainfall pattern changes and vegetation restoration feedback on water and carbon cycling.

Jingfeng Xiao is currently Research Associate Professor in Global Ecology at the Earth Systems Research Center, University of New Hampshire. Dr. Xiao's research interests include terrestrial carbon cycle, remote sensing, ecological modeling, vegetation dynamics, land cover/land use change, disturbances, and human–environment interactions. He is particularly interested in understanding the impacts of climate change, land cover/land use change, and disturbances in terrestrial carbon cycling as well as their feedback into the climate. He combines remote sensing, ecological modeling, field measurements (e.g., eddy covariance measurements), model–data fusion, and statistical analysis to answer important carbon cycle, climate change, and ecology-related questions. He has served on the review committee of the 2nd State of Carbon Cycle Report (SOCCR-2) of the National Academies of Sciences, and was a Co-Chair of the United States–China Carbon Consortium (USCCC). He earned his Ph.D. from University of North Carolina, Chapel Hill, in 2006, M.Sc. from Peking University in 2000, and B.Sc. from Lanzhou University in 1997.

Ge Sun is a Research Hydrologist with the Eastern Forest Environmental Threat Assessment Center, USDA Forest Service Southern Research Station, and an Adjunct Professor at North Carolina State University. Dr. Sun has conducted forest hydrological research on various ecosystems, from Florida's cypress swamps in the humid southeastern United States to northern China's Loess Plateau drylands. Currently Dr. Sun's research focuses on the effects of climate change, land use change, and wildland fires on water and carbon resources at multiple scales. Dr. Sun has authored more than 250 journal articles and book chapters. Dr. Sun has received several distinguished awards, including Fellow of the American Water Resources Association and the US Forest Service Chief's Distinguished Science Awards. He is a co-founder of the US–China Carbon Consortium and has served as a forestry expert for the Forest Service International Programs mission in Asia, Africa, and Mexico. He received degrees in forest hydrology from Beijing Forestry University (B.Sc. in 1985 and M.Sc. in 1988) and the University of Florida (Ph.D. in 1995).

Lu Hao is Professor at the College of Applied Meteorology, Nanjing University of Information Science and Technology. Dr. Hao obtained both her Ph.D. and Master degrees in Physical Geography from Beijing Normal University. She was a visiting scholar at CSIRO (Commonwealth Scientific and Industrial Research Organization), Australia, in 2013, and worked as a senior meteorological engineer at the Inner Mongolia Meteorological Bureau, China, from 1992 to 2009. Dr. Hao has authored more than 70 journal articles and book chapters. Her current research interests focus on the effects of climate change and land use and land cover change on evapotranspiration, and water and energy balance, and applications of ecohydrological simulation models, GIS, and remote sensing to assess water supply and demand relations at basin and regional scales. Her current projects study the impacts of urbanization on evapotranspiration and associated ecohydrological processes in a paddy wetland-dominated basin, and the coupling of hydrology and carbon cycling in terrestrial ecosystems in response to climate change and land use management.

Zhiqiang Zhang is Professor of Ecohydrology and currently serving as Dean at the College of Soil and Water Conservation at Beijing Forestry University, China. He received his Bachelor, Master, and Ph.D. degrees in forestry from Beijing Forestry University in 1988, 1991, and 1999, respectively. Dr. Zhang's research interests include ecohydrological processes and modeling, water yield and land use change, carbon and water flux coupling of managed forest ecosystems, and urban forestry. Dr. Zhang has led a number of research projects supported by the Ministry of Science and Technology, National Science Foundation of China, the State Forestry and Grassland Administration, as well as International Cooperation Projects. Dr. Zhang has authored over 180 journal articles and book chapters and received several national, provincial, and ministry awards for his academic achievements. He is a co-founder of the US–China Carbon Consortium.

Preface to "Afforestation and Reforestation: Drivers, Dynamics, and Impacts"

Afforestation and reforestation have been implemented worldwide as an effective measure towards sustainable ecosystem services. The conversion of grasslands, croplands, shrublands, or bare lands to forests can dramatically alter forest water, energy, and carbon cycles, e.g., through carbon sequestration, soil erosion control, and water quality improvement. Large-scale afforestation/ reforestation is typically driven by policies and, in turn, can also have substantial socioeconomic impacts. To enable success, forestation endeavors require novel approaches that involve a series of complex processes and interdisciplinary sciences. For example, exotic or fast-growing tree species are often used to improve the soil conditions of degraded lands or maximize productivity, and it often takes a long time to understand and quantify the consequences of such practices at watershed or regional scales. Maintaining the sustainability of man-made forests is becoming increasingly challenging under a changing environment and disturbance regime changes such as wildland fires, urbanization, drought, air pollution, climate change, and socioeconomic change.

This book is a collection of 11 papers published as a Special Issue of Forests in 2018. These studies provide an in-depth understanding of the ecosystem changes and driving forces following afforestation and explore how plants regulate water consumption, energy utilization, and carbon exchange in different phenophases and climatic conditions, in addition to proposing new quantitative models and appraisal approaches for sustaining forests under a changing environment. We have organized the 11 papers into four major themes that address emerging issues about afforestation/reforestation in several unique regions. The first theme provides an overview of the grand challenges and opportunities facing vegetation greening in the Loess Plateau region of northern China and Yangtze River basin of southern China [1–3]. The second theme presents process-based studies on the effects of vegetation cover change on ecohydrological processes, including evapotranspiration [4] and net primary productivity [5] by remote sensing, transpiration by Granier's sap flow [6], surface runoff and base flow [7], and chlorophyll fluorescence [8]. The third theme is represented by a case study that uses modeling to demonstrate land cover change and vegetation dynamics under environmental changes [9]. The fourth and last theme covers studies that focus on innovative research methodology that are being used in ecological engineering [10] and nursery production [11]. In summary, these studies provide an update on the scientific advances related to forestation and show that human-dominated vegetation restoration activities will further exacerbate vegetation water demand, resulting in contrasting effects on ecosystem services in both magnitude and direction. We recommend that future ecological restoration programs pay more attention to maintaining the balance between ecosystem restoration and water resource demand to maximize the benefits of human activities and ensure that vegetation restoration is ecologically sustainable. We hope that the information provided by this book is timely and helpful for land managers and policy makers to better manage forest resources under rapidly changing environments. We would also like to thank the authors for sharing their research, and the reviewers and editors for their dedication, which were vital to the success of this Forests Special Issue.

References

1. Zhang, D.; Liu, X.; Bai, P. Different Influences of Vegetation Greening on Regional Water-Energy Balance under Different Climatic Conditions. *Forests* **2018**, *9*, 412.

2. Zhao, F.; Wang, J.; Zhang, L.; Ren, C.; Han, X.; Yang, G.; Doughty, R.; Deng, J. Understory Plants Regulate Soil Respiration through Changes in Soil Enzyme Activity and Microbial C, N, and P Stoichiometry Following Afforestation. *Forests* **2018**, *9*, 436.

3. Wei, X.; Bi, H.; Liang, W.; Hou, G.; Kong, L.; Zhou, Q. Relationship between Soil Characteristics and Stand Structure of Robinia pseudoacacia L. and Pinus tabulaeformis Carr. Mixed Plantations in the Caijiachuan Watershed: An Application of Structural Equation Modeling. *Forests* **2018**, *9*, 124.

4. Wang, Y.; Liu, Y.; Jin, J. Contrast Effects of Vegetation Cover Change on Evapotranspiration during a Revegetation Period in the Poyang Lake Basin, China. *Forests* **2018**, *9*, 217.

5. Wang, H.; Liu, G.; Li, Z.; Wang, P.; Wang, Z. Assessing the Driving Forces in Vegetation Dynamics Using Net Primary Productivity as the Indicator: A Case Study in Jinghe River Basin in the Loess Plateau. *Forests* **2018**, *9*, 374.

6. Schwärzel, K.; Zhang, L.; Strecker, A.; Podlasly, C. Improved Water Consumption Estimates of Black Locust Plantations in China's Loess Plateau. *Forests* **2018**, *9*, 201.

7. Ouyang, L.; Liu, S.; Ye, J.; Liu, Z.; Sheng, F.; Wang, R.; Lu, Z. Quantitative Assessment of Surface Runoff and Base Flow Response to Multiple Factors in Pengchongjian Small Watershed. *Forests* **2018**, *9*, 553.

8. Ren, C.; Wu, Y.; Zha, T.; Jia, X.; Tian, Y.; Bai, Y.; Bourque, C.; Ma, J.; Feng, W. Seasonal Changes in Photosynthetic Energy Utilization in a Desert Shrub (Artemisia ordosica Krasch.) during Its Different Phenophases. *Forests* **2018**, *9*, 176.

9. Wang, J.; Ning, S.; Khujanazarov, T. Modeling Hydrological Appraisal of Potential Land Cover Change and Vegetation Dynamics under Environmental Changes in a Forest Basin. *Forests* **2018**, *9*, 451.

10. Zhang, D.; Cheng, J.; Liu, Y.; Zhang, H.; Ma, L.; Mei, X.; Sun, Y. Spatio-Temporal Dynamic Architecture of Living Brush Mattress: Root System and Soil Shear Strength in Riverbanks. *Forests* **2018**, *9*, 493.

11. González-Orozco, M.; Prieto-Ruíz, J.; Aldrete, A.; Hernández-Díaz, J.; Chávez-Simental, J.; Rodríguez-Laguna, R. Nursery Production of Pinus engelmannii Carr. with Substrates Based on Fresh Sawdust. *Forests* **2018**, *9*, 678.

Gang Dong, Jingfeng Xiao, Ge Sun, Lu Hao, Zhiqiang Zhang
Special Issue Editors

Article

Relationship between Soil Characteristics and Stand Structure of *Robinia pseudoacacia* L. and *Pinus tabulaeformis* Carr. Mixed Plantations in the Caijiachuan Watershed: An Application of Structural Equation Modeling

Xi Wei [1], Huaxing Bi [1,2,3,4,*], Wenjun Liang [5], Guirong Hou [1], Lingxiao Kong [1] and Qiaozhi Zhou [1]

[1] School of Soil and Water Conservation, Beijing Forestry University, Beijing 100083, China; weixi860826@163.com (X.W.); hgralex@163.com (G.H.); konglx958 @163.com (L.K.); Gracezqz@163.com (Q.Z.)
[2] Beijing Collaborative Innovation Center for Eco-Environmental Improvement with Forestry and Fruit Trees, Beijing 102206, China
[3] Ji County Station, Chinese National Ecosystem Research Network (CNERN), Beijing 100083, China
[4] Beijing Forestry University, Key Laboratory of State Forestry Administration on Soil and Water Conservation, Beijing 100083, China
[5] College of Forestry, Shanxi Agricultural University, Taigu 030801, China; liangwenjun123@163.com
*   Correspondence: bhx@bjfu.edu.cn; Tel.: +86-10-6233-6756

Received: 20 December 2017; Accepted: 3 March 2018; Published: 6 March 2018

Abstract: In order to study the multi-factor coupling relationships between typical *Robinia pseudoacacia* L. and *Pinus tabulaeformis* Carr. mixed plantations in the Caijiachuan basin of the Loess Plateau of Shanxi Province, West China, 136 sample plots were selected for building a structural equation model (SEM) of three potential variables: terrain, stand structure, and soil characteristics. Additionally, the indicators (also known as observed variables) were studied in this paper, including slope, altitude, diameter at breast height (DBH), tree height (TH), tree crown area, canopy density, stand density, leaf area index (LAI), soil moisture content, soil maximum water holding capacity (WHC), soil organic matter (SOM), total nitrogen (TN), total phosphorus (TP), ammonia-nitrogen (NH_3-N), nitrate-nitrogen (NO_3-N), and available phosphorus (AP). The results showed that terrain was the most important factor influencing soil moisture and nutrients, with a total impact coefficient of 1.303 and a direct path coefficient of 0.03, which represented mainly positive impacts; while correspondingly stand structure had a smaller negative impact on soil characteristics, with a total impact coefficient of −0.585 and a direct path coefficient of −0.01. The terrain also had a positive impact on the stand structure, with a total impact coefficient of 0.487 and a direct path coefficient of 0.63, indicating that the topography factors were more suitable for site conditions and both the stand structure and the soil moisture and nutrient conditions were relatively superior. By affecting the stand structure, terrain could restrict some soil, water, and nutrient functions of soil and water conservation. The influence coefficients of the four observed variables of DBH, stand density, soil water content, and organic matter, and potential variable topography reached 0.686, −0.119, 1.117, and 0.732, respectively; and the influence coefficients of soil moisture, organic matter and stand structure were −0.502 and −0.329, respectively. Therefore, besides observing the corresponding latent variables, the observed variables had a considerable indirect influence on other related latent variables. These relationships showed that the measures, such as changing micro-topography and adjusting stand density, should effectively maintain or enhance soil moisture and nutrient content so as to achieve improved soil and water conservation benefits in the ecologically important Loess Area.

Keywords: structural equation model; *Robinia pseudoacacia* L. and *Pinus tabulaeformis* Carr. mixed plantations; stand structure; soil characteristics; soil and water conservation function

---

1. Introduction

Large-scale afforestation in the Western Shanxi Loess Area started in the early 1990s, and the area has been covered by larger-sized plantations over the past 20 years. Mixed afforestation methods—with plantations consisting of two or more tree species—are often used. In the mixed plantations, tree species other than the main tree species represent over 20% of the plantation, in terms of number of trees, cross-sectional area, or volume. The impact of vegetation construction on soil and water resources has also aroused great attention, both at home and abroad, particularly in the Loess Plateau—an area known for soil and water loss, a lack of water resources, a fragile ecological environment, and a lack of a strong conservation ethics. The main reasons for the creation of large-scale plantations in this region, and their impacts on soil and water conservation, water resources security, and regional sustainable development, have been a particular cause for concern. The relationships between the stand structures of the plantations and soil and water conservation have also gradually become the foci of academic research.

Many studies on the stand structure and soil and water conservation functions of the Loess Plateau, as well as other ecologically sensitive areas, have been carried out, usually aimed at one, or several, dimensions. For example, Bi Huaxing et al. utilized the principle of water balance to establish a suitable coverage calculation model based on spatial and temporal differentiation of soil moisture and water consumption [1]; Brzostek et al. proposed that chronic water stress could reduce the tree growth of forests, and also considered the extent to which forests ameliorate climate warming [2]; and Panagos et al. presented an assessment of soil loss due to water erosion in Europe, and also suggested some policy measures that should be targeted [3]. Other scholars have also presented research results on some factors related to the stand structure, soil moisture [4,5], and soil nutrients [6] in the Loess Plateau. However, research on the multi-factor coupling relationships between stand structure and soil moisture and nutrients [7] is relatively lacking. Traditionally, the main functions of water and soil conservation in gully areas of this type are regarded as water resource conservation and erosion reduction [8]. Water conservation has manifested as soil water storage capacity and soil conservation often incorporates the preservation, storage, recycling, conversion, and acquisition of soil organic matter, nitrogen, phosphorus, and other nutrients [9].

Structural equation modeling (SEM) has become increasingly precise and is now widely used in ecological studies [10–13], mainly for the purpose of quantifying the relationships between multiple factors. Essentially, SEM aims to generate strong and distinct links between theoretical and experimental ideas [14]. The ability to disentangle causal relationships and to test competing models and theories (as opposed to null hypotheses) are key strengths of SEM methods [15]. Due to their statistical strength and applicability, SEM approaches have been employed in a wide range of environmental and ecological studies [16–19]. For example, SEM has been applied to evaluate the effect of grazing on ecosystem processes [20,21]; the relationships between fire and edaphic factors and woody vegetation structure and composition [22]; the sensitivity of soil respiration to environmental factors [23]; the impacts of land uses on stream integrity [24]; the factors that affect plant richness in recovering forests [25,26]; the relationships associated with the decline in species richness, as natural landscapes undergo conversion to human-dominated landscapes [27]; and both the direct and indirect association of plant species richness to landscape conditions and local environmental factors [28,29]. However, to our knowledge, SEM methods have not been applied to study stand structure impacts on soil and water conservation.

In this paper, the covariance SEM is used to quantify the multi-factor coupling relationship between soil moisture and nutrients in typical plantations. Soil moisture content and soil maximum water holding capacity are taken as water conservation indicators, and organic matter, nitrogen, and phosphorus are used as conservation soil indicators. These indicators are all used to study soil characteristics and their relationships with the topography and stand structure in order to reveal the role of stand structure on the conservation of water sources and soil function mechanisms, and to further provide references of control technology regarding the practical and suitable slope stand structural adjustments in the Loess Plateau.

2. Materials and Methods

2.1. Site Description

The Caijiachuan Watershed served as the study site; it is located on the Loess Plateau in Ji County, Shanxi Province, China (35°53′–36°21′ N, 110°27′–111°7′ E; elevation 904–1592 m), and is a typical gully area. Meteorological records indicate that the long-term mean annual air temperature is 10.2 °C and the frost-free period is 172 days. The average annual precipitation is 571 mm, with an uneven distribution. The average annual potential evapotranspiration (PET) is 1724 mm, which far exceeds the rainfall. Inside this area, the type of soil is mainly Haplic Luvisols (Soil classification of the Food and Agriculture Organization of the United Nations) and is mostly alkaline. There are mainly artificial shelterbelts of black locust (*Robinia pseudoacacia* L.) and Chinese pine (*Pinus tabulaeformis* Carr.) in the nested watershed, with an area of 38 km^2 and a forest cover rate of 72%. The main shrubs under the forests are periploca (*Periploca sepium* Bunge), yellow rose (*Rosa xanthine* Lindl.), sophora viciifolia (*Sophora davidii* (Franch.) Skeels), meadowsweet (*Spiraea salicifolia* L.), lilac (*Syringa* linn.), elaeagnus umbellata (*Elaeagnus pungens* Thunb.), etc. Through the investigation of forestland, shrubland, and grassland in this area, dominant species of artificial *R. pseudoacacia* and *P. tabulaeformis* forests with different slopes, aspects, and altitudes were selected as objects and studied.

2.2. Data Acquisition and Processing Methods

Thirty-four standard plots of 20 m × 20 m were set up at the plantations and 136 sample plots of 10 m × 10 m were set up with shady, semi-shady, sunny, and semi-sunny aspects (one standard plot was divided into four equal sample plots to reduce the heterogeneity) The slopes of the plots ranged from 15° to 45°, and were distributed at an average elevation of 1133.5 m above sea level (Table 1). The mixture of plantation species consisted of *R. pseudoacacia* and *P. tabulaeformis* at a ratio of 8:2, as *R. pseudoacacia* was the dominant species. Using individual field measurements in these plots, the varieties of trees, diameter at breast height (DBH), tree height (TH), and tree crown area were measured, and then the canopy density and stand density were calculated. The leaf area indices of the quadrats were determined using a LAI-2000 (LI-COR Company, Lincoln, NE, USA) vegetation canopy analyzer. According to the trophic classification scheme for functions of soil and water conservation [8], indicators of water resources and soil protection were confirmed. Mixed soil samples, 0–60 cm, were collected using the cutting ring method and were representative of forest soils in this area; soil moisture content was determined using the drying method; and water holding capacity (WHC) was measured using the soil infiltration method [30]. After air-dried soil was sieved (0.15 mm sieve), indoor experiments were conducted. The contents of soil organic matter (SOM), total nitrogen (TN), total phosphorus (TP), ammonia-nitrogen (NH_3-N), nitrate-nitrogen (NO_3-N), and available phosphorus (AP) were measured with a SmartChem-200 (AMS/Alliance Instruments, Paris, France) discrete wet chemistry analyzer. The major geographical and biological characteristics of the investigated plots are summarized in Table 2.

Table 1. The distribution of aspects, slopes, and altitudes of the sample plots in the Caijiachuan Watershed.

Aspect	Shady	Semi-Shady	Sunny	Semi-Sunny
Sample quantity	7	11	6	10

Slope/°	$\leq$15	16–25	26–35	$\geq$36
Sample quantity	2	15	15	2

Altitude/m	900–1000	1000–1100	1100–1150	1150–1200	1200–1300	>1300
Sample quantity	2	6	15	8	3	0

Notes: (1) The distribution of aspect tends to be mostly homogeneous. (2) The lands featuring gentle slopes ($\leq$15) are usually cropland and those that are dangerously steep ($\geq$36) are difficult sites to access, as such, there are very few of either type of these sites for afforestation, whereas many sites with deep slopes (16–35) are used for afforestation in order to restore vegetation and improve the environment in China. (3) In the watershed, low-altitude (900–1000 m) areas are mainly agricultural lands or areas where people are living, so there are very few low-altitude sites for afforestation; high-altitude (>1300 m) areas are mainly distributed with natural forests, so there are no high-altitude plantations; the mid-altitude areas (1000–1300 m) are the main afforestation areas of mixed plantations, and can therefore be regarded as being representative of the mixed plantations in the region.

Table 2. The survey of the species of *Robinia pseudoacacia* L. and *Pinus tabulaeformis* Carr. mixed plantations in the Caijiachuan Watershed, Shanxi Province, West China. DBH, diameter at breast height; LAI, leaf area index; WHC, water holding capacity; SOM, soil organic matter; TN, total nitrogen; TP, total phosphorus; NH3-N, ammonia-nitrogen; NO3-N, nitrate-nitrogen; AP, available phosphorus.

Stands and Soil Characteristics	Maximum	Minimum	Average
Slope (°)	45	15	26.50
Altitude (m)	1220	960	1133.53
DBH (cm)	18.54	6.37	10.85
Tree height (m)	13.4	3.0	8.2
Crown area (m^2)	16.74	2.80	8.12
Canopy density	0.88	0.38	0.64
Stand density (trees·hectare^{-1})	4400	500	1679
LAI	4.50	0.88	2.06
Soil moisture content (%)	40.03	5.66	13.71
WHC (%)	122.88	25.54	50.41
SOM (g·kg^{-1})	122.55	1.31	16.08
TN (g·kg^{-1})	4.65	0.01	0.69
TP (g·kg^{-1})	7.60	0.03	0.66
NH$_3$-N (mg·kg^{-1})	66.84	2.79	20.99
NO$_3$-N (mg·kg^{-1})	88.40	0.12	10.27
AP (mg·kg^{-1})	117.64	0.16	36.00

2.3. Structural Equation Modeling

SEMs (also known as path analyses) are statistical multivariate models that are used to estimate causality and direct or indirect relationships between multiple variables [31]. These models are less restrictive than regression models in that some variables may play the role of predictor variable and dependent variable simultaneously [32,33]. SEM starts by constructing an a priori schema: an analytical model that represents all hypothetical causal links between the predictors and the response variables, based on previous knowledge of the ecological system [34].

Thus, SEM in ecology is a method to test ecosystem structure and function [35], which is closely related to (and is actually a more general form of) several types of statistical analyses, including regression, principal components analysis (PCA), and path analysis [36]. Furthermore, it can explore the relationships between observed variables, latent variables, and residuals to quantitatively describe the influence of independent variables on dependent variables, including direct, indirect, and total impacts [13]. However, in contrast to some other methods, SEM provides a means to evaluate both the structure of the model as well as a specific parameterization of the model structure using data [37].

By "model structure", we mean the pattern of relationships among variables (correlations, direct, and indirect relationships among variables). The structural equation model includes two parts of the measurement model and the structural model; the formula is as follows [38–40]:

$$X = \Lambda x\xi + \delta \tag{1}$$

$$Y = \Lambda y\eta + \varepsilon \tag{2}$$

$$\eta = \eta + \Gamma\xi + \zeta \tag{3}$$

Equations (1) and (2) are measurement models, used to describe the relationship between latent and observed variables, where X is the exogenous observation variable vector, Y is the endogenous observation variable vector, Λx and Λy are the factor loadings of the indicator variables (X, Y), δ and ε are the measurements of the exogenous observation variables and the endogenous observation variables, ξ is the exogenous latent variable, and η is the endogenous latent variable.

Equation (3) is a structural model that can reflect the relationships between the potential variables, where is the structural coefficient matrix of the relationship between endogenous latent variables, Γ is the structural coefficient matrix of the relationships between endogenous latent variables and exogenous latent variables, and ζ is the interference factor or residual value of the structural model. The initial model with the aid of a path map reflects the relationships among the variables in the structural model, and the path coefficients represent the extent of the relationship between the variables. After the path diagram is established, the path coefficients of all the paths are usually calculated using the maximum likelihood method [13], which is also used in this study.

The climate, hydrology, and other environmental conditions in the area were basically the same, and the different topographical factors (ξ_1) were used as potential exogenous variables. The slope, aspect, slope position, and altitude were taken into account in the modeling. In consideration of the aspect and slope position being subject to qualitative description and random selection in the survey, the slope (x_1) and elevation (x_2) were determined to be the index of the initial model belonging to the exogenous observation variables, because they are the indicators of accurate measurement. Their corresponding errors are δ_1 and δ_2, respectively.

The differences between quadrats were also reflected in the stand structure (ξ_2). The measuring indices affecting the stand structure mainly included DBH (x_3), TH (x_4), tree crown area (x_5), canopy density (x_6), stand density (x_7), and LAI (x_8), which were also exogenous observation variables as well. The corresponding errors were δ_3, δ_4, δ_5, δ_6, δ_7, and δ_8. Soil properties (η) were taken as potential endogenous variables, and their corresponding indices, such as soil moisture content (y_1), WHC (y_2), SOM (y_3), TN (y_4), TP (y_5), NH$_3$-N (y_6), NO$_3$-N (y_7), and AP (y_8), were determined as endogenous observation variables, of which the corresponding errors were ε_1, ε_2, ε_3, ε_4, ε_5, ε_6, ε_7, and ε_8. In addition, the modeling process needed to consider the residuals of three potential variables as ζ_1, ζ_2, and ζ_3.

After creating an initial model based on previous knowledge, site information, and background data, a chi-square value (χ^2) test was then conducted to examine whether the covariance structure suggested by the model satisfactorily fits the covariance structures [19]. The χ^2, degree of freedom (df, $0 \leq \chi^2 / df \leq 3$), probability level ($p > 0.05$), root mean square error of approximation ($0 \leq$ root meant square error of approximation (RMSEA) ≤ 0.05) [40], and comparative fit index ($0.9 \leq$ comparative fit index (CFI) ≤ 1.00) were given to determine the "best" model that has the highest predictive performance, while the comparative fit index ($0.7 \leq$ CFI ≤ 0.90) was "tolerable" [41]. If the parameters exhibit beyond the proper range after model running the model, we should use two methods for model correction: "modification index" and "critical ratio (CR) for difference", provided by the Amos 22.0 software (IBM/International Business Machines Corporation, Armonk, NY, USA) package. In the model diagram, the double arrow ("<->") section was the covariance correction index between the residual variables, and the single arrow ("->") section was a regression weight correction index between variables, indicating that if an arrow is added between two variables, at least the chi-squared value of the model will be reduced. In addition, the CR statistic followed a normal distribution, so its value can be used to judge whether there is a significant difference between the two estimated parameters. In this study, the former method was selected to modify the model, according to the characteristics of the investigated measurable variables.

3. Results

3.1. Model Construction and Correction

SEM has been shown to be verifiable, therefore, the basis for the study of the coupling relationships between stand structure and soil properties was taken into consideration when building the model. The data in this research conformed to a multivariate normal distribution, and maximum likelihood estimation was used to quantitatively analyze potential and observed variables. SPSS 19.0 software (IBM/International Business Machines Corporation, Armonk, NY, USA) was used for data exploratory analysis. Using Amos 22.0 software (IBM/International Business Machines Corporation, Armonk, NY, USA) for drawing the path map and parameter estimation, the path coefficient between each variable, the factor load of each variable, the measurement error of observed variables, and the residual of potential variables could all be obtained.

The initial model was constructed based on generally-known experience. After running in Amos, the chi-square (χ^2) test statistic value was 414.592 with 101 df, and a significant probability (p) value of 0.0001 (<0.05) in the model (Figure 1), which resulted in the rejection of the null hypothesis. The root meant square error of approximation (RMSEA) was 0.152, under the null hypothesis of "close fit" (i.e., RMSEA is no greater than 0.05), as such, the adaptability of the hypothetical model to the observed data should be modified (Table 3).

Table 3. The fitting parameters describing the coupling relationship between stand structure and soil properties of artificial mixed forests in the Caijiachuan Watershed.

Index Name	Evaluation Criterion	Initial Model	Modified Model
The chi-square (χ^2)	The smaller the better.	414.592	247.554
The ratio of chi-square and freedom (χ^2/df)	1~3. When the ratio is less than 1, the model is over adapted; when the ratio is between 1 and 3, the model is well adapted; when the ratio is greater than 3, the model is poorly fitted.	4.105	2.782
Significant probability (p)	>0.05	0.000	0.078
Normative fit index (NFI)	0~1. A value greater than 0.7 is acceptable, the closer to 1 the better.	0.421	0.754
Incremental fit index (IFI)	0~1. A value greater than 0.7 is acceptable, the closer to 1 the better.	0.490	0.747
Comparative fit index (CFI)	0~1. A value greater than 0.7 is acceptable, the closer to 1 the better.	0.474	0.734
The root meant square error of approximation (RMSEA)	<0.05. The smaller the better.	0.152	0.045
Akaike information criterion (AIC)	The smaller the better.	516.592	373.554
Bayes criterion (BCC)	The smaller the better.	531.287	391.707

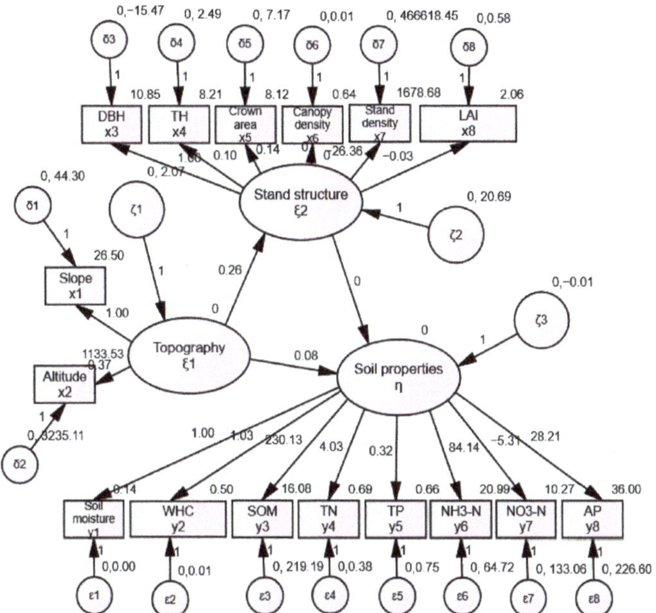

Figure 1. The initial structural equation model (SEM) used in the study. Note: The hypothesized initial model used for predicting topography, stand structure, and soil properties is based on soil and water conservation science. A rectangular box is used for each observed variable, with a measurement error, and the numbers correspond to the standardized path coefficients of the initial model on the single arrows in operation. A value outside of a rectangular box is the mean of the indicator, and a value outside of a round box is the residual error before modification. In the figure, DBH is the abbreviation for diameter at breast height; TH is the acronym for height of tree; LAI is the acronym for leaf area index; WHC is the abbreviation for soil maximum water holding capacity; SOM is the acronym for soil organic matter; TN is the is the acronym for total nitrogen; TP is the acronym for total phosphorus; NH_3-N is the acronym for ammonia-nitrogen, NO_3-N is the acronym for nitrate-nitrogen; and AP is the acronym for available phosphorus.

According to the current theoretical research and qualitative analyses of the conclusions, the model correction was completed using the "Modification Indices" hints of the Amos software. Principally, adding double arrows could express the correlation between the residuals of each variable, so that the parameters of the model were within an allowable range. After being modified, the chi-square (χ^2) test statistic value became 247.554 and the degree of freedom (df) reduced to 89, with significant probability (p) value of 0.078 (>0.05) in the SEM (Figure 2), which accepted the null hypothesis. The value of RMSEA was 0.045 (<0.05), while normative fit index (NFI), incremental fit index (IFI), and CFI were each greater than 0.7, inside the acceptable range, thus the model could be tolerated. Therefore, the test of the fittest indicators also met the standard, indicating that the fit of the model and the observed data were better after correction (Table 3).

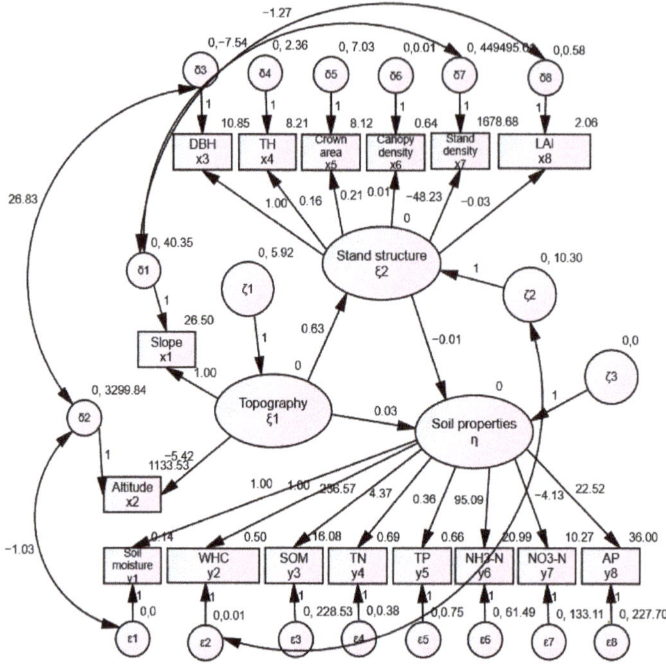

Figure 2. The modified model. Note: The numbers correspond to the standardized path coefficients on the single arrows, and to correlation coefficients on the double arrows. A value outside of a rectangular box is the mean of the indicator, and a value outside of a round box is the residual error after modification. In the figure, DBH is the abbreviation for diameter at breast height; TH is the acronym for height of tree; LAI is the acronym for leaf area index; WHC is the abbreviation for soil maximum water holding capacity; SOM is the acronym for soil organic matter; TN is the is the acronym for total nitrogen; TP is the acronym for total phosphorus; NH_3-N is the acronym for ammonia-nitrogen, NO_3-N is the acronym for nitrate-nitrogen; and AP is the acronym for available phosphorus.

3.2. Model Explanation

3.2.1. Relationship between Latent Variables

The topography had positive effects on the stand structure and soil characteristics (Figure 2), and the path coefficients were 0.63 and 0.03, respectively. Numerically, the effect of topography on the stand structure was far greater than its impact on the soil. The stand structure had a negative impact on soil properties, with a path coefficient of −0.01. Standardized influence coefficients characterized the effects of the latent variables, which were calculated using the SEM method (Table 4). The impact coefficient of topography on soil characteristics was 1.303, with a direct impact of 1.589 and an indirect impact of −0.285; the total impact and direct impact coefficients of stand structure on soil properties were both −0.585; and the total impact and direct impact coefficients of topography on stand structure were both 0.487.

This showed that, after optimizing the stand structure, soil moisture and nutrients should be slightly influenced. In practice, the stand structure should be optimized as much as possible so as to increase the ecological function of stand. Simultaneously, the moderate stand structure adjustments should also be observed in order to avoid heavier constraints for the soil water resources and fertility when the negative impacts accumulate to a certain extent.

Table 4. Standardized influence coefficients of latent variables in the structural equation model of *Robinia pseudoacacia* and *Pinus tabulaeformis* mixed plantations in the Caijiachuan Watershed.

Effect Type		Influences	
		Topography	Stand Structure
Standardized total impact	Stand structure	0.487	
	Soil characteristics	1.303	−0.585
Standardized direct impact	Stand structure	0.487	
	Soil characteristics	1.589	−0.585
Standardized indirect impact	Stand structure		
	Soil characteristics	−0.285	

3.2.2. Relationship between Latent and Observed Variables

The extent and effect of the influence between latent variables and observed variables were also reflected in the calculation of the path coefficient of the fit model (Figure 2). First, among the observed variables affecting the topographic factors, the slope showed a positive effect, but the altitude showed an opposite effect, which had a great influence on the topographic factors from a numerical point of view. Second, in the observed variables of influencing the stand structure, DBH, TH, crown width, and canopy density showed positive effects, whereas stand density and LAI both showed negative effects. The effect of stand density on the stand structure was much greater than the other factors. Third, among all of the observed variables that affected soil properties, all of them showed positive effects, except NO_3-N, and the order of their impacts on soil characteristics were: SOM > NH_3-N > AP > TN > soil moisture > WHC > TP (Figure 2).

4. Discussion

The concept of SEM has been applied to many multivariable coupling studies in the natural sciences and has achieved good results in forest ecology studies. Application studies in forests [12,14,42,43], shrublands [13], wetlands [35], and other ecosystems [15] have proved the scientific basis and reliability of the model. Its importance is that it allows for the detection of topographic, stand structural, and soil characteristic factors that are responsible for structural and functional changes. However, the correct interpretation of these models requires a strong supply of ecological data and a good understanding of the characteristics of the topography–stand–soil system.

4.1. Topography Mainly Impacted Stand Structure

Topographical indicators commonly incorporate slope, aspect, altitude, slope length, and slope position, while the basal impacting factors of stand structure were DBH, TH, crown area, canopy density, stand density, forests diversity, LAI, and so on. Different topographic factors influenced forest diversity, structure, and dynamic change by their respective influencing mechanisms as supported by previous research that has demonstrated some aspects of these observations. For instance, tree height had a remarkable dissimilarity based on the slope and aspect of sites [44]; there was significant interaction between aspect and elevation in influencing forest structure [45]; and the vegetation should be adapted to various elevations according to relevant temperatures and other conditions [46].

Examining the data of the mixed plantation, a path coefficient of 0.63 illustrated that the effect of topography on the stand structure was the most important positive factor in the complex relationships. Standardized influence coefficients characterized the effects of the latent variables between topography and stand structure (Table 4). The total impact coefficient of the topography on the stand structure was 0.487, all of which was attributed to direct impacts. Both of the results above demonstrated that if the conditions of the topography in an area became more advantageous, the stand structure can grow more favorably.

In addition to representing the corresponding latent variables, observational variables indirectly affected other related latent variables and further revealed the indirect relationship between the latent variables [42]. The interactive relationship between the observed variables could preferably show the focus of the interaction among the systems represented by the latent variables and explain the current relationship of the latent variables more deeply. Observed variables such as DBH, TH, crown width, canopy density, stand density, and LAI, could not only characterize the stand structure, with the impact coefficients of 1.407, 0.424, 0.305, 0.234, −0.243, and −0.070, but also relate to the topographical factors, with the impact coefficients reaching 0.686, 0.277, 0.144, 0.114, −0.119, and −0.034 (Table 5). As such, a deeper relationship between the indicators of the stand structure and the topographic factors could be clearly distinguished among them. The results showed that if the values of DBH, TH, crown area and canopy density were greater then the stand structure was more favorable; conversely, the direct influence of stand density and LAI on the stand structure was negative. Analogously, the indirect influence coefficients between the DBH, TH, crown area, canopy density, and topography were positive, but the others were negative.

4.2. Topography Significantly Influenced Soil by Stand Structure Indirectly

The soil characteristics studied in this paper were soil moisture content, WHC, SOM, TN, TP, NH_3-N, NO_3-N, and AP. The relation of one or several dimensions of these soil characteristics to topography have been found in past research. Slope has been shown to influence antecedent soil moisture which can potentially lead to either an increase or decrease in soil erosion, and is therefore a crucial consideration for recommending appropriate measures to protect soils [47]; there were some significant effects of altitudinal zone and slope aspect on the vertical distribution of soil organic carbon (close to SOM) density [48]. Topography usually has some form of direct impact on soil and, here, topography influenced soil indirectly via stand structure.

Topography had positive effects on the soil characteristics, with a path coefficient of 0.03 (Figure 2), which was a small absolute value. The most influential factor on soil characteristics was topography, with a total impact coefficient of 1.303; the direct impact being positive, and the indirect impact being negative (Table 4). This indicated that the majority of its impact on soil properties was through an indirect impact by influencing the stand structure. Through the stand structure adjustment, the topographic factors might change the formation of soil moisture and nutrients.

In addition to the representation of topography, the observed indicators, such as slope and elevation, were also related to soil characterization, and every factor of the soil characteristics had a positive impact on the topography, except for NO_3-N. The absolute value of the influence coefficient of soil water content on the topographical factors was 1.117, indicating that the impact of the factor on soil moisture was quite high. The values of TP and AP were only 0.052 and 0.043, respectively, showing that the impact of phosphorus was weaker than the other factors' impacts on the topography. The results of this study align with other recent research that suggests that soil moisture content was the key factor for study site conditions and the status of stand structure. In the future, soil nutrient-related variables, such as organic matter, nitrogen, and phosphorus, should also be given special consideration.

Table 5. Standardized influence coefficients between observed variables and latent variables in the structural equation model of *Robinia pseudoacacia* and *Pinus tabulaeformis* mixed plantations in the Caijiachuan Watershed.

Observed Variables	Influences								
	Standardized Total Impact			Standardized Direct Impact			Standardized Indirect Impact		
	Topography	Stand Structure	Soil Characteristics	Topography	Stand Structure	Soil Characteristics	Topography	Stand Structure	Soil Characteristics
Slope	0.342			0.342					
Altitude	−0.317			−0.317					
DBH	0.686	1.407			1.407		0.686		
Tree height	0.207	0.424			0.424		0.207		
Tree crown area	0.149	0.305			0.305		0.149		
Canopy density	0.114	0.234			0.234		0.114		
Stand density	−0.119	−0.243			−0.243		−0.119		
LAI	−0.034	−0.070			−0.070		−0.034		
Soil moisture content	1.117	−0.502	0.857			0.357	1.117	−0.502	
WHC	0.546	−0.245	0.419			0.419	0.546	−0.245	
SOM	0.732	−0.329	0.561			0.561	0.732	−0.329	
TN	0.306	−0.138	0.235			0.235	0.306	−0.138	
TP	0.052	−0.023	0.040			0.040	0.052	−0.023	
NH$_3$-N	0.547	−0.245	0.419			0.419	0.547	−0.245	
NO$_3$-N	−0.088	0.040	−0.068			−0.068	−0.088	0.040	
AP	0.043	−0.019	0.033			0.033	0.043	−0.019	

Note: In the table, DBH is the abbreviation for diameter at breast height; LAI is the acronym for leaf area index; WHC is the abbreviation for soil maximum water holding capacity; SOM is the acronym for soil organic matter; TN is the is the acronym for total nitrogen; TP is the acronym for total phosphorus; NH$_3$-N is the acronym for ammonia-nitrogen, NO$_3$-N is the acronym for nitrate-nitrogen; and AP is the acronym for available phosphorus.

4.3. Stand Structure Impacted Soil Properties to a Comparatively Smaller Degree

The relationships of the stand structure and soil properties were the essential issues. There have been many studies on these factors in the field of ecological systems and soil and water conservation. Soil variables have been observed to differ in terms of soil TN across the different canopy types [49]. Forest structure has effects on microbiological soil properties and nutrient content [50]; however, a great many of these effects still remain in the realm of single-factor impact research. Additionally, using the SEM method to explore the multi-factors, we found that the stand structure impacted soil properties with a path coefficient of −0.01. The total impact coefficient of stand structure on soil properties was −0.585, all of which was attributed to direct impacts. The results indicated that, under the suitable conditions of topographic factors, the stand might negatively impact soil moisture and nutrient conditions.

The effects of the observed variables on latent variables are also described in Table 5. Soil moisture content, WHC, SOM, TN, TP, NH_3-N, NO_3-N, and AP were related to stand structure, apart from the representation of soil characterization. Here, stand structure had a positive relationship to NO_3-N, but it had negative influence coefficients with all of the other soil factors, along with a small absolute value. Soil moisture content was most sensitive, with an influence coefficient of −0.502 on the stand structure. Adversely, the values of TP and AP were −0.023 and −0.019, demonstrating that phosphorus was weakly influenced by the stand structure. These results verify that the benefits of studying stand and soil moisture in an arid area are great, and that the soil nutrients also play a significant and crucial role in forest stand research.

5. Conclusions

The soil characteristics of *R. pseudoacacia* and *P. tabulaeformis* mixed plantations in the Caijiachuan Watershed were mainly limited by factors such as the regional topography and the stand structure; SEM satisfactorily quantified the relationships between these factors. The modeling indicated that if the topography was more suitable and the stand structure was more favorable then soil moisture and fertility conditions would be better. This finding was consistent with the initial empirical assumption [1–7,47,50], with the results of this study offering a more statistically accurate expression than previously determined. The stand structure had little effect on soil properties, all of which were negatively and directly affected, indicating that it imposed certain restrictions on the retained moisture and nutrients in the soil. Observed variables also correlated well with the latent variables. The mutual indirect influences between the observed variables and the two latent variables of topography and stand structure were comparatively high.

Based on all the above, the quantification process should provide new insights into the management and conservation of forests. In line with the quantified effective values of impact between these interrelated factors, some measures may in fact adjust terrain and stand structure to effectively maintain or increase the soil moisture and nutrient content in reality, such as changing the gradient of micro-topography, or reducing the stand density. These measures could enhance the functions of soil and water conservation. In the Loess Plateau, the conservation and forestry staff can achieve this positive outcome by excavating level steps and fish-scale pits to lessen the slopes or reduce runoff, or by artificial tree-tending in order to change the stand structure according to the path coefficient of the modified model. The results can serve as a reference for determining care and management measures suitable for the *R. pseudoacacia* and *P. tabulaeformis* plantations, as well as for controlling the stand structure on the Loess Plateau and improving the region's soil and water conservation functions.

Acknowledgments: This study was supported by grants from the National Key R & D Program of China (No. 2016YFC0501704), the Natural Science Fund (31500523), the Fund for Introduced Talents for Shanxi Agricultural University (2014YJ19), and the Beijing Municipal Education Commission Beijing Collaborative Innovation Center for Eco-Environmental Improvement with Forestry and Fruit Trees (PXM2017_014207_000024). We would like to thank our colleagues for their comments on this paper before submission.

Author Contributions: H.B. and X.W. designed the experiments. X.W. and W.L. wrote the paper and reviewed the manuscript. All authors conducted the experiments.

Conflicts of Interest: The authors declare no conflict of interest.

References

1. Bi, H.X.; Li, X.Y.; Li, J.; Guo, M.X.; Liu, X. Study on suitable vegetation cover on Loess area based on soil water balance. *Sci. Silvae Sin.* **2007**, *43*, 17–24.
2. Brzostek, E.R.; Dragoni, D.; Schmid, H.P.; Rahman, A.F.; Sims, D.; Wayson, C.A.; Johnson, D.J.; Phillips, R.P. Chronic water stress reduces tree growth and the carbon sink of deciduous hardwood forests. *Glob. Chang. Biol.* **2014**, *28*, 2531–2539. [CrossRef] [PubMed]
3. Panagos, P.; Borrelli, P.; Poesen, J.; Ballabio, C.; Lugato, E.; Meusburger, K.; Montanarella, K.; Alewell, C. The new assessment of soil loss by water erosion in Europe. *Environ. Sci. Policy* **2015**, *54*, 438–447. [CrossRef]
4. Li, S.; Liang, W.; Fu, B.J.; Lv, Y.H.; Fu, S.Y.; Wang, S.; Su, H.M. Vegetation changes in recent large-scale ecological restoration projects and subsequent impact on water resources in China's Loess Plateau. *Sci. Total Environ.* **2016**, *569*, 1032–1039. [CrossRef] [PubMed]
5. Zhang, X.; Zhao, W.W.; Liu, Y.X.; Fang, X.N.; Feng, Q. The relationships between grasslands and soil moisture on the Loess Plateau of China: A review. *Catena* **2016**, *145*, 56–67. [CrossRef]
6. Shi, W.Y.; Du, S.; Joseph, C.M.; Guan, J.H.; Wang, K.B.; Ma, M.G.; Norikazu, Y.; Ryunosuke, T. Physical and biogeochemical controls on soil respiration along a topographical gradient in a semiarid forest. *Agric. For. Meteorol.* **2017**, *247*, 1–11. [CrossRef]
7. Wang, Y.H.; Yu, P.T.; Wang, J.Z.; Xu, L.H.; Karl, H.F.; Xiong, W. Multifunctional forestry on the Loess Plateau. *Earth Environ. Sci.* **2017**, *4*, 79–107.
8. Zhang, C.; Wang, Z.G.; Ling, F.; Ji, Q.; Meng, F.B. Function evaluation of soil and water conservation and its application in soil and water conservation regionalization. *Sci. Soil Water Conserv.* **2016**, *14*, 90–99.
9. Pimentel, D.; Harvey, C.; Resosudarmo, P.; Sinclair, K.; Kurz, D.; McNair, M.; Crist, S.; Shpritz, L.; Fitton, L.; Saffouri, R.; et al. Environmental and Economic Costs of Soil Erosion and Conservation Benefits. *Science* **1995**, *267*, 1117–1123. [CrossRef] [PubMed]
10. Grace, J.B. The factors controlling species density in herbaceous plant communities: An assessment. *Perspect. Plant Ecol. Evol. Syst.* **1999**, *2*, 1–28. [CrossRef]
11. Grace, J.B.; Anderson, T.M.; Smith, M.D.; Seabloom, E.; Andelman, S.J.; Meche, G.; Weiher, E.; Allain, L.K.; Jutila, H.; Sankaran, M.; et al. Does species diversity limit productivity in natural grassland communities? *Ecol. Lett.* **2007**, *10*, 680–689. [CrossRef] [PubMed]
12. Miao, S.L.; Carstenn, S.; Nungesser, M. *Real World Ecology: Large-Scale and Long-Term Case Studies and Methods*; Springer: New York, NY, USA, 2009.
13. Wang, S.L.; Liang, X.J.; Ma, C.; Zhou, J.P. Coupling relationship between *Hedysarum mongdicum* shrub plantation and sand soil based on structural equation model. *J. Beijing For. Univ.* **2017**, *39*, 1–8.
14. Grace, J.B.; Anderson, T.M.; Olff, H.; Scheiner, S.M. On the specification of structural equation models for ecological systems. *Ecol. Monogr.* **2010**, *80*, 67–87. [CrossRef]
15. Erin, M.M.; Samiran, B.; Edward, W.B.; Linda, M.H.; Donald, D.H. Structural equation modeling reveals complex relationships in mixed forage swards. *Crop Prot.* **2015**, *78*, 106–113.
16. Shipley, B. *Cause and Correlation in Biology*; Cambridge University Press: Cambridge, UK, 2000.
17. Jonsson, M.; Wardle, D.A. Structural equation modelling reveals plantcommunity drivers of carbon storage in boreal forest ecosystems. *Biol. Lett.* **2010**, *6*, 116–119. [CrossRef] [PubMed]
18. Lamb, E.G.; Kennedy, N.; Siciliano, S.D. Effects of plant species richness and evenness on soil microbial community diversity and function. *Plant Soil* **2011**, *338*, 483–495. [CrossRef]
19. Lamb, E.; Shirtliffe, S.; May, W. Structural equation modelling in the plant sciences: An example using yield components in oat. *Can. J. Plant Sci.* **2011**, *91*, 603–619. [CrossRef]
20. Laliberté, E.; Tylianakis, J.M. Cascading effects of long-term land-use changes on plant traits and ecosystem functioning. *Ecology* **2012**, *93*, 145–155. [CrossRef] [PubMed]
21. Chen, D.; Zheng, S.; Shan, Y.; Taube, F.; Bai, Y. Vertebrate herbivore-induced changes in plants and soils: Linkages to ecosystem functioning in a semi-arid steppe. *Funct. Ecol.* **2013**, *27*, 273–281. [CrossRef]

22. Diouf, A.; Barbier, N.; Lykke, A.M.; Couteron, P.; Deblauwe, V.; Mahamane, A.; Bogaert, J. Relationships between fire history, edaphic factors andwoody vegetation structure and composition in a semi-arid savanna landscape (Niger, West Africa). *Appl. Veg. Sci.* **2012**, *15*, 488–500. [CrossRef]

23. Matías, L.; Castro, J.; Zamora, R. Effect of simulated climate change on soil respiration in a Mediterranean-type ecosystem: Rainfall and habitat type are more important than temperature or the soil carbon pool. *Ecosystems* **2012**, *15*, 299–310. [CrossRef]

24. Riseng, C.M.; Wiley, M.J.; Black, R.W.; Munn, M.D. Impacts of agricultural land use on biological integrity: A causal analysis. *Ecol. Appl.* **2011**, *21*, 3128–3146. [CrossRef]

25. Leithead, M.; Anand, M.; Duarte, L.D.S.; Pillar, V.D. Causal effects of latitude, disturbance and dispersal limitation on richness in a recovering temperate, subtropical and tropical forest. *J. Veg. Sci.* **2012**, *23*, 339–351. [CrossRef]

26. Virginia, C.; Madhur, A. Assessing ecological integrity: A multi-scale structural and functional approach using Structural Equation Modeling. *Ecol. Indic.* **2016**, *71*, 258–269.

27. Desrochers, R.E.; Kerr, J.T.; Currie, D.J. How, and how much, natural cover loss increases species richness. *Glob. Ecol. Biogeogr.* **2011**, *20*, 857–867. [CrossRef]

28. Gazol, A.; Tamme, R.; Takkis, K.; Kasari, L.; Saar, L.; Helm, A.; Pärtel, M. Landscape- and small-scale determinants of grassland species diversity: Directand indirect influences. *Ecography* **2012**, *35*, 944–951. [CrossRef]

29. Santibáñez-Andrade, G.; Castillo-Argüero, S.; Vega-Peña, E.V.; Lindig-Cisnerosc, R.; Zavala-Hurtado, J.A. Structural equation modeling as a tool to develop conservation strategies using environmental indicators: The case of the forests of the Magdalena river basin in Mexico City. *Ecol. Indic.* **2015**, *54*, 124–136. [CrossRef]

30. Xu, Z. Study on Ecological Function Evaluation and Biodiversity of Typical Forest Stands in Jinxi of the Loess Plateau. Master's Thesis, Beijing Forestry University, Beijing, China, 2014.

31. Quinn, G.P.; Keough, M.J. *Experimental Design and Data Analysis for Biologists*; Cambridge University Press: New York, NY, USA, 2002.

32. Ruiz, M.A.; Pardo, A.; San Martín, R. Structural equation models. *Psychol. Roles* **2010**, *31*, 34–45.

33. Schumacker, R.E.; Lomax, R.G. *A Beginner's Guide to Structural Equation Modeling*; Taylor and Francis Group, LLC: Mahwah, NJ, USA, 2004.

34. Valdés, A.; García, D. Direct and indirect effects of landscape change on the reproduction of a temperate perennial herb. *J. Appl. Ecol.* **2011**, *48*, 1422–1431. [CrossRef]

35. Sutton-Grier, A.E.; Kenney, M.A.; Richardson, C.J. Examining the relationship between ecosystem structure and function using structural equation modelling: A case study examining denitrification potential in restored wetland soils. *Ecol. Model.* **2010**, *221*, 761–768. [CrossRef]

36. McCune, B.; Grace, J.B. *Analysis of Ecological Communities*; MjM Software Design: Gleneden Beach, OR, USA, 2002.

37. Bollen, K.A. *Structural Equations with Latent Variables*; John Wiley & Sons: New York, NY, USA, 1989.

38. Joreskog, K.G.; Sorbom, D. *LISREL 8 User's Reference Guide*; Scientific Software International: Chicago, IL, USA, 1993.

39. Hau, K.T.; Cheng, Z.J. Application and analytical strategies of structural equation modeling. *Explor. Psychol.* **1999**, *19*, 54–59.

40. Schermelleh, K.; Moosbrugger, H. Evaluating the fit of structural equation models: Tests of significance and descriptive goodness-of-fit measures. *Methods Psychol. Res.* **2003**, *8*, 23–74.

41. Wu, M.L. *Structural Equation Model—The Operation and Application of AMOS*; Chongqing University Press: Chongqing, China, 2010.

42. Li, H.; Wang, J.K.; Pei, J.B.; Li, S.Y. Equilibrium relationships of soil organic carbon in the main croplands of northeast China based on structural equation modeling. *Acta Ecol. Sin.* **2015**, *35*, 517–525. [CrossRef]

43. Brahim, N.; Blavet, D.; Gallali, T.; Bernoux, M. Application of structural equation modeling for assessing relationships between organic carbon and soil properties in semiarid Mediterranean region. *Int. J. Environ. Sci. Technol.* **2011**, *8*, 305–320. [CrossRef]

44. Hanieh, S.; Lalit, K.; Russell, T.; Christine, S. Airborne LiDAR derived canopy height model reveals a significant difference in radiata pine (*Pinus radiata* D. Don) heights based on slope and aspect of sites. *Trees* **2014**, *28*, 733–744.

45. Eshetu, Y.; Mike, S.; Mesele, N.; Fantaw, Y. Influence of topographic aspect on floristic diversity, structure and treeline of afromontane cloud forests in the Bale Mountains, Ethiopia. *J. For. Res.* **2015**, *26*, 919–931.
46. Koichi, T.; Satomi, M. Morphological variations of the *Solidago virgaurea* L. complex along an elevational gradient on Mt Norikura, central Japan. *Plant Species Biol.* **2017**, *32*, 238–246.
47. Ziadat, F.M.; Taimeh, A.Y. Effect of rainfall intensity, slope, land use and antecedent soil moisture on soil erosion in an arid environment. *Land Degrad. Dev.* **2013**, *24*, 582–590. [CrossRef]
48. Tshering, D.; Inakwu, O.A.O.; Damien, J.F. Vertical distribution of soil organic carbon density in relation to land use/cover, altitude and slope aspect in the eastern Himalayas. *Land* **2014**, *3*, 1232–1250.
49. Ou, Z.Y.; Cao, J.Z.; Shen, W.H.; Tan, Y.B.; He, Q.F.; Peng, Y.H. Understory flora in relation to canopy structure, soil nutrients, and gap light regime: A case study in southern China. *Pol. J. Environ. Stud.* **2015**, *24*, 2559–2568. [CrossRef]
50. Lucas-Borja, M.E.; Hedo, J.; Cerdá, A.; Candel-Pérez, D.; Viñegla, B. Unravelling the importance of forest age stand and forest structure driving microbiological soil properties, enzymatic activities and soil nutrients content in Mediterranean Spanish black pine (*Pinus nigra Ar.* ssp. *salzmannii*) Forest. *Sci. Total Environ.* **2016**, *562*, 145–154. [CrossRef] [PubMed]

Article

Seasonal Changes in Photosynthetic Energy Utilization in a Desert Shrub (*Artemisia ordosica* Krasch.) during Its Different Phenophases

Cai Ren [1,2], Yajuan Wu [1,2], Tianshan Zha [1,2,*], Xin Jia [1,2], Yun Tian [2], Yujie Bai [2], Charles P.-A. Bourque [2,3], Jingyong Ma [2] and Wei Feng [1,2]

[1] Yanchi Research Station, School of Soil and Water Conservation, Beijing Forestry University, Beijing 100083, China; cairenbjfu@foxmail.com (C.R.); yajuanwu2015@sina.com (Y.W.); xinjia@bjfu.edu.cn (X.J.); weifeng@bjfu.edu.cn (W.F.)

[2] Key Laboratory for Soil and Water Conservation, State Forestry Administration, Beijing Forestry University, Beijing 100083, China; tianyun@bjfu.edu.cn (Y.T.); bloweer@bjfu.edu.cn (Y.B.); cbourque@unb.ca (C.P.-A.B.); majy2015@bjfu.edu.cn (J.M.)

[3] Faculty of Forestry and Environmental Management, 28 Dineen Drive, P.O. Box 4400, University of New Brunswick, Fredericton, NB E3B5A3, Canada

* Correspondence: tianshanzha@bjfu.edu.cn; Tel.: +86-10-6233-6608

Received: 26 February 2018; Accepted: 26 March 2018; Published: 29 March 2018

Abstract: Our understanding of the mechanisms of plant response to environment fluctuations during plants' phenological phases (phenophases) remains incomplete. Continuous chlorophyll fluorescence (ChlF) measurements were acquired from the field to quantify the responses in a desert shrub species (i.e., *Artemesia ordosica* Krasch. (*A. ordosica*)) to environmental factors by assessing variation in several ChlF-linked parameters and to understand plant acclimation to environmental stresses. Maximal quantum yield of PSII photochemistry (F_v/F_m) was shown to be reduced by environmental stressors and to be positively correlated to air temperature (T_a) during the early and late plant-growing stages, indicating a low temperature-induced inhibition during the leaf expansion and coloration phases. Effective quantum yield of PSII photochemistry (Φ_{PSII}) was negatively correlated to incident photosynthetically active radiation (PAR) irrespective of phenophase, suggesting excessive radiation-induced inhibition at all phenophases. The main mechanism for acclimating to environmental stress was the regulatory thermal dissipation (Φ_{NPQ}) and the long-term regulation of relative changes in Chl *a* to Chl *b*. The relative changes in photosynthetic energy utilization and dissipation in energy partitioning meant *A. ordosica* could acclimatize dynamically to environmental changes. This mechanism may enable plants in arid and semi-arid environments to acclimatize to increasingly extreme environmental conditions under future projected climate change.

Keywords: *Artemisia ordosica*; chlorophyll fluorescence; energy partitioning; phenophase; photoprotection

---

1. Introduction

Photosynthesis is a fundamental mechanism in plant metabolism that converts light energy into biochemical energy [1]. Excessive incident energy under high illumination usually leads to an accumulation of excitation pressure, which can impair photosystem II (PSII) and its associated light harvesting proteins, causing subsequent loss of PSII electron-transfer activity and finally inducing Photoinhibition [2–4].

Various tolerance and/or acclimation mechanisms exist in naturally grown plants to prevent damage to the photosynthetic apparatus from Photoinhibition [5]. For example, plants may (i) reduce the absorbed incident radiation; (ii) regulate photo-protective thermal dissipation of absorbed energy by means of the xanthophyll cycle; and/or (iii) modulate the chlorophyll content and the ratio of chlorophyll *a* to chlorophyll *b* [6,7]. The assessment of Photoinhibition and its tolerance and/or acclimation can be accurately determined by a chlorophyll fluorescence (ChlF) measurement on a long time scale using mostly PAM fluorometers [8,9] and by a measurement of the OJIP fluorescence rise as reviewed, e.g., by Lazár [10]. A vast number of studies have shown that the activity of PSII and its associated light harvesting proteins gradually declined during the entire time of exposure to high illumination and that non-photochemical quenching (NPQ) increased in response to excessive radiation energy [11–13]. However, such inhibition of photosynthesis can not only arise from exposure to high light in the absence of other stresses, but also from too much light in the presence of other stresses that limit photosynthesis and thus, plant growth [14]. Studies on extreme temperatures suggested that heat stress significantly inhibited the maximal (F_v/F_m) and effective quantum yield of PSII photochemistry (Φ_{PSII}), and decreased the relative leaf chlorophyll content [15–17]. When facing low temperature stress, however, plants showed a sustained NPQ and improved photo-protection, which is associated with the reorganization of the light harvesting complex into aggregates [18,19]. Studies on water-deficit conditions showed a significant decrease in leaf chlorophyll concentration, whereas decreasing F_v/F_m was only observed during the more advanced stages of dehydration [20,21]. For natural conditions, both processes of absorption and utilization of photosynthetic energy are complicated when extreme temperatures, water status, and other stress factors become enmeshed [22]. On the other hand, acclimation of plants to a stressor (e.g., to high temperature) might cause some resistance against acute action of the stressor—see, e.g., Lazár et al. [23]. It is commonly observed that multiple stress factors in naturally growing plants synergistically reduce the photochemical rate and cause greater stress than what would have resulted from a single factor [24,25]. Further understanding of the mechanisms associated with photosynthetic tolerance and/or acclimation to multiple stressors is needed for naturally growing plants, particularly for desert plants.

Deserts cover ~15% of the terrestrial surface and have been rapidly expanding due to climate change and human practices [26,27]. Cold-desert ecosystems are characterized by harsh environments and strong hydrometeorological gradients [28]. The absorption and utilization of photosynthetic energy by cold-desert plants are usually inhibited by high illumination, extreme temperature, water deficits, and other climatic irregularities [29]. Thus, plants in cold-desert ecosystems may produce lower amounts of photosynthate and tend to exhibit lower resilience and resistance to disturbances of similar severity than plants in environmentally-moderate ecosystems [30]. Most of the existing studies have focused on plants of the Mediterranean region and North America, where ecosystems are characterized by cold–wet winters and hot–dry summers [31,32]. The cold-desert regions in semi-arid China, however, differed climatically by having cold–dry winters and hot–wet summers due to the impact of the monsoon season, which could lead to different stress conditions. Spring and summer droughts are mostly common stresses in this region [33,34]. Cold-desert plant responses to environmental stress would provide additional insight into the mechanistic understanding of photosynthetic energy utilization by terrestrial plants.

Artemisia ordosica Krasch. (hereafter, *A. ordosica*), a long-lived, deciduous woody shrub species, is naturally dispersed throughout the Mu Us desert [35]. It is the main species providing wind and sand protection in arid- and semi-arid areas of northwest China. *A. ordosica* usually buds in early April and accumulates its greatest biomass in June, slowing its growth from July to defoliating in October [36]. The functional responses of *A. ordosica* are known to vary with plant phenological phases (or phenophases [37]) as diverse environmental conditions induce differential photochemical responses in photosynthetic apparatus [38]. We hypothesized that photosynthetic energy utilization of a cold-desert plant in response to environmental stress differs seasonally and varies as a function of phenophase. To test our hypothesis, ChlF of *A. ordosica* was measured continuously in situ during the growing season of 2014. The specific objectives of the study were to: (1) examine temporal dynamics of several ChlF-associated parameters and their responses to environmental factors during different phenophases; and (2) to understand the mechanisms associated with defining the species' tolerance and acclimation to environmental stresses in relation to its phenophases.

2. Materials and Methods

2.1. Site Characteristics

The experimental site is located in the Yanchi Desert Ecosystem Research Station (37.68° N to 37.73° N, 107.20° E to 107.26° E, 1570 m a.s.l.), northwestern China. The site is in a transitional region between the southern Mu Us desert and northern Loess plateau. It belongs to a mid-temperate, semi-arid region with a continental monsoon climate. From 1954 to 2004, the region's mean annual air temperature was 8.1 °C. The mean T_a from spring to autumn was 9.3, 21.3, and 7.7 °C, respectively [39]. Mean annual precipitation (PPT) was 287 mm, showing large seasonal (~80% falling during the June–September period) and inter-annual variation (133–572 mm year^{-1}). The area's soil type is classified as a sierozem, with >70% fine sand (0.02–0.20 mm). Mean annual potential evapotranspiration is >PPT. Subsurface water levels lie between 8–10 m from the ground surface and is currently decreasing. The dominant native shrub species of the region is *A. ordosica*.

2.2. Environmental Measurements

Environmental factors were measured simultaneously over the growing season from 12 April to 12 October 2014, including (1) photosynthetically active radiation (PAR) with a quantum sensor (PAR-LITE, Kipp and Zonen, Delft, The Netherlands); air temperature (T_a) and relative humidity (RH) with a thermohygrometer (HMP155A, Vaisala, Vantaa, Finland); soil water content (SWC) with ECH2O-5TE sensors (Decagon Devices, Pullman, WA, USA) set at depths of 10, 30, and 70 cm into the soil; and PPT with a tipping bucket rain gauge (TE525WS, Campbell Scientific Inc., Logan, UT, USA). All meteorological variables were measured every 10 s, and averaged or summed to generate 30-min values before being stored on data loggers (CR200X for PPT, CR3000 for all others, Campbell Scientific Inc.). Half-hourly values of vapor pressure deficit (VPD), representing air dryness, were calculated as defined by Wilhelm et al. [40]

$$VPD = e - e \times \frac{RH}{100} \tag{1}$$

$$e = 0.611 \times \exp^{\frac{17.27\,T_a}{237.3+T_a}} \tag{2}$$

where *e* is the partial water vapor pressure at the point of saturation.

Phenological observations were made using photographs taken twice each week. Phenophases include three obvious phases, i.e., (i) leaf-expanding phase during day of year (DOY) 102–147; (ii) leaf-expanded phase during DOY 148–247; and (iii) leaf-coloring phase during DOY 238–285. For details concerning phenological observations at the site, consult Chen et al. [41].

Chlorophyll *a* and *b* content were measured approximately once per week using spectrophotometry, UV-2100 (UNICO, China). Fresh south-facing leaves were collected from nearby plants similar to those being sampled. Two grams of clean leaves were weighed, ground into a powder with a small amount of quartz sand and calcium carbonate added. The powder was dissolved into a turbid fluid with dehydrated ethanol. The turbid fluid was subsequently filtered and calibrated to a volume of 200 mL. Finally, aqueous-chlorophyll extract was subsequently injected into a spectrophotometer using the energy spectrum from 645 to 663 nm to assess chlorophyll content. The values of chlorophyll content (ChlC, mg g^{-1}) and the ratio of chlorophyll *a* to chlorophyll *b* content (Chl *a/b*) were calculated with a modified Arnon equation set [42,43], i.e.,

$$Chl\ a = (12.71A_{663} - 2.59A_{645}) \times \frac{V}{1000 \times W} \tag{3}$$

$$\mathrm{Chl}\,b = (22.88A_{645} - 4.67A_{663}) \times \frac{V}{1000 \times W} \tag{4}$$

$$\mathrm{ChlC} = \mathrm{Chl}\,a + \mathrm{Chl}\,b \tag{5}$$

$$\mathrm{Chl}\,a/b = \mathrm{Chl}\,a\,/\,\mathrm{Chl}\,b \tag{6}$$

where A_{663} and A_{645} are the light absorption at 663 and 645 nm, respectively, V (mL) is the volume of the extract, and W (g) is the weight of fresh leaves.

2.3. ChlF Measurements, Data Treatment, and Analysis

Field measurements of ChlF parameters were made from three fully-grown *A. ordosica* plants, which averaged 50 cm in height and 80 × 60 cm in canopy size. Three Monitoring-PAM detectors (MONI-head; Walz, Effeltrich, Germany) were individually set on a fascicle of randomly chosen leaves (~12 leaves) in the middle of south-facing stems on each of three sampled shrubs. Three detector heads represent three replicates of the same measurement. ChlF-associated parameters (i.e., effective and maximum fluorescence values, F' and $F_{m'}$, respectively) were recorded with modulated light pulses. The actinic light driving photosynthesis was the light from the Sun. The time interval between measurements was 10 min. Measurements were stored in memory of the PAM analyzer (MONI-DA). The PAM system was powered by a solar panel.

From the two basic fluorescence parameters (i.e., F' and $F_{m'}$), defined by the mean of three replicate, simultaneous measurements, other parameters can be calculated [44]. Effective quantum yield of PSII photochemistry (Φ_{PSII}, Equation (7)) is a light-adapted parameter that reflects the effective portion of absorbed quanta used in PSII. Maximal quantum yield of PSII photochemistry (F_v/F_m, Equation (8)) is a dark-adapted parameter, which estimates the maximum portion of absorbed quanta used in PSII. Stern–Volmer non-photochemical fluorescence quenching (NPQ, Equation (9)) is a parameter that reflects heat-dissipation of excitation energy [45]. The three ChlF-associated parameters were calculated as

$$\Phi_{PSII} = \frac{F_{m'} - F'}{F_{m'}} \tag{7}$$

$$F_V/F_m = \frac{F_m - F_o}{F_m} \tag{8}$$

$$NPQ = \frac{F_m - F_{m'}}{F_{m'}} \tag{9}$$

where F' is the steady-state fluorescence in ambient light and $F_{m'}$ is the maximum fluorescence in ambient light following a saturating light pulse. The intensity of the measuring light and the saturating pulses were 0.9 µmol m^{-2} s^{-1} and 3500 µmol m^{-2} s^{-1}, respectively. Daytime F' and $F_{m'}$ become nighttime F_o and F_m. Nighttime was defined as the period with PAR < 5 µmol m^{-2} s^{-1} [46,47].

The approach of energy partitioning was introduced and developed by researchers (i.e., Genty, Cailly, Kramer, Hendrickson, etc.; [48–51]) to discriminate different types of non-photochemical losses [52]. Parameter Φ_{NPQ} is the quantum yield of regulatory (light-induced) non-photochemical quenching that represents thermal dissipation related to all photo-protective mechanisms, whereas Φ_{NO} is the quantum yield of constitutive non-regulatory (basal or dark) energy dissipation, which represents all other components of non-photochemical quenching that are not photo-protective [52], i.e.,

$$\Phi_{NO} = \frac{1}{NPQ + 1 + qL\left(\frac{F_m}{F_o} - 1\right)} \tag{10}$$

$$\Phi_{NPQ} = 1 - \Phi_{PSII} - \Phi_{NO} \tag{11}$$

where qL is coefficient of photochemical quenching in the lake model, which is taken to describe the energetic connectivity in terrestrial plants, which is a better estimate than that provided in the puddle model, as reviewed by Lazár [53]. Parameter $F_{o'}$ is the minimal fluorescence in light-adapted conditions, representing F_o with non-photochemical quenching [54]. Both parameters were calculated as

$$qL = \frac{F_{m'} - F'}{F_{m'} - F_{o'}} \left(\frac{F_{o'}}{F'} \right) \tag{12}$$

$$F_{o'} = \frac{F_o}{\frac{F_v}{F_m} + \frac{F_o}{F_{m'}}} \tag{13}$$

Rainy days were removed from the analysis to avoid rainfall-pulse disturbances. A Pearson correlation analysis over the entire growing season was conducted in MATLAB (MathWorks, Natick, MA, USA), largely to quantify correlations between ChlF-associated parameters and meteorological factors. Generalized linear regression was used to quantify response sensitivity during each phenophase [46]. Multivariate stepwise regression was used to further analyze interaction between individual meteorological factors on ChlF-associated parameters [32].

3. Results

3.1. Phenological Dynamics in Biophysical Factors

Daily-integrated PPT, daily means of PAR, T_a, RH, and SWC at three soil depths showed distinct patterns for the different phenophases (Figure 1). Daily mean PAR increased from 108.0 μmol m^{-2} s^{-1} (DOY 108) during the early leaf-expanding phase to a maximum of 591.0 μmol m^{-2} s^{-1} (DOY 195) during the leaf-expanded phase, and then decreased to 98.6 μmol m^{-2} s^{-1} (DOY 284) during the late leaf-coloring phase (Figure 1a). Daily mean T_a increased from 8.6 °C (DOY 108) during the early leaf-expanding phase to a maximum of 25.8 °C (DOY 214) during the leaf-expanded phase, and then decreased to 3.5 °C (DOY 284) during the late leaf-coloring phase (Figure 1b). Daily mean RH decreased from 49.9% (DOY 102) during the early leaf-expanding phase to a minimum of 13.8% (DOY 147) during the late leaf-expanding phase, and increased to 80% during the leaf-expanded and coloring phases (Figure 1b). Daily mean VPD varied in a manner similarly observed to that of T_a (Figure 1c). Total PPT during the observation period was 295.9 mm. Rainy days occurred about 24% of the time (11 of 45 days), 40% (31 of 77 days), and 48% (23 of 48 days) during the leaf-expanding, expanded, and coloring phases, respectively. Concomitant PPT-amounts during the three phenophases accounted for 9.3, 52.0, and 38.7% of the total PPT falling during the growing season (Figure 1d). Soil water content at a soil depth of 10 and 30 cm responded instantly to rain pulses exceeding 5 mm. The response decreased with an increase in soil depth during the growing season, and showed nominal fluctuation (<0.01 m^3 m^{-3}) at depths $\geq$70 cm (Figure 1d). In general, the annual mean of each of meteorological parameters was basically close to its multi-year average. The leaf-expanding phase occurred during a dry and cold period, and the leaf-expanded phase occurred during a high radiation period. Leaf coloring, in contrast, occurred during rainy and cold conditions.

3.2. Phenological Variations in ChlF Parameters

Figure 2 shows the seasonal dynamics of ChlF-associated parameters in vivo. Daily mean F_v/F_m increased from a minimum of 0.66 (DOY 107) to 0.78 during the leaf-expanding phase, then decreased to 0.67 (DOY 190), and fluctuated at 0.76 subsequently (Figure 2a), averaging 0.71 $\pm$ 0.08, 0.76 $\pm$ 0.03, 0.76 $\pm$ 0.03 ($\pm$standard deviation) during the leaf-expanding, expanded, and coloring phases, respectively. Daily mean Φ_{PSII} demonstrated a similar trend to that observed with F_v/F_m, with low values occurring in the early leaf-expanding and leaf-expanded phases (Figure 2b). The seasonal pattern in ChlC was consistent with the pattern observed in F_v/F_m; Chl a/b varied in an opposite direction from that of ChlC (Figure 2c,d). A drastic change of ChlC and Chl a/b occurred from DOY 210 to 240, after which Chl a/b increased to a maximum of 6.0 (DOY 284) during the late leaf-coloring phase (Figure 2c,d).

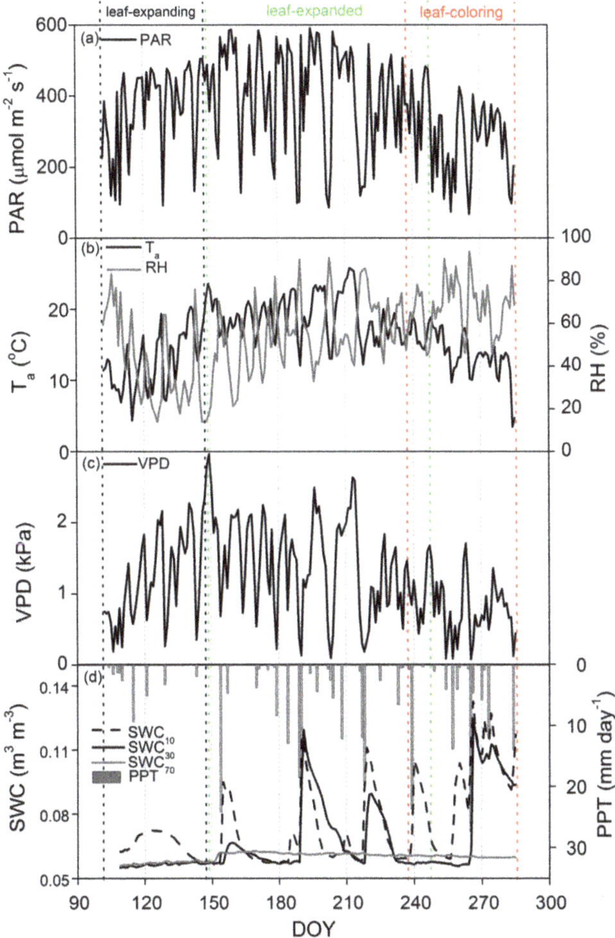

Figure 1. Time series of meteorological variables: daily mean incident photosynthetically active radiation (PAR, (**a**)), air temperature (T_a, (**b**)), relative humidity (RH, (**b**)), atmospheric vapor pressure deficit (VPD, (**c**)), soil water content (SWC, (**d**)) at three soil depths (10, 30, and 70 cm), and daily-cumulated rainfall (PPT, (**d**)) during the growing season (DOY 102 to 285) of 2014. Black, green, and red vertically dotted lines represent leaf-expanding, expanded, and coloring phenophases.

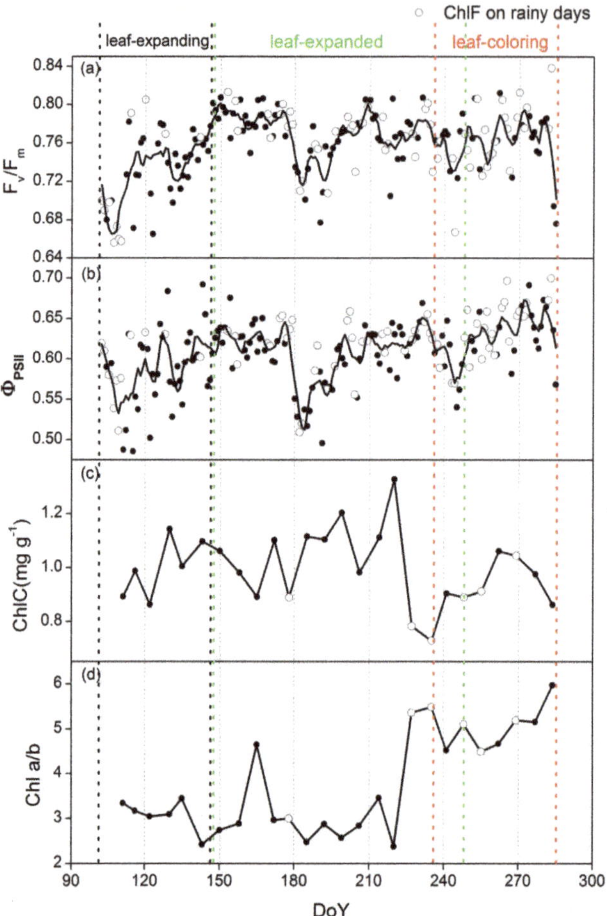

Figure 2. Time series of ChlF-associated parameters, i.e., maximal quantum yield of PSII photochemistry (F_v/F_m, (**a**)), daily mean effective quantum yield of PSII photochemistry (Φ_{PSII}, (**b**)), sum content of chlorophyll a and b (ChlC, (**c**)), and ratio of chlorophyll a to chlorophyll b (Chl a/b, (**d**)) during the growing season (DOY 102 to 285) of 2014. Open circles represent ChlF-associated parameter values during rainy days. Major trend lines for F_v/F_m and Φ_{PSII} were modified according to the Savitzky-Golay method. Black, green, and red vertically dotted lines represent leaf-expanding, expanded, and coloring phenophases.

3.3. Proportion of Energy Partitioning and Response of ChlF to Abiotic Factors

Daily response of ChlF was modified by different meteorological factors. Generally, F_v/F_m increased with increasing daily mean of T_a during the leaf-expanding and coloring phases (Figure 3a). The slopes of F_v/F_m-to-T_a regressions for the leaf-expanding phase were similar to those observed for the leaf-coloring phase (Table 1). Multivariate regressions confirmed T_a as the main controlling factor of F_v/F_m during both phenophases (Table 2). Regression of ChlF with ChlC as independent variable showed that ChlC was positively correlated to F_v/F_m during the leaf-expanding and coloring phases (Figure 3b).

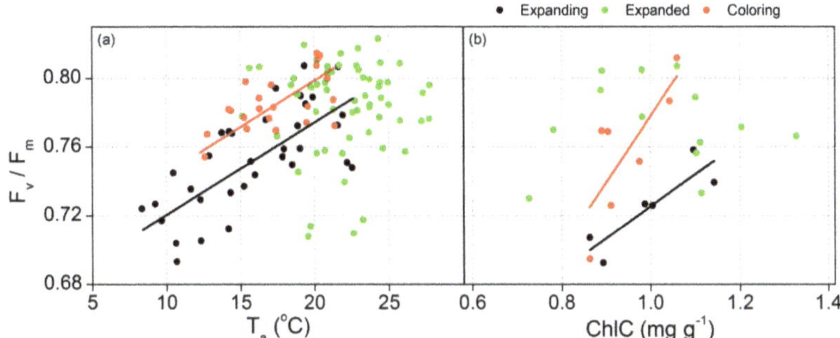

Figure 3. Relationship between maximal quantum yield of PSII photochemistry (F_v/F_m) and daily mean air temperature (T_a, (**a**)), and the sum content of chlorophyll *a* and chlorophyll *b* (ChlC, (**b**)) during each phenophase for the 2014 growing season. Black, green, and red colored circles and lines coincide with values and fitting functions (Table 1) for the leaf-expanding, expanded, and coloring phases, respectively. Data from rainy days were excluded from the analysis.

Table 1. Regression equations between meteorological variables and ChlF-associated parameters during the growing season (DOY 102 to 285) of 2014. Independent variables include daily mean incident photosynthetically active radiation (PAR), air temperature (T_a), relative humidity (RH), atmospheric vapor pressure deficit (VPD), sum content of chlorophyll *a* and chlorophyll *b* (ChlC), and ratio of chlorophyll *a* to chlorophyll *b* (i.e., Chl *a/b*). Dependent variables include maximal quantum yield of PSII photochemistry (F_v/F_m), daily mean effective quantum yield of PSII photochemistry (Φ_{PSII}), Stern–Volmer non-photochemical fluorescence quenching (NPQ), and quantum yield of regulatory light-induced non-photochemical quenching (Φ_{NPQ}). *p*-values represent significant level for the regression models. r^2 is the coefficient of determination.

	Factors	Phase	Equation	r^2	p
F_v/F_m	T_a	Expanding	$y = 0.67 + 53.7 \times 10^{-4}x$	0.51	<0.05
		Coloring	$y = 0.69 + 54.3 \times 10^{-4}x$	0.50	0.01
	ChlC	Expanding	$y = 0.54 + 0.19x$	0.71	0.02
		Coloring	$y = 0.39 + 0.38x$	0.52	0.04
Φ_{PSII}	T_a	Expanding	$y = 0.52 + 45.9 \times 10^{-4}x$	0.25	0.02
	PAR	Expanding	$y = 0.67 - 1.81 \times 10^{-4}x$	0.14	0.02
		Expanded	$y = 0.67 - 1.29 \times 10^{-4}x$	0.17	<0.05
		Coloring	$y = 0.72 - 2.45 \times 10^{-4}x$	0.44	<0.05
NPQ	Chl *a/b*	Expanding	$y = 2.70 - 0.68x$	0.78	0.01
		Expanded	$y = 1.06 - 0.12x$	0.30	0.02
Φ_{NPQ}	RH	Expanding	$y = 22.85 - 0.18x$	0.34	<0.05
		Expanded	$y = 23.73 - 0.09x$	0.14	<0.05
	VPD	Expanding	$y = 11.82 + 4.20x$	0.25	<0.05
		Expanded	$y = 15.99 + 2.44x$	0.15	<0.05
		Coloring	$y = 11.26 + 4.96x$	0.41	<0.05
	PAR	Expanding	$y = 4.23 + 0.03x$	0.48	<0.05
		Expanded	$y = 9.84 + 0.02x$	0.42	<0.05
		Coloring	$y = 6.87 + 0.03x$	0.58	<0.05

Table 2. Multi-variate linear regressions between meteorological variables and ChlF-associated parameters during the growing season (DOY 102 to 285) of 2014. Independent variables include daily mean incident photosynthetically active radiation (PAR), air temperature (T_a), relative humidity (RH), and atmospheric vapor pressure deficit (VPD). Dependent variables include maximal quantum yield of PSII photochemistry (F_v/F_m), daily mean effective quantum yield of PSII photochemistry (Φ_{PSII}), and quantum yield of regulatory light-induced non-photochemical quenching (Φ_{NPQ}). p-values represent significant level for the regression models. r^2 is the coefficient of determination.

	Factors	Phase	Equation	r^2	p
F_v/F_m	T_a	Expanding	$y = 56.77 \times 10^{-4}x + 0.66$	0.45	<0.01
		Expanded	-	-	
		Coloring	$y = 30.14 \times 10^{-4}x + 0.73$	0.26	0.01
Φ_{PSII}	PAR (x_1), T_a (x_2) VPD(x_3)	Expanding	$y = -3.54 \times 10^{-4}x_1 + 78.34 \times 10^{-4}x_2 + 0.61$ ($-0.70x_1 + 0.85x_2$ in standardized form)	0.67	<0.01
		Expanded	$y = -2.22 \times 10^{-4}x_1 + 314.6 \times 10^{-4}x_3 + 0.66$ ($-0.70x_1 + 0.49x_3$ in standardized form)	0.31	<0.01
		Coloring	-	-	
Φ_{NPQ}	PAR	Expanding	$y = 0.026x + 7.73$	0.32	<0.01
		Expanded	$y = 0.016x + 13.02$	0.28	<0.01
		Coloring	$y = 0.023x + 8.44$	0.46	<0.01

Daily mean Φ_{PSII} increased with increasing T_a during the leaf-expanding phase (Figure 4a), but decreased with increasing PAR irrespective of phenophase (Figure 4b). The slope of the Φ_{PSII}-to-PAR relationship was lowest during the leaf-expanded phase (Table 1). Multivariate regressions indicated that the relative effect of PAR on Φ_{PSII} was small than that of T_a during the leaf-expanding phase (standardized coefficient of -0.70 for PAR vs. 0.85 for T_a), but larger than that of VPD during the leaf-expanded phase (standardized coefficient of -0.70 for PAR vs. 0.49 for VPD; Table 2). Chl a/b was negatively correlated to NPQ during the leaf-expanding and expanded phases (Figure 5).

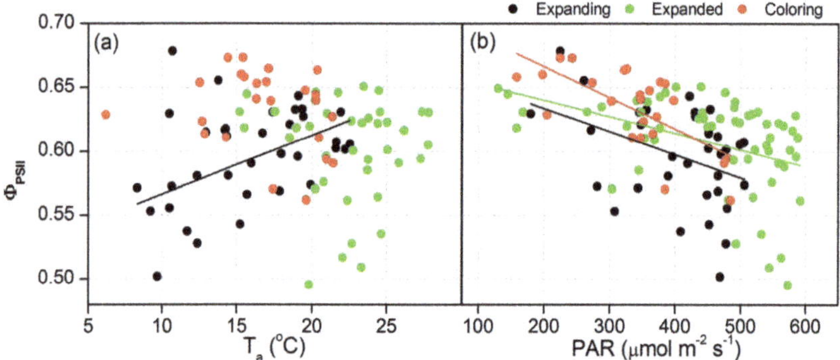

Figure 4. Relationship between daily mean effective quantum yield of PSII photochemistry (Φ_{PSII}), daily mean air temperature (T_a, (**a**)), and incident photosynthetically active radiation (PAR, (**b**)) during each phenophase during the 2014 growing season. Black, green, and red colored circles and lines coincide with values and fitting functions (Table 1) for the leaf-expanding, expanded, and coloring phases, respectively. Data from rainy days were excluded from the analysis.

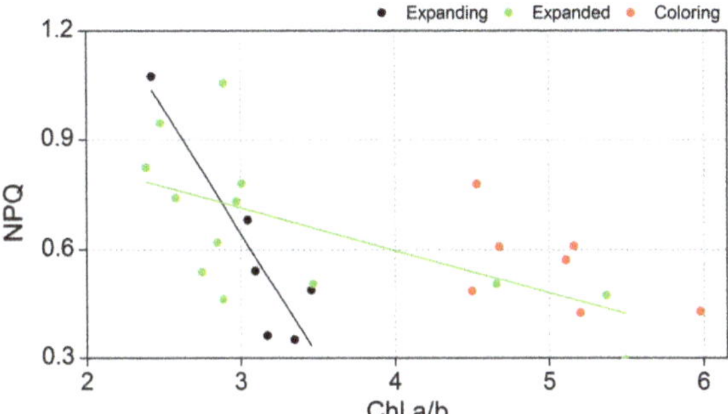

Figure 5. Relationship between Stern–Volmer non-photochemical fluorescence quenching (NPQ) and the ratio of chlorophyll a to chlorophyll b content (i.e., Chl a/b) in each phenophase during the 2014 growing season. Black, green, and red colored circles and lines coincide with values and fitting functions (Table 1) for the leaf expanding, expanded, and coloring phases, respectively. Data from rainy days were excluded from the analysis.

Figure 6 shows the dynamics of energy partitioning in PSII. The energy partitioning of Φ_{PSII} ranged from 48.6% to 71.4%, 49.6% to 67.6%, and 54.0% to 70.0% during the leaf-expanding, expanded and coloring phases, respectively, averaging $59.3 \pm 5.2\%$, $60.3 \pm 3.9\%$, and $63.0 \pm 3.8\%$ ($\pm$standard deviation). The proportion of Φ_{NPQ} increased from 10% (DOY 107) during the early leaf-expanding phase to a maximum level of 25.0% (DOY 190), and declined to 10% (DOY 284) during the late leaf-coloring phase, averaging $16.3 \pm 5.7\%$, $18.7 \pm 3.9\%$, and $15.4 \pm 3.7\%$ during the three phenophases, respectively. Parameter Φ_{NPQ} exhibited greater variance (i.e., larger standard deviations) during the leaf-expanding phase. Days with Φ_{NO} within the range of 20–30% accounted for 74% (34 of 46 days), 94.8% (73 of 77 days), and 95.8% (46 of 48 days) during the three phenophases.

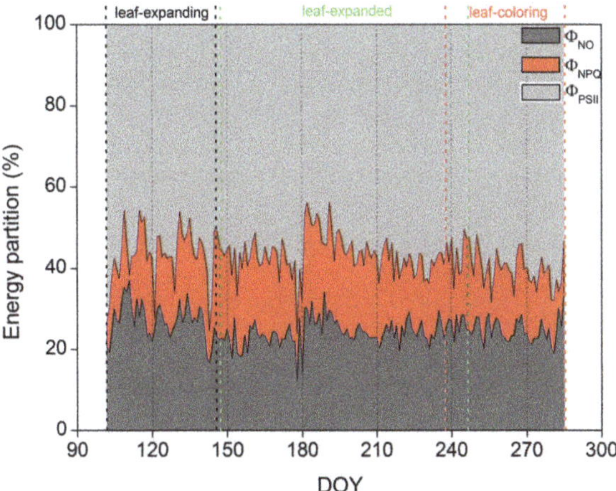

Figure 6. Timeseries of percentage of energy partitioning during the 2014 growing season. The red area is the quantum yield of regulatory (light-induced) energy dissipation (Φ_{NPQ}); the gray area is quantum yield of constitutive non-regulatory (basal or dark) energy dissipation (Φ_{NO}); and the light gray area is the quantum yield of light-induced photochemical quenching, which is equal to effective quantum yield of PSII photochemistry (Φ_{PSII}). Each phenophase is identified with a different color of dot lines.

Daily mean Φ_{NPQ} showed a correlation with PAR, T_a, and atmospheric water-related factors (i.e., RH and VPD; Figure 7). Daily mean Φ_{NPQ} decreased with increasing RH during the leaf-expanding and coloring phases (Figure 7a), whereas it increased with increasing VPD and PAR during the three phenophases (Figure 7b,c). Multivariate regressions showed PAR as the main controlling factor of Φ_{NPQ} during the three phenophases (Table 2).

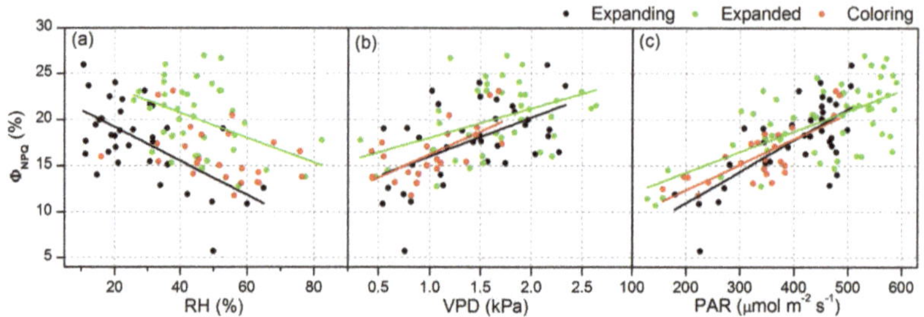

Figure 7. Response of quantum yield of regulatory light-induced non-photochemical quenching (Φ_{NPQ}) to daily mean air relative humidity (RH, (**a**)), atmospheric vapor pressure deficit (VPD, (**b**)), and incident photosynthetically active radiation (PAR, (**c**)) for each phenophase. Black, green, and red colored circles and lines coincide with values and fitting functions (Table 1) for the leaf-expanding, expanded, and coloring phases, respectively. Data from rainy days were excluded from the analysis.

4. Discussion

4.1. Effect of Environmental Stresses on F_v/F_m, Φ_{PSII}, and NPQ

Parameter F_v/F_m is commonly viewed as a reliable indicator of inhibition and of recovery from environmental stress. Variations in F_v/F_m can indicate photochemical regulation of PSII in response to stress [55]. Our results show that F_v/F_m fluctuated seasonally for 162 days of the 184 days of the growing season with a value below 0.80~0.83 in non-stressed environments [46,56–58]. Fluctuations coincided to inhibition by environmental stress and subsequent recovery as stress diminished (Figure 2). The inhibition occurred frequently in response to cold temperatures during the leaf-expanding and coloring phases and to summer drought during the leaf-expanded phase (Figures 1d and 2a). The shrubs were generally capable to recover to an optimal state when leaves were fully expanded (Figure 2). Presence of large fluctuations in F_v/F_m during spring implied greater inhibition during the leaf-expanding phase. A similar seasonal pattern in F_v/F_m to Φ_{PSII} indicated that stressful environmental conditions reduced both PSII photochemical capacity and efficiency. The averages of F_v/F_m for the three phenophases were 0.71 ± 0.08, 0.76 ± 0.03, and 0.76 ± 0.03, close to the non-inhibition range, indicating *A. ordosica* can generally recover from inhibition and well adapted to the prevailing desert conditions.

Air temperature is a crucial factor affecting F_v/F_m or photochemical capacity during the leaf-expanding and coloring phases (Table 2, Figure 3a). The result that F_v/F_m linearly increased with increasing daily mean T_a (Figure 3a), indicated that photochemical capacity of PSII was downregulated by cold temperature during the early and late growing phases, suggesting probable low-temperature stress and capability of recovery as temperature increased. A similar relationship between F_v/F_m and ChlC indicated that the synthesis of chlorophyll coincided with an increase in photochemical capacity (Figure 3b). It is thus assumed that cold temperature during the early and late growing season reduced the chlorophyll content, thus reducing photochemical capacity and making the plant acclimatize to low temperatures. Some studies also reported that the PSII photochemical capacity of new leaves was often inhibited by a photo-oxidative cold-shock in early spring [59,60]. It was noted that cold temperature only reduced photochemical efficiency during the leaf-expanding rather than leaf-coloring phase (Figures 3a and 4a). A reason for this may be due to physiological senescence around leaf coloration. The drastic changes of ChlC and Chl a/b was the result of the concomitant decrease of Chl a and b during the end of leaf expanded phase, with larger decrease in Chl b than Chl a, when PPT was high and temperature was low. These environmental conditions could activate the synthesis of abscisic acid [61] and induce the foliar defoliation process.

Variations in Φ_{PSII} reflected responsive sensitivity of the photosynthetic apparatus to abrupt ambient fluctuations [12]. Incident radiation has been shown to play an important role in modifying actual plant photochemical efficiency (Figure 4). Our results show that Φ_{PSII} decreased with increasing daily mean PAR during the three phenophases (Figure 4b), indicating that the reduction in photochemical efficiency was driven by excessive solar radiation during the growing season. Multivariate regression disclosed the changes in the relative role of PAR to T_a and VPD in affecting Φ_{PSII} during the leaf-expanding and expanded phases (Table 2), varying in opposing ways with PAR, T_a, and VPD controlling Φ_{PSII} during the leaf-expanding and expanded phases, respectively. Extreme temperatures were shown by some researchers to influence the fluidity and enzyme activity in and through the thylakoid membrane [62,63], thus dominating the acclimation of PSII to variable solar radiation during the leaf-expanding phase. Moreover, low temperature and high radiation could have a synergistic effect on photochemical efficiency when incident radiation increases beyond plant physiological needs on sunny days [64]. This can exacerbate reductions in photochemical efficiency compared to reductions associated with single stressors, leading to large reductions in photochemical capacity. Parameter NPQ decreased with increasing Chl a/b during the leaf-expanding and expanded phenophases (Figure 5), suggesting that the capacity of thermal dissipation through the consumption of excessive energy was regulated with changes in Chl a to Chl b content. This was because the majority of chlorophyll b is associated with light-harvesting complex proteins, which have an intrinsic ability to switch from an efficient light-harvesting state to a photo-protective state as a long-term strategy of acclimation to stressful conditions [65,66].

As a rule, low temperatures and high irradiance cause stress in plants during the leaf-expanding and expanded phenophases. However, plants are largely able to prevent and/or alleviate inhibition by thermal dissipation and regulation of changes in Chl a to Chl b content.

4.2. Acclimation of Energy Partitioning to Environmental Stresses

By dividing the Stern–Volmer NPQ parameter into quantum yield of regulatory and constitutive non-photochemical quenching, energy partitioning can provide quantitative information about the fraction of energy that is dissipated as heat [52,67]. The mean energy fraction of Φ_{PSII} in this study (i.e., 60%) was larger than that of the near 40% observed in some Mediterranean species [68,69], indicating a higher efficiency of energy utilization in *A. ordosica*. However, our result was slightly less than that of the approximately 65–73% for some plant species in mid-to-low latitude regions [51,70], due to the high inhibition observed at our site. Φ_{NO} values (Figure 6) in our study were generally within the range 20–30% reported for healthy green leaves [71], suggesting physiological resistance of *A. ordosica* during most of the growing season.

Energy partitioning was affected by the various abiotic factors. Low RH and high VPD was shown to produce high Φ_{NPQ} (Figure 7a,b), suggesting that the fraction of photo-protective thermal dissipation was increased as the air became drier. Water availability is reportedly one of the most common limitations to energy partitioning of incident radiation for arid and semi-arid plants [69]. The reason for water stress was presumably associated with low soil moisture induced stomata closure, increasing the trans-thylakoid proton gradient and mediating xanthophyll de-epoxidation, leading to increased protective thermal dissipation [72]. Specifically, during the leaf-expanding phase, RH was shown to explain more of the variation in Φ_{NPQ} than did VPD, assuming increasing photo-protection with dehydration stress. In contrast, due to the synergy of temperature and moisture, VPD explained more of the variation in Φ_{NPQ} than did RH during the leaf-coloring phase. We did not find significant correlations between soil water content and photosynthetic efficiency. This may be due to inadequate water gradients and synergistic regulation by other meteorological factor (i.e., incident radiation and temperature). Further study is suggested to see how soil water content affects the photosynthetic efficiency by control experiment.

Incident radiation served as a controlling factor in the process of photosynthetic energy partitioning (Table 2). Parameter Φ_{NPQ} increased with increasing PAR in all phenophases (Figure 7c), suggesting that the fraction of photo-protective thermal dissipation was enhanced by relieving excitation pressure under conditions of excessive light. Some studies suggested that excitation pressure can be relaxed by the xanthophyll cycle, in which violaxanthin was de-epoxidated to zeaxanthin [73,74]. Concomitantly, excessive energy was dissipated and thylakoid pH difference was regulated to a moderate level. Reversible conversion between violaxanthin and zeaxanthin in the xanthophyll cycle can be viewed as a rapid response of the PSII system to variable irradiation.

In summary, the fraction of photosynthetic energy utilization and dissipation were dynamically regulated by simultaneous changes in the environment. Photosynthetic energy conversion can switch to a beneficial or protective state by regulating the relative changes in photochemical energy and thermal dissipation [75]. Through this regulation, plants can optimize the energy partitioning in a dynamic balance between the cost of photosynthetic utilization and the risk of photo-oxidative damage [76]. Strategy of dynamic acclimation can improve resistance to environmental extremes.

5. Conclusions

Inhibition happened throughout the growing season, being largely induced by cold air temperature (T_a) during the leaf-expanding and coloring phases and PAR in all phenophases. *A. ordosica* acclimated to stressful environmental conditions by way of non-photochemical quenching (NPQ) that was associated with reversible regulatory thermal dissipation and long-term regulation of relative decrease in Chl *a* to Chl *b* content. The fraction of thermal dissipation varied in response to PAR and atmospheric- and soil-water conditions. Thermal dissipation against excitation pressure fluctuated more greatly during the leaf-expanding phase relative to the leaf-expanded and coloring phases. *A. ordosica* was shown to undergo dynamic acclimation to changes in its external environment during most of its phenophases. Climate change may cause more extreme weather events in arid and semi-arid regions, leading to unfavorable environmental conditions for plant growth [77]. Our work added to the understanding of acclimation of desert species to changing climate. Further research is suggested to understand the acclamatory responses to various environmental stresses with different intensities and durations based on multi-year, continuous measurements of ChlF parameters.

Acknowledgments: We acknowledge the support obtained from National Natural Science Foundation of China (NSFC) (no. 31670710 and 31670708), the National Key Research and Development Program of China (no. 2016YFC0500905), and the Fundamental Research Funds for the Central Universities (no. 2015ZCQ-SB-02). We thank M. Y. Zhang, P. Liu, and S. J. Liu for their assistance with the field work. The U.S–China Carbon Consortium (USCCC) supported this work by way of helpful discussion and exchange of ideas. We greatly appreciate the valuable comments from all anonymous reviewers of this manuscript.

Author Contributions: T.Z. and X.J. designed the experiment; C.R. and Y.W. did the measurements, analyzed the data, and led the writing of the manuscript; C.P.-A.B., J.M., W.F., Y.B., and Y.T. discussed the results and contributed to the revision of the manuscript. All authors approved this manuscript for publication.

Conflicts of Interest: The authors declare no conflict of interest.

References

1. Ensminger, I.; Busch, F.; Huner, N. Photostasis and cold acclimation: Sensing low temperature through photosynthesis. *Physiol. Plant.* **2006**, *126*, 28–44. [CrossRef]
2. Yakushevska, A.E.; Jensen, P.E.; Keegstra, W.; van Roon, H.; Scheller, H.V.; Boekema, E.J.; Dekker, J.P. Supermolecular organization of photosystem II and its associated light-harvesting antenna in *Arabidopsis thaliana*. *FEBS. J.* **2001**, *268*, 6020–6028. [CrossRef]
3. Demmig-Adams, B.; Cohu, C.M.; Muller, O.; Adams, W.W. Modulation of photosynthetic energy conversion efficiency in nature: From seconds to seasons. *Photosynth. Res.* **2012**, 1–14. [CrossRef] [PubMed]

4. Ware, M.A.; Belgio, E.; Ruban, A.V. Photoprotective capacity of non-photochemical quenching in plants acclimated to different light intensities. *Photosynth. Res.* **2015**, *126*, 261–274. [CrossRef] [PubMed]

5. Zhou, Y.; Lam, H.M.; Zhang, J. Inhibition of photosynthesis and energy dissipation induced by water and high light stresses in rice. *J. Exp. Bot.* **2007**, *58*, 1207–1217. [CrossRef] [PubMed]

6. Müller, P.; Li, X.P.; Niyogi, K.K. Non-photochemical quenching. A response to excess light energy. *Plant Physiol.* **2001**, *125*, 1558–1566. [CrossRef] [PubMed]

7. Savitch, L.V.; Ivanov, A.G.; Gudynaite-Savitch, L.; Huner, N.P.; Simmonds, J. Effects of low temperature stress on excitation energy partitioning and photoprotection in Zea mays. *Funct. Plant Biol.* **2009**, *36*, 37–49. [CrossRef]

8. Porcar-castell, A.; Pfündel, E.; Korhonen, J.F.; Juurola, E. A new monitoring pam fluorometer (moni-pam) to study the short- and long-term acclimation of photosystem II in field conditions. *Photosynth. Res.* **2008**, *96*, 173–179. [CrossRef] [PubMed]

9. Durako, M.J. Using PAM fluorometry for landscape-level assessment of *Thalassia testudinum*: Can diurnal variation in photochemical efficiency be used as an ecoindicator of seagrass health? *Ecol. Indic.* **2012**, *18*, 243–251. [CrossRef]

10. Lazár, D. The polyphasic chlorophyll a fluorescence rise measured under high intensity of exciting light. *Funct. Plant Biol.* **2006**, *33*, 9–30. [CrossRef]

11. Harel, Y.; Ohad, I.; Kaplan, A. Activation of photosynthesis and resistance to photoinhibition in cyanobacteria within biological desert crust. *Plant Physiol.* **2004**, *136*, 3070–3079. [CrossRef] [PubMed]

12. Demmig-Adams, B.; Garab, G.; Adams, W., III; Govindjee. *Non-Photochemical Quenching and Energy Dissipation in Plants, Algae and Cyanobacteria*; Springer: Dordrecht, The Netherlands, 2014. [CrossRef]

13. Ruban, A.V. Nonphotochemical chlorophyll fluorescence quenching: Mechanism and effectiveness in protecting plants from photodamage. *Plant Physiol.* **2016**, *170*, 1903–1916. [CrossRef] [PubMed]

14. Adams, W.W., III; Muller, O.; Cohu, C.M.; Demmig-Adams, B. May photoinhibition be a consequence, rather than a cause, of limited plant productivity? *Photosynth. Res.* **2013**, *117*, 31–44. [CrossRef] [PubMed]

15. Haque, M.S.; Kjaer, K.H.; Rosenqvist, E.; Sharma, D.K.; Ottosen, C.O. Heat stress and recovery of photosystem II efficiency in wheat (*Triticum aestivum* L.) cultivars acclimated to different growth temperatures. *Environ. Exp. Bot.* **2014**, *99*, 1–8. [CrossRef]

16. Kalaji, H.M.; Jajoo, A.; Oukarroum, A.; Brestic, M.; Zivcak, M.; Samborska, I.A.; Cetner, M.D.; Łukasik, I.; Goltsev, V.; Ladle, R.J. Chlorophyll a fluorescence as a tool to monitor physiological status of plants under abiotic stress conditions. *Acta Physiol. Plant* **2016**, *38*, 102. [CrossRef]

17. Marias, D.E.; Meinzer, F.C.; Woodruff, D.R.; McCulloh, K.A. Thermo tolerance and heat stress responses of Douglas-fir and ponderosa pine seedling populations from contrasting climates. *Tree Physiol.* **2016**, *37*, 301–315. [CrossRef]

18. Öquist, G.; Huner, N.P. Photosynthesis of overwintering evergreen plants. *Annu. Rev. Plant Biol.* **2003**, *54*, 329–355. [CrossRef] [PubMed]

19. Liang, Y.; Chen, H.; Tang, M.J.; Yang, P.F.; Shen, S.H. Responses of *Jatropha curcas* seedlings to cold stress: Photosynthesis-related proteins and chlorophyll fluorescence characteristics. *Physiol. Plant.* **2007**, *131*, 508–517. [CrossRef] [PubMed]

20. Chaves, M.M.; Pereira, J.S.; Maroco, J.; Rodrigues, M.L.; Ricardo, C.P.P.; Osório, M.L.; Carvalho, I.; Faria, T.; Pinheiro, C. How plants cope with water stress in the field? Photosynthesis and growth. *Ann. Bot.* **2002**, *89*, 907–916. [CrossRef] [PubMed]

21. Ditmarová, L'.; Kurjak, D.; Palmroth, S.; Kmeť, J.; Střelcová, K. Physiological responses of Norway spruce (*Picea abies*) seedlings to drought stress. *Tree Physiol.* **2009**, *30*, 205–213. [CrossRef] [PubMed]

22. Suzuki, N.; Rivero, R.M.; Shulaev, V.; Blumwald, E.; Mittler, R. Abiotic and biotic stress combinations. *New Phytol.* **2014**, *203*, 32–43. [CrossRef] [PubMed]

23. Lazár, D.; Ilík, P.; Nauš, J. An appearance of K-peak in fluorescence induction depends on the acclimation of barley leaves to higher temperatures. *J. Lumin.* **1997**, *72*, 595–596. [CrossRef]

24. Niinemets, Ü. Responses of forest trees to single and multiple environmental stresses from seedlings to mature plants: Past stress history, stress interactions, tolerance and acclimation. *For. Ecol. Manag.* **2010**, *260*, 1623–1639. [CrossRef]

25. Gerganova, M.; Popova, A.V.; Stanoeva, D.; Velitchkova, M. Tomato plants acclimate better to elevated temperature and high light than to treatment with each factor separately. *Plant Physiol. Biochem.* **2016**, *104*, 234–241. [CrossRef] [PubMed]
26. Geist, H.J.; Lambin, E.F. Dynamic causal patterns of desertification. *AIBS Bull.* **2004**, *54*, 817–829. [CrossRef]
27. Ezcurra, E. *Global Deserts Outlook*; Scanprint: Aarhus, Denmark, 2006; ISBN 92-807-2722-2.
28. Lioubimtseva, E.; Henebry, G.M. Climate and environmental change in arid Central Asia: Impacts, vulnerability, and adaptations. *J. Arid. Environ.* **2009**, *73*, 963–977. [CrossRef]
29. Reynolds, J.F.; Smith, D.M.S.; Lambin, E.F.; Turner, B.L.; Mortimore, M.; Batterbury, S.P.; Huber-Sannwald, E. Global desertification: Building a science for dryland development. *Science* **2007**, *316*, 847–851. [CrossRef] [PubMed]
30. Chambers, J.C.; Bradley, B.A.; Brown, C.S.; D'Antonio, C.; Germino, M.J.; Grace, J.B.; Hardegree, S.P.; Miller, R.F.; Pyke, D.A. Resilience to stress and disturbance, and resistance to *Bromus tectorum* L. invasion in cold desert shrublands of western North America. *Ecosystems* **2014**, *17*, 360–375. [CrossRef]
31. Barker, D.H.; Adams, W.W.; Demmig-Adams, B.; Logan, B.A.; Verhoeven, A.S.; Smith, S.D. Nocturnally retained zeaxanthin does not remain engaged in a state primed for energy dissipation during the summer in two *Yucca* species growing in the Mojave Desert. *Plant Cell Environ.* **2002**, *25*, 95–103. [CrossRef]
32. Ogaya, R.; Penuelas, J.; Asensio, D.; Llusià, J. Chlorophyll fluorescence responses to temperature and water availability in two co-dominant Mediterranean shrub and tree species in a long-term field experiment simulating climate change. *Environ. Exp. Bot.* **2011**, *73*, 89–93. [CrossRef]
33. Xie, J.; Zha, T.; Jia, X.; Qian, D.; Wu, B.; Zhang, Y.; Bourque, C.P.-A.; Chen, J.; Sun, G.; Peltola, H. Irregular precipitation events in control of seasonal variations in CO_2 exchange in a cold desert-shrub ecosystem in northwest China. *J. Arid Environ.* **2015**, *120*, 33–41. [CrossRef]
34. Jia, X.; Zha, T.S.; Gong, J.N.; Wu, B.; Zhang, Y.Q.; Qin, S.G.; Chen, G.P.; Feng, W.; Kellomäki, S.; Peltola, H. Energy partitioning over a semi-arid shrubland in northern China. *Hydrol. Processes* **2016**, *30*, 972–985. [CrossRef]
35. Jiang, G.M.; Zhu, G.J. Different patterns of gas exchange and photochemical efficiency in three desert shrub species under two natural temperatures and irradiances in Mu Us Sandy Area of China. *Photosynthetica* **2001**, *39*, 257–262. [CrossRef]
36. Deng, W.H. Allelopathy of *Artemisia ordosica* Community in the Process of Plant Succession. Ph.D. Thesis. Beijing Forestry University, Beijing, China, 2016.
37. Porcar-Castell, A.; Juurola, E.; Ensminger, I.; Berninger, F.; Hari, P.; Nikinmaa, E. Seasonal acclimation of photosystem II in *Pinus sylvestris*. II. Using the rate constants of sustained thermal energy dissipation and photochemistry to study the effect of the light environment. *Tree. Physiol.* **2008**, *28*, 1483–1491. [CrossRef] [PubMed]
38. Repo, T.; Leinonen, I.; Ryyppö, A.; Finér, L. The effect of soil temperature on the bud phenology, chlorophyll fluorescence, carbohydrate content and cold hardiness of Norway spruce seedlings. *Physiol. Plant.* **2004**, *121*, 93–100. [CrossRef] [PubMed]
39. Qian, D.; Zha, T.S.; Wu, B.; Jia, X.; Qin, S.G. Spatio-temporal distribution characteristics of reference crop evapotranspiration in the Mu Us desert. *Acta Ecol. Sin.* **2017**, *37*, 1966–1974. [CrossRef]
40. Wilhelm, E.; Battino, R.; Wilcock, R.J. Low-pressure solubility of gases in liquid water. *Chem. Rev.* **1977**, *77*, 219–262. [CrossRef]
41. Chen, Z.H.; Zha, T.; Jia, X.; Wu, Y.; Wu, B.; Zhang, Y.; Guo, J.B.; Qin, S.G.; Chen, G.P.; Peltola, H. Leaf nitrogen is closely coupled to phenophases in a desert shrub ecosystem in China. *J. Arid. Environ.* **2015**, *122*, 124–131. [CrossRef]
42. Arnon, D.I. Copper enzymes in isolated chloroplasts. *Polyphenoloxidase* in Beta *vulgaris*. *Plant Physiol.* **1949**, *24*, 1. [PubMed]
43. Su, Z.S.; Zhang, X.Z. Comparision of determined methods of chlorophyll content. *Plant Physiol. J.* **1989**, 77–78. [CrossRef]
44. Schreiber, U.; Schliwa, U.; Bilger, W. Continuous recording of photochemical and non-photochemical chlorophyll fluorescence quenching with a new type of modulation fluorometer. *Photosynth. Res.* **1986**, *10*, 51–62. [CrossRef] [PubMed]
45. Roháček, K. Chlorophyll fluorescence parameters: The definitions, photosynthetic meaning, and mutual relationships. *Photosynthetica* **2002**, *40*, 13–29. [CrossRef]

46. Zha, T.S.; Wu, Y.J.; Jia, X.; Zhang, M.Y.; Bai, Y.J.; Liu, P.; Ma, J.Y.; Bourque, C.-P. A.; Peltola, H. Diurnal response of effective quantum yield of PSII photochemistry to irradiance as an indicator of photosynthetic acclimation to stressed environments revealed in a xerophytic species. *Ecol. Indic.* **2017**, *74*, 191–197. [CrossRef]

47. Porcar-Castell, A. A high-resolution portrait of the annual dynamics of photochemical and non-photochemical quenching in needles of *Pinus sylvestris*. *Physiol. Plant.* **2011**, *143*, 139–153. [CrossRef] [PubMed]

48. Genty, B.; Harbinson, J. Regulation of light utilization for photosynthetic electron transport. In *Photosynthesis and the Environment*; Springer: Dordrecht, The Netherlands, 1996; Volume 5, pp. 67–99.

49. Cailly, A.L.; Rizzal, F.; Genty, B.; Harbinson, J. Fate of excitation at PSII in leaves, the nonphotochemical side. *Plant Physiol. Biochem.* **1996**, *86*, 28.

50. Kramer, D.M.; Johnson, G.; Kiirats, O.; Edwards, G.E. New fluorescence parameters for the determination of Q_A redox state and excitation energy fluxes. *Photosynth. Res.* **2004**, *79*, 209–218. [CrossRef] [PubMed]

51. Hendrickson, L.; Furbank, R.T.; Chow, W.S. A simple alternative approach to assessing the fate of absorbed light energy using chlorophyll fluorescence. *Photosynth. Res.* **2004**, *82*, 73. [CrossRef] [PubMed]

52. Lazár, D. Parameters of photosynthetic energy partitioning. *J. Plant Physiol.* **2015**, *175*, 131–147. [CrossRef] [PubMed]

53. Lazár, D. Chlorophyll *a* fluorescence induction. *BBA-Bioenerg.* **1999**, *1412*, 1–28. [CrossRef]

54. Oxborough, K.; Baker, N.R. Resolving chlorophyll a fluorescence images of photosynthetic efficiency into photochemical and non-photochemical components–calculation of qP and F_v-/F_m without measuring F_o. *Photosynth. Res.* **1997**, *54*, 135–142. [CrossRef]

55. Maxwell, K.; Johnson, G.N. Chlorophyll fluorescence—A practical guide. *J. Exp. Bot.* **2000**, *51*, 659–668. [CrossRef] [PubMed]

56. Björkman, O.; Demmig, B. Photon yield of O_2 evolution and chlorophyll fluorescence characteristics at 77 K among vascular plants of diverse origins. *Planta* **1987**, *170*, 489–504. [CrossRef] [PubMed]

57. Lambers, H.; Chapin, F.S., III; Pons, T.L. *Plant Physiological Ecology*, 2nd ed.; Springer: Dordrecht, The Netherlands, 2008; pp. 39–40. [CrossRef]

58. Murchie, E.H.; Lawson, T. Chlorophyll fluorescence analysis: A guide to good practice and understanding some new applications. *J. Exp. Bot.* **2013**, *64*, 3983–3998. [CrossRef] [PubMed]

59. Kuk, Y.I.; Shin, J.S.; Burgos, N.R.; Hwang, T.E.; Han, O.; Cho, B.H.; Jung, S.; Guh, J.O. Antioxidative enzymes offer protection from chilling damage in rice plants. *Crop. Sci.* **2003**, *43*, 2109–2117. [CrossRef]

60. Pietrini, F.; Chaudhuri, D.; Thapliyal, A.P.; Massacci, A. Analysis of chlorophyll fluorescence transients in mandarin leaves during a photo-oxidative cold shock and recovery. *Agric. Ecosyst. Environ.* **2005**, *106*, 189–198. [CrossRef]

61. Baron, K.N.; Schroeder, D.F.; Stasolla, C. Transcriptional response of abscisic acid (ABA) metabolism and transport to cold and heat stress applied at the reproductive stage of development in *Arabidopsis thaliana*. *Plant Sci.* **2012**, *188–189*, 48–59. [CrossRef] [PubMed]

62. Ač, A.; Malenovský, Z.; Olejníčková, J.; Gallé, A.; Rascher, U.; Mohammed, G. Meta-analysis assessing potential of steady-state chlorophyll fluorescence for remote sensing detection of plant water, temperature and nitrogen stress. *Remote Sens. Environ.* **2015**, *168*, 420–436. [CrossRef]

63. Faik, A.; Popova, A.V.; Velitchkova, M. Effects of long-term action of high temperature and high light on the activity and energy interaction of both photosystems in tomato plants. *Photosynthetica* **2016**, *54*, 611–619. [CrossRef]

64. Wilhelm, C.; Selmar, D. Energy dissipation is an essential mechanism to sustain the viability of plants: The physiological limits of improved photosynthesis. *J. Plant Physiol.* **2011**, *168*, 79–87. [CrossRef] [PubMed]

65. Niyogi, K.K.; Truong, T.B. Evolution of flexible non-photochemical quenching mechanisms that regulate light harvesting in oxygenic photosynthesis. *Curr. Opin. Plant Biol.* **2013**, *16*, 307–314. [CrossRef] [PubMed]

66. Campos, H.; Trejo, C.; Peña-Valdivia, C.B.; García-Nava, R.; Conde-Martínez, F.V.; Cruz-Ortega, M.R. Photosynthetic acclimation to drought stress in *Agave salmiana Otto ex Salm-Dyck* seedlings is largely dependent on thermal dissipation and enhanced electron flux to photosystem I. *Photosynth. Res.* **2014**, *122*, 23–39. [CrossRef] [PubMed]

67. Bajkán, S.; Várkonyi, Z.; Lehoczki, E. Comparative study on energy partitioning in photosystem II of two *Arabidopsis thaliana* mutants with reduced non-photochemical quenching capacity. *Acta Physiol. Plant* **2012**, *34*, 1027–1034. [CrossRef]

68. Sperdouli, I.; Moustakas, M. Spatio-temporal heterogeneity in *Arabidopsis thaliana*, leaves under drought stress. *Plant Biol.* **2012**, *14*, 118–128. [CrossRef] [PubMed]
69. Osório, M.L.; Osório, J.; Romano, A. Photosynthesis, energy partitioning, and metabolic adjustments of the endangered Cistaceae species *Tuberaria major* under high temperature and drought. *Photosynthetica* **2013**, *51*, 75–84. [CrossRef]
70. Song, L.; Chow, W.S.; Sun, L.; Li, C.; Peng, C. Acclimation of photosystem II to high temperature in two *Wedelia* species from different geographical origins: Implications for biological invasions upon global warming. *J. Exp. Bot.* **2010**, *61*, 4087. [CrossRef] [PubMed]
71. Klughammer, C.; Schreiber, U. Complementary PSII quantum yields calculated from simple fluorescence parameters measured by PAM fluorometry and the saturation pulse method. *PAM Appl. Notes* **2008**, *1*, 27–35.
72. Flexas, J.; Bota, J.; Escalona, J.M.; Sampol, B.; Medrano, H. Effects of drought on photosynthesis in grapevines under field conditions: An evaluation of stomatal and mesophyll limitations. *Funct. Plant Biol.* **2002**, *29*, 461–471. [CrossRef]
73. Horton, P.; Johnson, M.P.; Perez-Bueno, M.L.; Kiss, A.Z.; Ruban, A.V. Photosynthetic acclimation: Does the dynamic structure and macro-organisation of photosystem II in higher plant grana membranes regulate light harvesting states? *FEBS. J.* **2008**, *275*, 1069–1079. [CrossRef] [PubMed]
74. Sacharz, J.; Giovagnetti, V.; Ungerer, P.; Mastroianni, G.; Ruban, A.V. The xanthophyll cycle affects reversible interactions between PsbS and light-harvesting complex II to control non-photochemical quenching. *Nat. Plants* **2017**, *3*, 16225. [CrossRef] [PubMed]
75. Terashima, I.; Hanba, Y.T.; Tazoe, Y.; Vyas, P.; Yano, S. Irradiance and phenotype: Comparative eco-development of sun and shade leaves in relation to photosynthetic CO_2 diffusion. *J. Exp. Bot.* **2005**, *57*, 343–354. [CrossRef] [PubMed]
76. Oguchi, R.; Hikosaka, K.; Hiura, T.; Hirose, T. Costs and benefits of photosynthetic light acclimation by tree seedlings in response to gap formation. *Oecologia* **2008**, *155*, 665. [CrossRef] [PubMed]
77. Donat, M.G.; Lowry, A.L.; Alexander, L.V.; O'Gorman, P.A.; Maher, N. More extreme precipitation in the world's dry and wet regions. *Nat. Clim. Chang.* **2016**, *6*, 508–513. [CrossRef]

 *forests*

Article

Improved Water Consumption Estimates of Black Locust Plantations in China's Loess Plateau

Kai Schwärzel [1,*], Lulu Zhang [1], Andreas Strecker [2] and Christian Podlasly [2]

[1] Institute for Integrated Management of Material Fluxes and of Resources (UNU-FLORES), United Nations University, 01067 Dresden, Germany; lzhang@unu.edu

[2] Institute of Soil Science and Site Ecology, Faculty of Environmental Sciences, Technische Universität Dresden, 01737 Tharandt, Germany; AndreasStrecker@gmx.net (A.S.); christian.podlasly@freenet.de (C.P.)

* Correspondence: schwaerzel@unu.edu; Tel.: +49-351-8921-9381

Received: 7 March 2018; Accepted: 11 April 2018; Published: 11 April 2018

Abstract: Black locust (*Robinia pseudoacacia* L.) is a major tree species in China's large-scale afforestation. Despite its significance, black locust is underrepresented in sap flow literature; moreover, the published water consumption data might be biased. We applied two field methods to estimate water consumption of black locust during the growing seasons in 2012 and 2013. The application of Granier's original sap flow method produced a very low transpiration rate (0.08 mm d^{-1}) while the soil water balance method yielded a much higher rate (1.4 mm d^{-1}). A dye experiment to determine the active sapwood area showed that only the outermost annual ring is responsible for conducting water, which was not considered in many previous studies. Moreover, an in situ calibration experiment was conducted to improve the reliability of Granier's method. Validation showed a good agreement in estimates of the transpiration rate between the different methods. It is known from many studies that black locust plantations contribute to the significant decline of discharge in the Yellow River basin. Our estimate of tree transpiration at stand scale confirms these results. This study provides a basis for and advances the argument for the development of more sustainable forest management strategies, which better balance forest-related ecosystem services such as soil conservation and water supply.

Keywords: afforestation; heat dissipation probes; in situ calibration; soil water balance; transpiration; dye tests; ring-porous trees

---

1. Introduction

Accurate estimates of tree transpiration are needed not only for the management of a forest and its water-related services, but also for understanding a forest's response to climate and other changing environmental conditions to identify risks and manage them accordingly. Such estimates are usually obtained by measuring the speed of xylem sap movement through the stem using heat sensors. Three different systems of heat sensors are commonly used [1]: (i) heat pulse velocity sensors; (ii) heat field deformation sensors; and (iii) thermal dissipation sensors. Vandegehuchte and Steppe [2] discussed the working principles and applicability of these sensors in detail, which we will not include in this work.

Previous sap flow studies have demonstrated that multiple issues can introduce errors in the measurement of sap flow and estimate of stand transpiration rates: ignoring the radial variability in sap flow [3], neglecting nocturnal fluxes [4–6], sensor installation into non-conducting sapwood [7,8], errors in scaling-up sap flow measurements within the tree stem and over the ecosystem [9], wounding of sapwood due to sensor installation [10], diel dynamics in stem water content [11,12], or using universal sensor calibration instead of species-specific calibration [13–15]. Research on the reliability of the sap flow technique suggests that most of these sources of error can probably be minimized [16]. However, the remaining uncertainty and thus confidence in the results need to be assessed [17].

Black locust (*Robinia pseudoacacia* L.) is one of the most frequently cultivated broad-leaved tree species in the world [18]. Black locust was one of the first North American tree species introduced to Europe in the early 17th century [19] and is one of the three most widely distributed non-native plant species in Europe [20]. Outside of Europe, black locust has been introduced to temperate Asia, Australia, and New Zealand. For instance, black locust plays a major role in China's large-scale afforestation programs, particularly in the drylands of the north-western and northern regions. These afforestation programs aim to control desertification and soil erosion [21]. At present, the area covered by black locust plantations is the largest among the planted forests in China's Loess Plateau region [22]. Despite this significance, very few studies have estimated the water budget of black locust plantations, which is crucial scientific knowledge for water resources management in dryland areas [23–25].

The transpiration of trees is commonly determined using thermal dissipation sensors (i.e., Granier's method) [26]. Granier's method measures the temperature differential between heated and unheated probes and relates the normalized temperature differential to the sap flow density through a universal calibration equation [10]. This equation was derived by Granier using three stem segments of *Pseudotsuga menziesii*, *Pinus nigra*, and *Quercus pedunuculata*. In these laboratory calibration experiments, water was forced through these samples under varying pressure, and the flow rate was determined simultaneously by weighing the exudates and by using the thermal dissipation sensor. It has yet to be acknowledged that the universal calibration equation substantially underestimates sap flow, at least when it is used for other ring-porous tree species [27].

Despite its widespread significance, black locust is underrepresented in the international sap flow literature in contrast to other tree species. An online search in databases, such as Scopus, Web of Science, and Google Scholar, produced merely eleven hits on the topic of sap flow measurements in black locust stands. In contrast, Scopus listed 37 results for sap flow measurements in spruce (*Picea abies*) and 29 results in beech (*Fagus sylvatica*) for a short period from 2010 to 2016. Only eight out of the abovementioned eleven studies provide estimates of black locust stand transpiration. These eight studies were carried out in different black locust stands across the Loess Plateau. With the exception of two studies [15,28], the different experimenters acknowledged that their daily transpiration rates were exceptionally low [23–25,29–31]. The latter group of experimenters reported daily transpiration averages of $\leq$0.5 mm d^{-1}. This is surprisingly lower than what is believed and contrary to the generally accepted fact that afforestation in the Loess Plateau region resulted in a significant discharge reduction through the elevated evapotranspiration [32–35]. This led to a mismatch of the scientific evidence between plot measurement/estimation and watershed observation/investigation.

Several authors [13,14] have recommended developing tree-specific calibration equations for Granier's method in ring-porous tree species to ensure a more accurate estimate of sap flow [27]. However, to our knowledge, only one group of the abovementioned experimenters [15,23–25,28–31] have heeded this advice when applying the Granier's method for measuring sap flow in black locust. Thus, we believe that black locust is not only understudied in international sap flow literature but also that the published values of black locust sap flow may be biased due to the lack of tree-specific calibration. Furthermore, we postulate that the underestimation in black locust sap flow is probably caused by overestimation of the conducting sapwood area. Black locust is a ring-porous tree species; it is generally accepted that trees falling into this classification have water conduction mainly through the large vessels of the outermost growth ring [36].

The most common method for tree-specific calibration of sap flow sensors is the laboratory calibration conducted on excised stems [13,14] or cut branches [8]. In most of these laboratory studies, water is pulled through stem segments or cut branches using a pump or by maintaining a constant head of water pressure on cut stem segments. A question may therefore arise: does calibration obtained under laboratory conditions deliver a reliable estimate of sap flow under field conditions? The cut tree technique was developed for in situ validation of sap flow measurement systems under natural conditions [37,38]. To our knowledge, this technique has not yet been applied for the calibration of thermal dissipation sensors.

In view of the identified knowledge gaps and the importance of accurate estimate of tree transpiration for managing forest and water resources as well as related services, the objectives of this paper are threefold: (i) estimate tree transpiration of a black locust stand using two different methods, namely, Granier's original calibration and soil water balance method, and compare the reliability of the two methods; (ii) identify the possible sources of error for the underestimated tree transpiration from Granier's original calibration based on analyses of other published studies and discuss their impacts; and (iii) improve the accuracy of tree transpiration estimation of Granier's method by applying a tree-specific calibration, for which an easy-to-install measurement setup for the in situ calibration of Granier's sensor system in living trees is developed and applied; additionally, the reliability of the obtained calibration is validated.

2. Materials and Methods

2.1. Study Site

In situ calibration and measurement of sap flow of black locust were carried out in the Zhonggou catchment (35°20′ N, 107°31′ E), which is a small catchment in the semi-humid gully region of the Loess Plateau in the upper reaches of the Jing River (Gansu Province, NW China). The catchment is 14 km^2 in size with an elevation ranging between 1000 and 1300 m a.s.l. According to the long-term records (1990–2007) of the nearby Jingchuan hydrological station, the average annual temperature is 10.2 °C, the average annual potential evapotranspiration is 1395 mm a^{-1}, and the mean annual precipitation is 590 mm a^{-1}. The catchment is covered by loess deposits with a thickness of 50–80 m. The soil is classified as Calcaric Regosols (IUSS working group WRB, 2006) with a silt loam texture [39]. Eighty-three per cent of the catchment is covered by forests, mainly by black locust. The measurement plot was established in 1990 by planting black locust saplings. This plot covers an area of 2050 m^2 and comprises 206 trees (1006 trees per ha) with a basal area of 2.2 m^2 ha^{-1}. The summer maximum leaf area index (LAI) of the stand varied between 2.4 and 2.8 (measured using the optical instrument LAI 2000-PCA, LI-COR Inc., USA). In 2012, the average tree height and diameter at breast height (DBH) were 11.8 m and 10.7 cm, respectively (see Table 1). The understory consists mainly of grasses (*Melilotus suaveolens, Astragalus adsurgens, Onobrychis taneitica, Equisetum arvense, Juncus bufonius, Pennisetum centrasiaticum, Phragmites communis, Setaria viridis,* and *Chenopodium album*) as well as some shrubs such as *Amorpha fruticosa, Prunus davidiana, Hippophae rhamnoides, Syringa persica, Caragana korshinskii,* and *Xanthocera sorbifolia.*

Table 1. Sap flow studies carried out in black locust stands across the Loess Plateau region. All the studies used Granier's original calibration for the calculation of the sap flow densities, i.e., Equation (3). MAT = mean annual temperature, MAP = mean annual precipitation, LAI = Leaf Area Index, DBH = diameter at breast height, A_S = sapwood area, A_G = stand ground surface, T_{BL} = transpiration of the black locust stand, N.S. = not specified.

Study Area	Understory	MAT [°C]	MAP [mm year^{-1}]	LAI_{max}	Mean DBH [cm]	Number of Trees [ha^{-1}]	A_S/A_G [m^{2}ha^{-1}]	Sapwood Thickness [mm]	T_{BL} [mm d^{-1}]	Relationship between DBH and A_S	Source
Mt. Gonglushan, Yan'an 36°25′24″ N 109°31′ 32″ E	Grass and a few scattered shrubs	10.6	498	2.89	9.3	3100	5.10	5–10 (mean 7.3)	0.41	$A_S = 0.546DBH^{1.508}$	[29]
Mt. Gonglushan, Yan'an 36°25′40″ N 109°31′ 53″ E	Small trees and shrubs	10.6	498	2.89	9.3	N.S.	N.S.	7.2–9.2	N.S.	N.S.	[40]
Caijiachuan catchment, Ji county 36°14′ 27″ to 36°18′23″ N 110°39′45″ to 110°47′45″ E	N.S.	10.0	579	N.S.	6.9	850	1.57	N.S.	<0.2	$A_S = 3.4DBH-8.7$	[30]
Mt. Gonglushan, Yan'an 36°25′24″ N 109°31′32″ E	Grass and a few scattered shrubs	10.6	498	2.73–3.14	9.3	3100	5.09	5–10	0.32 to 0.49	$A_S = 0.546DBH^{1.508}$	[23]
Yangjuangou, Yan'an 36°42′ N 109°31′ E	Patches of liana and herbs	9.8	531	2.32–2.98	9.9–10.8	1300	3.16 to 3.60	N.S.	0.14 to 0.23	$A_S = 0.61DBH^{1.55}$	[24]
Yangjuangou, Yan'an 36°42′ N 109°31′ E	N.S.	9.8	531	2.77 (12 years old) 2.38 (28 years old)	6.94 (12 years old) 8.93 (28 years old)	2500 (12 years old) 1200 (28 years old)	3.16 to 3.60	15.4 (12 years old) 8.4 (28 years old)	0.22 (12 years old) 0.39 (28 years old)	$A_S = 0.28DBH^{2.25}$ (12 years old) $A_S = 0.25DBH^{1.81}$ (28 years old)	[25]
Yeheshan forest reserve, Fufeng county 34°31′46″ N 107°54′40″ E	Grass	12.7	580	2.4–.8	6.0–11	2450 (15 years old)	5.13 to 5.30	10 ± 2	0.12−0.16 *	$A_S = 0.4024DBH^{1.9}$	[15]
Zhonggou, 35°20′ N 107°31′ E	Grass and a few scattered shrubs	10.2	509	2.8	10.7	1006 (33 years old)	0.61	3.43	0.08 *	$A_S = 0.0604DBH^{1.882}$	This study

* This estimate is based on Granier's original calibration but Ma et al. [15] developed a tree-specific calibration based on lab experiments.

2.2. Continuous Field Measurements

2.2.1. Weather Stations

An open-land weather station was installed in 2012, approximately 300 m east of the black locust stand. Wind direction and wind speed (Thies, Germany), global radiation and net radiation (Kipp & Zonen, Delft, The Netherlands), air temperature and relative humidity, soil temperature at different depths, and temperature at the soil surface were measured at 15-min intervals. A tipping bucket system with heating (Lambrecht, Göttingen, Germany) was used to measure the precipitation. An automatic weather station was also placed under the black locust canopy to measure air and soil temperature, relative humidity, as well as global and net radiation (Kipp & Zonen, The Netherlands). Two throughfall troughs (each 5 m long and 0.16 m wide and made of stainless steel) were installed beneath the black locust canopy. The troughs drain into a tipping bucket. The throughfall troughs operate only during the growing season from the end of March until the middle of October.

2.2.2. Quantifying Total Evapotranspiration

For the estimation of total evapotranspiration, ET_{total}, the soil water depletion method was applied. This method relies on simultaneous measurements of soil pressure head and soil water content as a function of time and depth [41]. For the measurement of soil water dynamics, a wireless sensor network (Forschungszentrum Jülich, Jülich, Germany) was installed. Details of this network can be found in Bogena et al. [42].

In our study, soil water content under the black locust canopy was measured using spade sensors (sceme.de GmbH, Horn-Bad Meinberg, Germany). The spade sensor is a time domain transmission sensor. The general operating principle of these sensors is similar to that of time domain reflectometry (TDR) sensors [43]. The spade sensor consists of a sensor head (8.0 cm long, 3.0 cm wide, 1.0 cm thick) and a transmission line embedded in a four-layer epoxy printed circuit board (14.0 cm long, 3.0 cm wide, 0.2 cm thick). With this sensor, not only the soil water content, but also the soil temperature, can be measured. The measurement resolution of the sensor is 0.01 [$m^3\ m^{-3}$].

We established soil water content measurements on a 420 cm long transect between two trees. The leaf canopies of these trees touch each other and partly overlap. Sensors at the start and the end of the transect were not installed next to the trunks but at a distance of about 100 cm from them. When mounting the spade sensors, hand shovels were used to dig narrow trenches (<10 cm wide, ca. 100 cm long, and up to a depth of 100 cm) perpendicular to the transect. To minimize the effects of sensor installation on root water uptake, we managed to avoid cutting roots with diameters larger than 0.2 cm during the excavations. To insert the sensors into the soil, holes for the sensor heads were pre-drilled into the soil profile wall. This procedure ensures a minimal soil disturbance and soil compaction.

In total, 36 sensors were placed at six depths with a horizontal distance of 60 cm between them. Sensors were placed at the depths of 5, 20, and 40 cm with eight replications and at the depths of 60, 80, and 100 cm with four replications. The depths of sensor installation and the corresponding replications were chosen according to the mapped root density distribution. Soil pressure head was measured at the depths of 20, 40, 60, and 80 cm using a pF-meter (ecoTech, Bonn, Germany). A pF-meter measures the heat capacity in a porous equilibrium body. Based on a sensor-specific calibration curve (provided by the manufacturer), the measured heat capacity can be converted to soil pressure head values (measurement range: pF 0.0 to pF 7.0). The pF-meter was installed vertically in the middle of the abovementioned transect. Soil water contents and pressure head values were automatically recorded every 30 min. Deriving the evapotranspiration rates from the soil water balance requires knowledge of the location of the zero-flux-plane (ZFP) within the soil profile. The location of the ZFP was determined by calculating the hydraulic gradients using the measured pressure head data as shown in Khalil et al. [44].

Total evapotranspiration (sum of understory and overstory water uptake of the plant roots), ET_{total} [mm d^{-1}], between times t_1 and t_2 was then calculated from the measured soil water content changes as follows [41]:

$$\int_{t_1}^{t_2} ET_{total}\,dt = \int_{t_1}^{t_2} P_{net}\,dt - \int_{0}^{W}\int_{t_1}^{t_2} \frac{\partial \theta}{\partial t}\,dt\,dz \tag{1}$$

where P_{net} is the net precipitation measured using the throughfall troughs [mm d^{-1}], W is the depth of the ZFP [m], θ is the soil water content measured by spade sensors (sceme.de GmbH, Germany) [m^3 m^{-3}], and z is the vertical coordinate [m]. For the calculation of ET_{total}, soil water depletion was first balanced on a daily scale for each sensor, then averaged for each soil depth, and finally summed with respect to the depth of the ZFP.

2.2.3. Quantifying Understory Transpiration

To quantify the transpiration of the understory (grass), a weighable lysimeter was installed under the black locust canopy. The lysimeter consists of a container (2.5 m high, 2.3 m long, 1.5 m wide) made of polyethylene (PE-HD), a lysimeter vessel made of stainless steel with a surface area of 1.0 m^2 and a length of 1.7 m filled with undisturbed soil, a weighing system, and a unit for the control of soil moisture and soil temperature at the lower bottom of the lysimeter. The container is water-tight and air-tight with two circular openings at the top; one opening served as an access hatch and the other opening was for the lysimeter vessel. Details of the installation and setup of the lysimeter can be found in Podlasly and Schwärzel [45].

At the lower bottom of the lysimeter, suction cups, tensiometers, temperature sensors, cooling coil, and hoses were embedded into the soil for the automatic regulation of soil temperature and soil pressure at the lower boundary of the lysimeter. A detailed description of the working principle of the automated control of the lower boundary of the lysimeter is given in Podlasly and Schwärzel [45]. Finally, six spade sensors (sceme.de GmbH, Horn-Bad Meinberg, Germany) for the automatic measurement of the soil water content were installed at the depths of 5, 15, 25, 50, 100, and 150 cm of the soil monolith. The spade sensors were described in greater detail above.

The measurement resolution of the weighing system of the lysimeter is 30 g, which corresponds to a water equivalent of 0.03 mm. The weight of the lysimeter vessel and the abovementioned soil water content readings were logged every 30 min. Decreases in the weight of the lysimeter are caused by transpiration of the vegetation and by evaporation from the soil. Thus, we use the term evapotranspiration of understory (ET_{us}) when referring to water consumption quantified by the weighable lysimeter.

Daily evapotranspiration rates of the understory, ET_{us}, [mm d^{-1}] were calculated as follows:

$$ET_{us} = P_{net} - S - \Delta W \pm \Delta T \tag{2}$$

where P_{net} is the net precipitation (measured by throughfall troughs) [mm d^{-1}], S is the seepage water collected at the bottom of the lysimeter [mm d^{-1}], ΔW is the change in the mass of the lysimeter [1 kg d^{-1} = 1 L^3 m^{-2} d^{-1} = 1 mm d^{-1}], and ΔT is the amount of water removed or added as a result of controlling the pressure head at the lower bottom of the lysimeter [1 kg d^{-1} = 1 L^3 m^{-2} d^{-1} = 1 mm d^{-1}].

2.2.4. Quantifying Overstory Transpiration

To quantify tree transpiration, Granier-style sap flow sensors were used. Probes were 20 mm long with a diameter of 1.5 mm (Ecomatik, Dachau, Germany). Altogether, fourteen trees (range of the DBH of the sample trees: 10 to 17 cm) were equipped with sap flow sensors. Twelve pairs of sensors were mounted on the northern side of the stem of twelve trees, and eight pairs of sensors were installed on the eastern and northern side of the two other trees. The vertical distance between the needles was 150 mm. The sensors were shielded from radiation and rain using Styrofoam covered with reflective material. The upper needles were heated by a constant current source (0.2 W, 84 mA); the lower needles were unheated. Temperature differences between the sensors were measured at 60 s intervals and stored as 60-min averages by the data logger.

Hourly sap flow densities, F_d [kg^3 m^{-2} h^{-1}], were calculated from observed temperature differences by Equation (3), according to Granier [26]:

$$F_d = 3600 \times a \times k^b = 3600 \times a \times \left(\frac{\Delta T_M - \Delta T}{\Delta T} \right)^b \tag{3}$$

where k is the dimensionless, normalized sensor output, ΔT [K] is the actual measured temperature difference between the two needles, ΔTM [K] is the value of ΔT when sap flux density is zero, the coefficient a (0.119 kg m^{-2} s^{-1}) and the exponent b (1.231) are Granier's original fitting parameters.

As aforementioned, installation of sensors in non-conducting sapwood [7] or ignoring the radial variability in sap flow [3] may result in significant underestimation of sap flow densities. The latter plays a minor role in ring-porous trees such as black locust. In ring-porous trees, only the outermost annual tree ring (i.e., active sapwood area) is responsible for conducting water [3]. To verify it, the tree-cutting technique (see Section 2.3) was applied to visualize water-conduction pathways in living black locust trees, in which we used dye—an aqueous solution of 0.1% acid fuchsin—similar to the procedure of Sano et al. [46]. To determine the active sapwood thickness, ten trees—with similar DBH to the trees used for the sap flow measurements—were harvested and the individual tree-ring width of the outermost ring was measured using stem disc at breast height. Based on these results, a regression between the DBH and the corresponding sapwood area was derived. This relationship was used to calculate the total sapwood area of the plot using the frequency of the tree's DBH. However, due to the very small active sapwood thickness, it is impossible to install the sensors without touching the non-conducting sapwood. To reduce the potential errors caused by sensor installation in non-conducting sapwood, a correction procedure following Clearwater et al. [7] was applied in preliminary tests. Despite the reduced errors from the correction, it may still have some uncertainties from an unknown source, as reported by Bush et al. [14].

To consider potential nocturnal fluxes due to both transpiration and recharge, ΔTM was determined using the software BaseLiner (developed by Ram Oren's C-H$_2$O Ecology Lab Group at the Nicholas School of the Environment, Duke University). BaseLiner helps to identify no-flow conditions and to set baseline points at these times. A linear interpolation is then applied to calculate the baseline between these points. In a next step, the hourly values of sap flow density of each tree were calculated using Equation (3). Finally, daily totals of sap flow density for each sample tree were computed from one midnight to the next midnight. No functional relationship between the daily totals of sap flow density and the corresponding DBH was found.

To estimate the daily transpiration, T_{BL}, of the black locust stand [mm d^{-1}], the daily sap flow densities of all sample trees were averaged. The averaged flux densities, J_{avg}, were then multiplied by the total sapwood area, A_{ST}, of the plot, and divided by the stand ground surface, A_G [47]:

$$T_{BL} = J_{avg} \times \frac{A_{ST}}{A_G} \tag{4}$$

Another method to estimate the transpiration of the black locust stand (hereinafter referred as $T_{residual}$) is to calculate the transpiration as a residual of the difference between total measured evapotranspiration Equation (1) and measured understory transpiration Equation (2) on a daily basis. These residuals together with the calculated grass reference evapotranspiration [48] were used to assess the reliability of the sap flow measurements.

2.3. Calibration Experiment and Cut Tree Technique

The calibration experiment is grounded in the fact that a tree can take up water from a water reservoir instead of from the soil. To conduct the calibration, the sapwood had to be cut from a tree (see below). The setup of the calibration experiment consists of three components: two collars and a Mariotte water supply (Figure 1).

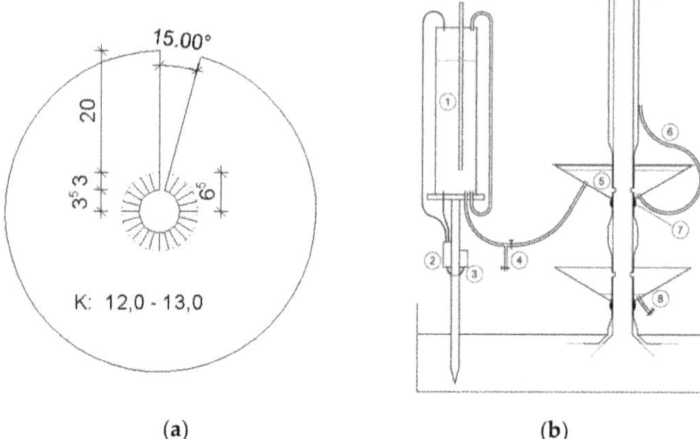

(a)   (b)

Figure 1. (a) Schematic of the cut tree experiment, 1 = Mariotte water supply, 2 = Differential pressure sensor, 3 = Data logger, 4 = Valve, 5 = Collar, 6 = Pipe for water level control, 7 = Lashing belt, 8 = Collar; (b) Prototype of a collar, K = Diameter, all dimensions in cm.

The funnel-shaped collar is made of a zinc sheet (Figure 1b). With a thickness of 0.5 mm, the collar is sufficiently sturdy but also easy to bend. A tree with a diameter of 14.3 cm and a height of 13.0 m was selected for cutting sapwood. After smoothing the bark, the collars were placed around the stem at the height of 40 cm and 90 cm above the soil surface. The inner metal cut strips were then bent downwards and attached to the stem by a lashing belt. Afterwards, both ends of the upper collar were joined by a waterproofing steam-tight tape (Siga Rissan, Russwil, Switzerland) whereas the ends overlapped. Finally, the gap between collar and stem was sealed with silicon adhesive. An S-form chisel (Kirschen, Remscheid, Germany) was used to cut a 1-cm-deep notch around the circumference of the black locust stem. The cut of the sapwood was done underwater to avoid air entry to the xylem. The same procedure was applied to the lower collar.

During the calibration, the upper collar serves as the tree's water reservoir. For this purpose, the upper collar is connected to a conventional Mariotte water supply (90.0 cm long, 18.7 cm in diameter, made of PVC). The latter has an adjustable tube that controls the outflow. It ensures a constant water level inside the collar throughout the experimental period and prevents the freshly cut sapwood from drying out. The function of the lower collar is to collect water dripping from the upper one. The collected water was periodically taken from the collar via a valve and stored in a bottle with a cap. The drip rate was determined by weighing. These data were used for calculations. To prevent evaporation of water from the collar during the experiment, two semi-circular lids were placed on the collars. Changes in the Mariotte water supply correspond to water uptake by the tree; these changes were monitored using a differential pressure sensor (PD-23, Fa. Keller, Winterthur, Switzerland). The data were logged every 60 s (EASYLOG 40NS K-0-10V, Fa. Greisinger, Regenstauf, Germany). At the same time, sap flow was measured using two pairs of Granier-style sap flow sensors, mounted at breast height on the eastern and northern side of the sample tree.

3. Results and Discussion

3.1. Diurnal Cycle of Transpiration Estimated Using Granier's Original Calibration and Soil Water Balance Method

In this section, we present the estimated transpiration of the black locust stand using two different methods. By comparing the two results, the impacts of potential errors on the measured sap flow and estimated stand transpiration rates based on Granier's original calibration are discussed. A warm period with low rainfall was selected for cross-checking the transpiration estimates. Figure 2 shows the daily rainfall and daily mean values of soil pressure head at different soil depths for the period from 1 May to 30 June 2012. By this time, the canopy of the black locust stand was fully developed. As the fast increase of soil pressure head data at the depths of 20 and 40 cm reveals (Figure 2), the soil became drier. This was caused by soil water depletion due to lack of rainfall (Figure 2). As outlined in the methods section, the measured soil pressure head data were used to identify the location of the zero-flux-plane (ZFP) by calculating the hydraulic gradients. Results of these calculations are also shown in Figure 2. Note that negative hydraulic gradients indicate upward water flow and positive hydraulic gradients indicate downward water flow. In May and June, a ZFP was located within the root zone but below the main root zone at a depth range of between 50 and 70 cm.

For the dry period from 2 June to 24 June soil moisture changes above the ZFP were balanced according to Equation (1). Although the ZFP was located within the root zone it can be assumed that soil moisture changes below the ZFP were mainly due to the downward movement of water and only to a minor extent because of root water extraction. This assumption is supported by our observation that (i) 90% of the fine roots were within the upper 50 cm of the soil; (ii) the rooting density at soil depths >50 cm was very low (1 to 2 fine roots per dm^2); and (iii) soil moisture changes at the soil depths of 80 and 100 cm were less than 3 vol % during the balanced period. Similar soil water depletion patterns of black locust were observed by other studies [30,40]. Chen et al. [30] measured soil moisture of black locust at the depth of 0–25 and 75–100 cm. They found insignificant soil moisture variations in the deeper soil during the vegetation season. Du et al. [40] concluded based on their soil moisture measurements at different soil depths that short-term changes of soil moisture below a depth of 1 m were small. Moreover, their plotted soil moisture variations along a vertical profile before and after a rain event revealed that about 90% of the soil moisture variations occurred within the upper 40 cm of the soil profile.

With this fact, we assume that neglecting root water extraction from below the ZFP will not result in a significant underestimation of the total evapotranspiration. For the dry period from 2 to 24 June, the totals of soil water depletion due to root water uptake and evaporation Equation (1) were contrasted with the totals of understory evapotranspiration (based on lysimeter measurements, Equation (4)) and black locust stand transpiration (based on sap flow measurements, Equation (2)) (Figure 2). The total evapotranspiration, understory evapotranspiration, and stand transpiration amounted to 82.0 mm (3.6 mm d^{-1}), 51.0 mm (2.2 mm d^{-1}), and 1.8 mm (0.08 mm d^{-1}). The latter was calculated based on Granier's original calibration parameters, but if calculated using soil water balance method (total evapotranspiration minus understory evapotranspiration), transpiration was 31.0 mm (1.4 mm d^{-1}). A large difference (ca. 30 mm) appears between the transpiration indirectly estimated as a "residual" of the water budget and the transpiration directly derived from Granier's original calibration.

So far only a few studies have balanced the total evapotranspiration of black locust stands in dryland China. Wang et al. [49] studied the water budget of black locust afforestation in the Loess Plateau Region by combining throughfall, stemflow, and soil water content measurements with soil hydrological modelling. Their calculated total evapotranspiration (ET_{total}) of the investigated black locust stand ranged from about 90 to 120 mm per month (or from 3.0 to 4.0 mm d^{-1}) during the summer months (June to August). Jian et al. [28] investigated the effect of different trees and shrubs on soil water storage and water balance in the semi-arid Loess Plateau area. In this study, evaporation from soil under black locust was measured using a micro lysimeter, while stand transpiration was derived from sap flow measurements using a modified Granier sensor system (SF-L, Ecomatik, Germany) from May to September during 2009–2013. Sums of evaporation from soil and stand transpiration amounted to between 56 and 84 mm (or from 1.9 to 2.7 mm d^{-1}) [28]. The mean potential evapotranspiration of study areas of Wang et al. [49] and Jian et al. [28] are about 1.6 and 1.5 mm a^{-1}, slightly higher than our study area (1.4 mm a^{-1}). Moreover, the mean annual precipitation values of the two study areas are 540 and 420 mm a^{-1}, smaller than the long-term average of rainfall in our study area (590 mm a^{-1}). Despite the differences in long-term potential evapotranspiration and rainfall, the total evapotranspiration rates are in the same range as our measured results based on the water balance method. This makes us confident that the soil water balance method delivered more reliable results than the application of Granier's original calibration.

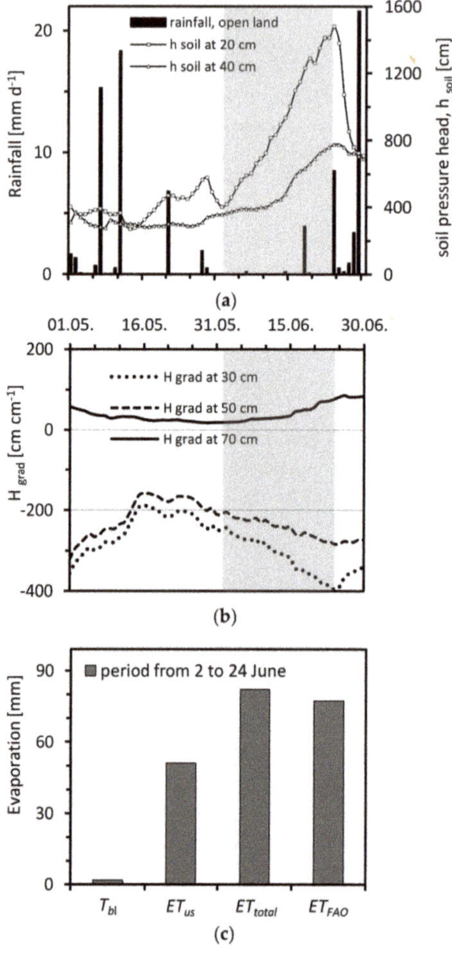

Figure 2. (**a**) Daily rainfall and daily averages of the soil pressure head at the depths of 20 and 40 cm of black locust (upper) and (**b**) soil hydraulic gradients (middle) for the period from 1 May to 30 June 2012 as well as (**c**) sums of tree transpiration (T_{bl}), understory evapotranspiration (ET_{us}), total evapotranspiration (ET_{total}), and FAO grass reference evapotranspiration for the period from 2 to 24 June 2012 (bottom).

Yet, what are the causes for the mismatch between the results of the soil water balance method and the combination of the lysimeter/sap flow method? Some of the possible issues, such as variability in sap flow or scaling-up of the measurements to the stand level, can be better understood if basic data in high temporal resolution are analyzed. For this purpose, daily courses of global radiation, as well as vapor pressure deficit and diurnal courses of sap flow density of trees with different tree diameter at breast height (DBH) on sunny and warm days during the abovementioned balance period, are examined (Figure 3). Night time sap flow was observed on selected nights and considered in this study. For the balance period from 2 to 24 June, consideration of nocturnal fluxes resulted in a 40% increase of stand transpiration in comparison to the estimation without consideration of night flow (not shown). Figure 3 reveals that there is no clear dependency of sap flow density on tree DBH of the measured black locust trees. This is in line with the findings of Wang et al. [29] and confirmed our presumption, as McCulloh et al. [50] have shown for black locust and other investigated ring-porous tree species, i.e., that their stem conductance per unit sapwood was independent of tree size.

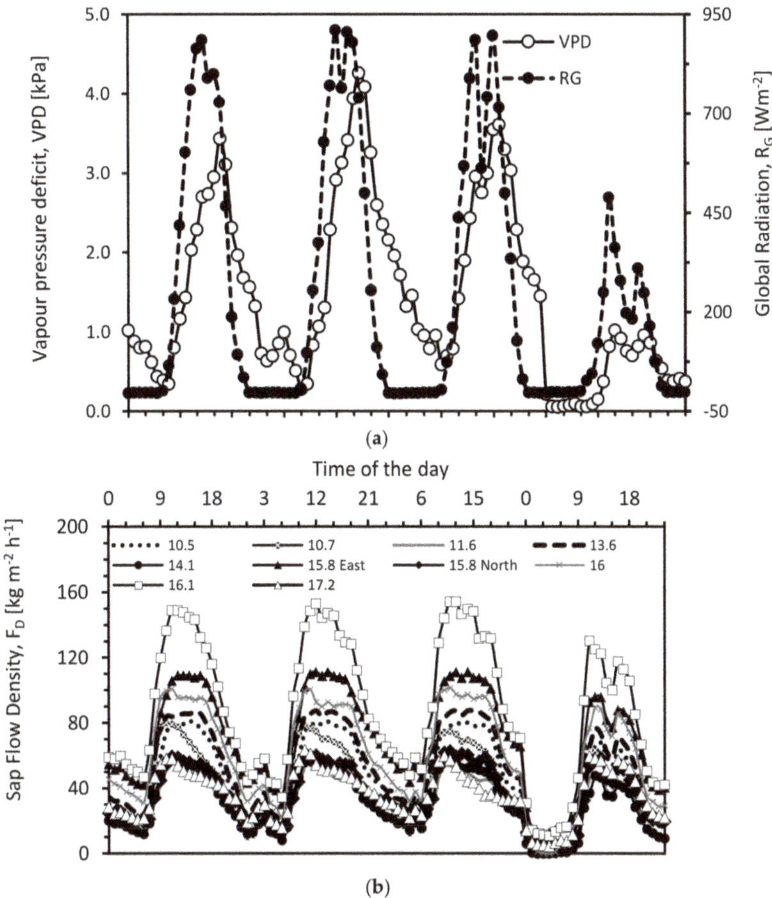

Figure 3. Time series of (**a**) global radiation, vapor pressure deficit (upper) and (**b**) sap flow density of individual trees with different diameter at breast height (bottom) for the period 21 to 24 June. The legend in the bottom figure provides the diameter (in cm) at breast height of the individual trees.

In our balance period, the maximum sap flow densities of the 33-year-old stand varied from about 60 kg m^{-2} h^{-1} (DBH = 17.1 cm, DBH = 15.8 cm, DBH = 12.1 cm) to about 150 kg m^{-2} h^{-1} (DBH = 16.1 cm). Jiao et al. [25] compared the sap flow densities and stand estimates of a 12-year-old and a 28-year-old black locust stand in the semi-arid Loess Plateau. Their maximum sap flow densities in the younger stand were between 45 and 75 kg m^{-2} h^{-1} and in the older stand between 80 and 180 kg m^{-2} h^{-1} on sunny and hot days. These values are in line with our results. In contrast, Wang et al. [29] found higher maximum flow densities (their values ranged from 105 to 408 kg m^{-2} h^{-1}) with a larger variation in another 30-year-old black locust stand in the semi-arid zone of the Loess Plateau. However, our measurement results still fall within the same order of magnitude and indicate that the values are representative for the black locust stands of the Loess Plateau region.

It is often discussed that ignoring circumferential variations in sap flow density may introduce significant errors when stand transpiration estimates are derived from sap flow measurements [9,51]. Thus, in individual cases, two pairs of sap flow sensors were installed at sampled trees (East, North). In our study, circumferential variations in sap flow/within-tree spatial variations were lower than the tree-to-tree variations as an example of circumferential variations in sap flow density (Figure 3); the maximum values were about 60 kg m^{-2} h^{-1} and 110 kg m^{-2} h^{-1} for the north- and east-facing sensors. Circumferential variations tend to be accompanied by slight asymmetries in the crown; we observed that often thicker branches were located above the sensor with larger sap flow densities. However, this issue was not systematically examined.

Kume et al. [51] studied the radial and circumferential variations in sap flow velocity in several directions and depths of tree trunks of black locust stands on the Loess Plateau. They found that the sap flow density was almost zero at the depth of 15 mm and most of the sap flow took place at the outermost area of the sapwood ($\leq$5-mm depth). Moreover, omitting radial variations and circumferential variations in sap flow density affected the stand transpiration estimation by 33% and 22%. This was lower than the effect of omitting tree-to-tree variation in sap flow density (error of 52%). We showed that radial, circumferential, and tree-to-tree variations may have introduced some errors when scaling up sap flow densities to the stand scale. However, the mismatch (1.3 mm d^{-1}) between the two applied methods cannot be sourced from these potential errors since the magnitude of these errors is too small to bridge the huge gap.

Another source of error may stem from the determination of the active sapwood area. The latter is needed for scaling-up the sap flow measurements to the stand level. In our study, we considered that the sap flow took place only in the outermost annual ring of black locust, which is supported by the results of dye injection experiments in ring-porous tree species [14,52] and our own dye tracer experiments in living black locust trees (Figure 4). The result showed that only the outermost growth ring was active for xylem flow. Moreover, isolated larger pores are often noticeable due to more intense coloring. A measurement of the maximum dye ascent velocities using three sample trees (with a height of 11.5, 13 and 14.5 m) revealed a range of 19–25.5 m h^{-1}. This falls within a similar range of values as measured by Zimmermann and Brown [53] and Huber and Schmidt [54].

A mean annual tree-ring width of 3.43 mm was found for our study area, ranging from 1.89 to 4.93 mm in an individual year. Based on the site-specific relationship between DBH and active sapwood area, a value of 0.125 m^2 was determined for the active sapwood area of our black locust stand. Table 1 shows that the ratio of total sapwood area to the stand ground surface in our study was much lower than other studies for black locust stands in the Loess Plateau region. This is due to the significantly different sapwood thicknesses that were taken into account for upscaling the stand transpiration in the different studies [25,29]. Other experimenters estimated a significantly larger sapwood thickness (ranging from 5.0 to 15.4 mm) suggesting possibly no limitation to the outermost growth, which may lead to uncertainties. Moreover, they did not give further details about how the sapwood thickness was determined in their studies. It is thus unclear if their assumed sapwood thickness corresponds to the active sapwood area.

Figure 4. Dye-stained area of the sample tree. The dye penetrated only the outermost growth ring. This area represents the hydrologically active sapwood.

To explore the possible impact of sapwood thickness on the estimated value of stand transpiration, we applied the relationships between DBH and active sapwood area of other studies (cf. Table 1) to estimate our stand transpiration, and the results are shown in Figure 5. It is evident that errors in sapwood thickness determination can have a significant effect on stand transpiration estimates (Figure 5). According to Equation (4), an error in determining the total sapwood area can lead to the same error in estimated stand transpiration.

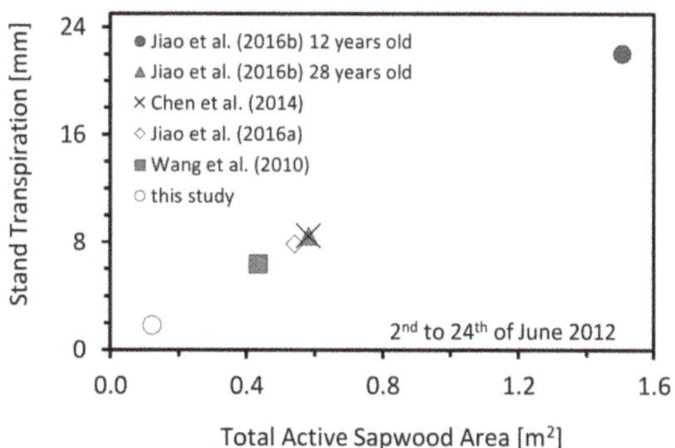

Figure 5. The possible impact of total active sapwood area on stand transpiration estimate.

3.2. Tree-Specific Calibration of Granier's Sensor System

As shown above, Granier's original calibration has a tendency of underestimating tree transpiration if it is applied directly without determining the active sapwood area for conducting water. To improve the accuracy of an estimate, an in situ experiment for the sap flow sensor calibration was developed and implemented in June 2013. In this section, the results of the black locust-specific calibration of the Granier sensor system are contrasted with the results of previous calibration studies carried out in other ring-porous tree species. Potential benefits and disadvantages regarding laboratory and field calibrations of sap flow sensor systems are elaborated.

The relationship between the outflow from the reservoir and the normalized sensor output is shown in Figure 6. The data were fitted to a power function Equation (3) of the same type as that proposed by Granier [26]. A closer inspection of Figure 6 reveals that the fitted function slightly underestimates the corrected outflow at higher k values (when k $\geq$ 0.25 in the north and k $\geq$ 0.3 in the east). The coefficient a, obtained from the fit, varied between 2.70 and 3.89 kg m^{-2} s^{-1}, and presents a significant disparity from the value (0.119 kg m^{-2} s^{-1}) found by Granier [26]. For coefficient b, we found a very similar value to the original one (1.231) as reported by Granier [26]. The k values (cf. Equation (3)) in calibration mainly ranged between 0.10 and 0.35; only 4% of the k values fell between 0.35 and 0.45 and 0.1% of them were larger than 0.5. So, extrapolation of the new calibration equation to $k > 0.35$ may induce errors in the estimation of transpiration.

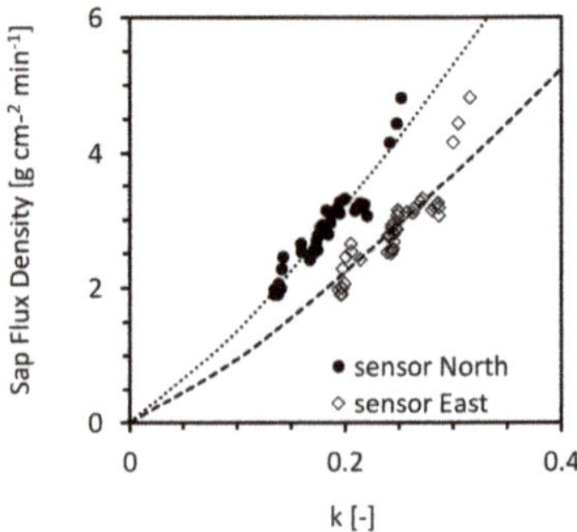

Figure 6. Relationship between the sap flux density and the normalized sensor parameter (k in Equation (3)).

A compilation of Granier's sensor calibration for different ring-porous trees is presented in Table 2. In most of the cases, calibration data differ significantly from Granier's original equation, particularly the coefficient a. It is worth noting that the values of the coefficient a vary considerably between different tree species. Possible reasons for discrepancies between calibration results of different studies have been reviewed by Sun et al. [16]. They discussed the impact of the following factors on the calibration of sap flow sensors: (i) differences in the construction of sensors; (ii) influence of the realized rate of water movement; (iii) occurrence of natural temperature gradients; (iv) errors in estimating zero-flux (cf. Equation (3); as well as (v) within- and between-tree variation. Another issue is likely important but has received surprisingly little attention so far, namely, the differences in calibration setup. The experimental setup is summarized for each case in Table 2. They differ in three aspects: size of plants (branches, stem segments, or trees), flow conditions (transient or steady state), and realization of the water flow through branches/stems (transpiration or application of external pressure). Cases showed that sensor calibration using external pressure to pull water through the stem delivered significantly larger values for the coefficient a than the calibration using xylem water movement driven by transpiration. Generally, laboratory calibration experiments on branches or stem segments using external pressures are quick, precise, and can be repeated at any time. However, it is often the case that the applied pressure was high and exceeded the in situ condition. Bush et al. [14] found that increases in pressure gradients can result in increased values of the coefficients a and b. It is thus likely that applying high pressures in calibration experiments may push water through conduits, which would not happen under natural flow conditions [16].

Moreover, the length of the segment may also play a role for a large value of a. The vessels are possibly open at both ends of a short segment. Under such conditions, sap flow densities would be overestimated [14]. Plant segments with a length ≤ 1 m were used for many calibrations of Granier's sap flow sensors [13–15,55–57]. According to Bush et al. [14], using relatively long plant segments can minimize the risk that both ends of earlywood vessels are open. Although a global analysis of xylem vessel length generally supports this view, it should be noted that the maximum vessel lengths exceeding 1 m were occasionally reported [58,59]. For this reason, it is suggested that ring-porous trees would require long stem segments for experiment [59]. Additionally, we noticed that one of the laboratory calibrations [15] also used a dye experiment to determine the active sapwood area in black locust. In this experiment, it took one hour until the dye was pulled through the 1-m long stem segments, which is twice as long as our experiment time (30 min) for a much longer length (>10 m). This can increase the possibility of error in determining active sapwood area resulted from the capillary rise. If an external pressure is applied during the dye test, the error is likely to be amplified. As a result, an extremely low value of parameter a may be obtained because of a relatively large active sapwood area (Table 2).

In contrast to laboratory calibrations, field calibration experiments on trees, in which the water flow through the xylem is driven by transpiration, have some advantages. They allow us to understand flow processes under a natural environmental condition. Without considering the limitations of the laboratory facilities, large trees can be examined. It is thus expected that xylem flow takes place mainly in the large earlywood pores of ring-porous trees in field calibration under natural flow conditions, as verified by Nagata et al. [60]. So, it is highly possible that the calibration data obtained from field experiments are more representative than the data obtained from laboratory calibration. Despite these discrepancies in sensor calibration, it can be concluded based on this study and previous research [13–15,61] that applying a universal calibration equation to ring-porous tree species will result in inaccurate estimates of tree transpiration. We highly recommend tree-specific calibration for sap flow measurement.

3.3. Validate the Accuracy of the New Calibration

To test the reliability of the sap flow estimates based on the newly derived calibration, we applied it in other selected periods of measurement and compared it with transpiration results calculated based on the soil water balance method. Periods of high evaporation demand and low rain are particularly suitable for cross-checking transpiration estimates [41], which are also dependent on the availability of datasets. Occasionally, sap flow and soil moisture measurements produced noises or failed. Data gaps were yet to be closed because gap filling could introduce additional uncertainties, and thus affect the validation. Transpiration estimates obtained from the two different methods are listed in Table 3 for periods with little or no rainfall in 2013. Moreover, transpiration estimates for the dry and warm period in 2012 (presented in Section 3.1) are also listed.

The results from both methods showed good agreement, despite that the data for soil water balance method, lysimeter method, and sap flow method were measured in different spatial extent and temporal resolution. For example, soil moisture measurements along the transect were 50 m away from the lysimeter. Small-scale variations in LAI might be another possible reason for the observed differences since LAI affects soil moisture by altering inputs of energy and precipitation. The reasonable agreement between the results makes us confident that the new calibration produces a much more reliable estimation of sap flow than Granier's original calibration.

As shown in Table 3, the mean daily stand transpiration ranged between 1.1 and 1.4 mm d^{-1}. These values are much larger than the ones ($\leq$0.5 mm d^{-1}) reported by some previous sap flow studies (Table 1) on black locust plantations [23–25,28–30]. A few studies quantified the water consumption of black locust stands in the Loess Plateau region with other methods. Wang [63] estimated ET using a water balance approach and reported that the annual evapotranspiration (ET) of a 14-year-old black locust stand, located in the semi-humid area of the Loess Plateau, ranged between 403 and 632 mm. In a more recent study of Wu et al. [64], the response of black locust seedlings to soil water availability was examined. The seedlings transpired up to about 1.4 mm d^{-1} when there was no water shortage. Ma et al. [15] and Jian et al. [28] used modified Granier methods to estimate the transpiration of black locust stands during growing season. The average values of Ma et al. [15] ranged between 1.6–2.1 mm d^{-1}, while the estimate of Jian et al. [28] ranged from 0.5 to 1.9 mm d^{-1}. Their results showed a good consistency with the findings of our study.

Table 2. Summary of the calibration of Granier's sensors for different ring-porous tree species. The coefficients a and b are fitting parameters for converting the measured temperature differences into sap flow densities according to Equation (3).

Tree Species	Characteristics of the Tree and Stem Segments	a [kg m^{-2} s^{-1}]	b [–]	Calibration Set-Up	Source
Elaeagnus Angustifolia (Russian Olive)	Diameter 4.36 cm, water-conducting sapwood area 1.70 cm^2	9.30	1.65	Lab experiments on cut stem segments: water was pulled through stem by applying varying pressure gradients, whereby the pressure was increased in several smaller steps ranging from 0.001 to 0.14 MPa m^{-1}	[14]
Gleditsia Triacanthos (Honey Locust)	Diameter 5.06 cm, water-conducting sapwood area 0.73 cm^2	30.7	1.40		
Qercus Gambelii (Gambel Oak)	Diameter 4.37 cm, water-conducting sapwood area 0.35 cm^2	58.1	1.88		
Sophora Japonica (Japanese Pagoda)	Diameter of stem segments 4.47 cm, water-conducting sapwood area 0.51 cm^2	11.90	1.24		
Quercus Pedunuculata (English Oak)	Diameter of stem segments between 4 and 5 cm	0.119	1.231	Lab experiment on cut stem segments: water was forced through stem segments under pressure, flow rate was varied by adjusting the water pressure	[26]
Quercus Petraea (Sessile Oak)	Diameter of the stem 13.7 cm	0.119	1.231	Cut tree experiment: water flow through the tree due to transpiration	[62]
Quercus Alba (White Oak)	Juvenile trees, diameter between 6.5 and 9 cm	0.128	1.47	Potometer experiments: juvenile trees were cut and placed in containers, water flow through the cut stems due to transpiration	[16]
Ulmus Americana (American Elm)	Juvenile trees, diameter between 6.5 and 9 cm	0.272	2.57		
Qercus Gambelii (Gambel Oak)	4–7 years old, diameter of the stem segment was between 5 and 6 cm	9.99 ± 6.18 (ranging from 2.38 to 18.1)	1.24 ± 0.19 (ranging from 1.05 to 1.50)	Lab experiments on cut stem segments: water was pulled through stem by applying varying pressure gradients, whereby the pressure increased in several smaller steps	[55]
Robinia Pseudoacacia (Black Locust)	15 years old stand, diameter 6–10 cm, 97–102 cm long, sapwood area 12.6–31.5 cm^2	0.051	1.18	Lab experiments on cut stem segments: water was pulled through stem by applying a series of pressure 0.005–0.04 MPa, whereby each pressure was achieved by varying the height of the reservoir and held for 30 min	[15]
Robinia Pseudoacacia (Black Locust)	33 years old stand, diameter 14.3 cm, 13 m long, active sapwood area 7.16 cm^2	3.29 (ranging from 2.70 to 3.89)	1.231	Cut tree experiment, water flow through the tree due to transpiration	this study

Table 3. Water balance components of a black locust stand with understory determined in periods with little or no rainfall. ET_{total} was determined using the zero-flux-plane method (soil water balance, Equation (1)), ET_{us} was measured using a weighable lysimeter (Equation (2)), T_{BL} was determined using the heat dissipation method (Equations (3) and (4)), and $T_{residual}$ was calculated from the difference between ET_{total} and ET_{us}. ET_{FAO} = grass reference evapotranspiration, ET = evapotranspiration, T = transpiration, US = understory, BL = black locust.

Period	Days	Rainfall (Open Land)	ET_{FAO}	ET_{total} (Soil Water Balance)	ET_{us} (Lysimeter)	T_{BL} (Sap Flow)	$T_{residual}$ ($ET_{total} - ET_{us}$)	Difference between the Two Transpiration Estimates ($T_{BL} - T_{residual}$)
				[mm]				
2 June 2012–24 June 2012	23	4.4	77.3	82.2	51.0	27.7 (1.2 *)	31.2	11%
29 May 2013–7 June 2013	10	2.2	39.0	34.4	24.7	13.9 (1.4)	9.7	30%
12 June 2013–18 June 2013	8	0.7	25.2	23.2	17.4	7.0 (1.1)	5.8	26%
23 June 2013–30 June 2013	8	1.1	39.0	33.2	23.5	10.4 (1.3)	9.7	7%
28 July 2013–6 August 2013	10	0.3	46.2	39.7	30.6	12.1 (1.2)	9.1	33%
12 August 2013–23 August 2013	12	0.0	51.7	41.8	29.5	14.0 (1.2)	12.3	14%

(*) The transpiration rates of the black locust stand in mm d^{-1} are shown in parentheses.

4. Summary and Conclusions

We found a large difference (ca. 30 mm) between the estimates of tree transpiration using two different methods—soil water balance and sap flow measurements based on Granier's original calibration—during a dry and warm period in June 2012. The estimate by applying Granier's original calibration was extremely low despite a similar range with other studies using Granier's original calibration on the Loess Plateau, while the estimate from soil water balance showed a reasonable result. A series of thorough analyses of the possible sources of error and exploration of datasets from our measurements indicate that the extremely low value of Granier's method probably stemmed from the application of a universal calibration and the inaccurate determination of the active sapwood area, which can lead to a significant error during scaling-up of stand transpiration. Our dye tracer test showed that most of the sap flow took place in larger earlywood vessels of the outer ring. Installation of sap flow sensors in non-active sapwood is another possible source of uncertainty.

An easy-to-install measurement setup for an in situ calibration of Granier's sensor system in living trees was developed and applied for black locust. The results showed that Granier's original calibration significantly underestimated the stand transpiration of black locust stands. A new calibration of Granier's sap flow sensor system for black locust can improve the accuracy of estimated sap flow and stand transpiration. Tree-specific calibration is recommended for future research on the use of Granier's method in ring-porous trees. Moreover, the use of short sensors (5 or 10 mm) can reduce the uncertainties due to sensor access to the non-active sapwood in ring-porous trees. Many researchers concluded that the black locust plantation is one of the main contributors to the significant decline of river discharge in the middle reaches of the Yellow River, according to their analyses of the long-term data on climate, streamflow trends, and land-use change. Our work confirms these conclusions, shown by a higher transpiration rate of black locust plantations on the Loess Plateau. More studies and evidence are required to clarify the role of afforestation and tree species to address regional environmental issues and their potential impacts on ecosystem services, which are directly linked to regional soil, water, and food security. In particular, the contribution of understory—which is an important feature of these plantation forests—to the total water consumption of such ecosystems needs to be understood. This will provide the basis for the development of forest management strategies that better balance forest-related ecosystem services such as soil conservation and water supply.

Acknowledgments: This study was funded by the German Research Foundation (DFG Schw1448-3/1). We would like to thank Atiqah Fairuz Binte Md Salleh for language editing.

Author Contributions: Kai Schwärzel, Andreas Strecker and Christian Podlasly conceived, designed and performed the experiments; Kai Schwärzel and Andreas Strecker analyzed the data; Kai Schwärzel and Lulu Zhang wrote the paper.

Conflicts of Interest: The authors declare no conflict of interest.

References

1. Kim, H.K.; Park, J.; Hwang, I. Investigating water transport through the xylem network in vascular plants. *J. Exp. Bot.* **2014**, *65*, 1895–1904. [CrossRef] [PubMed]
2. Vandegehuchte, M.W.; Steppe, K. Sap-flux density measurement methods: Working principles and applicability. *Funct. Plant Biol.* **2013**, *40*, 213–223. [CrossRef]
3. Köstner, B.; Granier, A.; Cermák, J. Sapflow measurements in forest stands: Methods and uncertainties. In *Annales des Sciences Forestières*; EDP Sciences: Les Ulis, France, 1998; Volume 55, pp. 13–27.
4. Regalado, C.M.; Ritter, A. An alternative method to estimate zero flow temperature differences for Granier's thermal dissipation technique. *Tree Physiol.* **2007**, *27*, 1093–1102. [CrossRef] [PubMed]
5. Rabbel, I.; Diekkrüger, B.; Voigt, H.; Neuwirth, B. Comparing Tmax Determination Approaches for Granier-Based Sapflow Estimations. *Sensors* **2016**, *16*, 2042. [CrossRef] [PubMed]
6. Oren, R.; Phillips, N.; Ewers, B.E.; Pataki, D.E.; Megonigal, J.P. Sap-flux-scaled transpiration responses to light, vapor pressure deficit, and leaf area reduction in a flooded *Taxodium distichum* forest. *Tree Physiol.* **1999**, *19*, 337–347. [CrossRef] [PubMed]

7. Clearwater, M.J.; Meinzer, F.C.; Andrade, J.L.; Goldstein, G.; Holbrook, N.M. Potential errors in measurement of nonuniform sap flow using heat dissipation probes. *Tree Physiol.* **1999**, *19*, 681–687. [CrossRef] [PubMed]

8. Paudel, I.; Kanety, T.; Cohen, S. Inactive xylem can explain differences in calibration factors for thermal dissipation probe sap flow measurements. *Tree Physiol.* **2013**, *33*, 986–1001. [CrossRef] [PubMed]

9. Ford, C.R.; Hubbard, R.M.; Kloeppel, B.D.; Vose, J.M. A comparison of sap flux-based evapotranspiration estimates with catchment-scale water balance. *Agric. For. Meteorol.* **2007**, *145*, 176–185. [CrossRef]

10. Wullschleger, S.D.; Childs, K.W.; King, A.W.; Hanson, P.J. A model of heat transfer in sapwood and implications for sap flux density measurements using thermal dissipation probes. *Tree Physiol.* **2011**, *31*, 669–679. [CrossRef] [PubMed]

11. Vergeynst, L.L.; Vandegehuchte, M.W.; McGuire, M.A.; Teskey, R.O.; Steppe, K. Changes in stem water content influence sap flux density measurements with thermal dissipation probes. *Trees* **2014**, *28*, 949–955. [CrossRef]

12. Hölttä, T.; Linkosalo, T.; Riikonen, A.; Sevanto, S.; Nikinmaa, E. An analysis of Granier sap flow method, its sensitivity to heat storage and a new approach to improve its time dynamics. *Agric. For. Meteorol.* **2015**, *211*, 2–12. [CrossRef]

13. Steppe, K.; De Pauw, D.J.; Doody, T.M.; Teskey, R.O. A comparison of sap flux density using thermal dissipation, heat pulse velocity and heat field deformation methods. *Agric. For. Meteorol.* **2010**, *150*, 1046–1056. [CrossRef]

14. Bush, S.E.; Hultine, K.R.; Sperry, J.S.; Ehleringer, J.R. Calibration of thermal dissipation sap flow probes for ring-and diffuse-porous trees. *Tree Physiol.* **2010**, *30*, 1545–1554. [CrossRef] [PubMed]

15. Ma, C.; Luo, Y.; Shao, M.; Li, X.; Sun, L.; Jia, X. Environmental controls on sap flow in black locust forest in Loess Plateau, China. *Sci. Rep.* **2017**, *7*, 13160. [CrossRef] [PubMed]

16. Sun, H.; Aubrey, D.P.; Teskey, R.O. A simple calibration improved the accuracy of the thermal dissipation technique for sap flow measurements in juvenile trees of six species. *Trees* **2012**, *26*, 631–640. [CrossRef]

17. Domec, J.-C.; Sun, G.; Noormets, A.; Gavazzi, M.J.; Treasure, E.A.; Cohen, E.; Swenson, J.J.; McNulty, S.G.; King, J.S. A comparison of three methods to estimate evapotranspiration in two contrasting loblolly pine plantations: Age-related changes in water use and drought sensitivity of evapotranspiration components. *For. Sci.* **2012**, *58*, 497–512. [CrossRef]

18. Schütt, P.; Weisgerber, H.; Schuck, H.J.; Lang, U.M.; Stimm, B.; Roloff, A. *Enzyklopädie der Laubbäume: Die große Enzyklopädie*; Nikol Verlag Barkhausenweg: Hamburg, Germany, 2006; ISBN 978-3-937872-39-1.

19. Vítková, M.; Müllerová, J.; Sádlo, J.; Pergl, J.; Pyšek, P. Black locust (*Robinia pseudoacacia*) beloved and despised: A story of an invasive tree in Central Europe. *For. Ecol. Manag.* **2017**, *384*, 287–302. [CrossRef]

20. Lambdon, P.W.; Pyšek, P.; Basnou, C.; Hejda, M.; Arianoutsou, M.; Essl, F.; Jarošík, V.; Pergl, J.; Winter, M.; Anastasiu, P. Alien flora of Europe: Species diversity, temporal trends, geographical patterns and research needs. *Preslia* **2008**, *80*, 101–149.

21. Cao, S.; Chen, L.; Shankman, D.; Wang, C.; Wang, X.; Zhang, H. Excessive reliance on afforestation in China's arid and semi-arid regions: Lessons in ecological restoration. *Earth-Sci. Rev.* **2011**, *104*, 240–245. [CrossRef]

22. Wang, J.J.; Hu, C.X.; Bai, J.; Gong, C.M. Carbon sequestration of mature black locust stands on the Loess Plateau, China. *Plant Soil Environ.* **2015**, *61*, 116–121.

23. Zhang, J.-G.; Guan, J.-H.; Shi, W.-Y.; Yamanaka, N.; Du, S. Interannual variation in stand transpiration estimated by sap flow measurement in a semi-arid black locust plantation, Loess Plateau, China. *Ecohydrology* **2015**, *8*, 137–147. [CrossRef]

24. Jiao, L.; Lu, N.; Sun, G.; Ward, E.J.; Fu, B. Biophysical controls on canopy transpiration in a black locust (*Robinia pseudoacacia*) plantation on the semi-arid Loess Plateau, China. *Ecohydrology* **2016**, *9*, 1068–1081. [CrossRef]

25. Jiao, L.; Lu, N.; Fu, B.; Gao, G.; Wang, S.; Jin, T.; Zhang, L.; Liu, J.; Zhang, D. Comparison of transpiration between different aged black locust (*Robinia pseudoacacia*) trees on the semi-arid Loess Plateau, China. *J. Arid Land* **2016**, *8*, 604–617. [CrossRef]

26. Granier, A. Une nouvelle méthode pour la mesure du flux de sève brute dans le tronc des arbres. In *Annales des Sciences Forestières*; EDP Sciences: Les Ulis, France, 1985; Volume 42, pp. 193–200.

27. Renninger, H.J.; Schäfer, K.V.R. Comparison of tissue heat balance-and thermal dissipation-derived sap flow measurements in ring-porous oaks and a pine. *Front. Plant Sci.* **2012**, *3*, 103. [CrossRef] [PubMed]

28. Jian, S.; Zhao, C.; Fang, S.; Yu, K. Effects of different vegetation restoration on soil water storage and water balance in the Chinese Loess Plateau. *Agric. For. Meteorol.* **2015**, *206*, 85–96. [CrossRef]

29. Wang, Y.-L.; Liu, G.-B.; Kume, T.; Otsuki, K.; Yamanaka, N.; Du, S. Estimating water use of a black locust plantation by the thermal dissipation probe method in the semiarid region of Loess Plateau, China. *J. For. Res.* **2010**, *15*, 241–251. [CrossRef]

30. Chen, L.; Zhang, Z.; Zeppel, M.; Liu, C.; Guo, J.; Zhu, J.; Zhang, X.; Zhang, J.; Zha, T. Response of transpiration to rain pulses for two tree species in a semiarid plantation. *Int. J. Biometeorol.* **2014**, *58*, 1569–1581. [CrossRef] [PubMed]

31. Wu, Y.; Zhang, Y.; An, J.; Liu, Q.; Lang, Y. Sap flow of black locust in response to environmental factors in two soils developed from different parent materials in the lithoid mountainous area of North China. *Trees* **2018**, 1–14. [CrossRef]

32. Chen, Y.; Wang, K.; Lin, Y.; Shi, W.; Song, Y.; He, X. Balancing green and grain trade. *Nat. Geosci.* **2015**, *8*, 739–741. [CrossRef]

33. Zhang, L.; Podlasly, C.; Feger, K.-H.; Wang, Y.; Schwärzel, K. Different land management measures and climate change impacts on the runoff–A simple empirical method derived in a mesoscale catchment on the Loess Plateau. *J. Arid Environ.* **2015**, *120*, 42–50. [CrossRef]

34. Sun, G.; Zhou, G.; Zhang, Z.; Wei, X.; McNulty, S.G.; Vose, J.M. Potential water yield reduction due to forestation across China. *J. Hydrol.* **2006**, *328*, 548–558. [CrossRef]

35. Feng, X.; Fu, B.; Piao, S.; Wang, S.; Ciais, P.; Zeng, Z.; Lü, Y.; Zeng, Y.; Li, Y.; Jiang, X. Revegetation in China's Loess Plateau is approaching sustainable water resource limits. *Nat. Clim. Chang.* **2016**, *6*, 1019–1022. [CrossRef]

36. Zimmermann, M.H. Hydraulic architecture of some diffuse-porous trees. *Can. J. Bot.* **1978**, *56*, 2286–2295. [CrossRef]

37. Vertessy, R.A.; Hatton, T.J.; Reece, P.; O'sullivan, S.K.; Benyon, R.G. Estimating stand water use of large mountain ash trees and validation of the sap flow measurement technique. *Tree Physiol.* **1997**, *17*, 747–756. [CrossRef] [PubMed]

38. Braun, P.; Schmid, J. Sap flow measurements in grapevines (*Vitis vinifera* L.) 2. Granier measurements. *Plant Soil* **1999**, *215*, 47–55. [CrossRef]

39. Yu, M.; Zhang, L.; Xu, X.; Feger, K.-H.; Wang, Y.; Liu, W.; Schwärzel, K. Impact of land-use changes on soil hydraulic properties of Calcaric Regosols on the Loess Plateau, NW China. *J. Plant Nutr. Soil Sci.* **2015**, *178*, 486–498. [CrossRef]

40. Du, S.; Wang, Y.-L.; Kume, T.; Zhang, J.-G.; Otsuki, K.; Yamanaka, N.; Liu, G.-B. Sapflow characteristics and climatic responses in three forest species in the semiarid Loess Plateau region of China. *Agric. For. Meteorol.* **2011**, *151*, 1–10. [CrossRef]

41. Schwärzel, K.; Menzer, A.; Clausnitzer, F.; Spank, U.; Häntzschel, J.; Grünwald, T.; Köstner, B.; Bernhofer, C.; Feger, K.-H. Soil water content measurements deliver reliable estimates of water fluxes: A comparative study in a beech and a spruce stand in the Tharandt forest (Saxony, Germany). *Agric. For. Meteorol.* **2009**, *149*, 1994–2006. [CrossRef]

42. Bogena, H.R.; Herbst, M.; Huisman, J.A.; Rosenbaum, U.; Weuthen, A.; Vereecken, H. Potential of wireless sensor networks for measuring soil water content variability. *Vadose Zone J.* **2010**, *9*, 1002–1013. [CrossRef]

43. Qu, W.; Bogena, H.R.; Huisman, J.A.; Vereecken, H. Calibration of a novel low-cost soil water content sensor based on a ring oscillator. *Vadose Zone J.* **2013**, *12*. [CrossRef]

44. Khalil, M.; Sakai, M.; Mizoguchi, M.; Miyazaki, T. Current and prospective applications of zero flux plane (ZFP) method. *J. Jpn. Soc. Soil Phys.* **2003**, *95*, 75–90.

45. Podlasly, C.; Schwärzel, K. Development of a continuous closed pipe system for controlling soil temperature at the lower boundary of weighing field lysimeters. *Soil Sci. Soc. Am. J.* **2013**, *77*, 2157–2163. [CrossRef]

46. Sano, Y.; Okamura, Y.; Utsumi, Y. Visualizing water-conduction pathways of living trees: Selection of dyes and tissue preparation methods. *Tree Physiol.* **2005**, *25*, 269–275. [CrossRef] [PubMed]

47. Clausnitzer, F.; Köstner, B.; Schwärzel, K.; Bernhofer, C. Relationships between canopy transpiration, atmospheric conditions and soil water availability—Analyses of long-term sap-flow measurements in an old Norway spruce forest at the Ore Mountains/Germany. *Agric. For. Meteorol.* **2011**, *151*, 1023–1034. [CrossRef]

48. Allen, R.G.; Pereira, L.S.; Raes, D.; Smith, M. Crop evapotranspiration-Guidelines for computing crop water requirements-FAO Irrigation and drainage paper 56. *FAO Rome* **1998**, *300*, D05109.

49. Wang, L.; Wei, S.; Horton, R.; Shao, M. Effects of vegetation and slope aspect on water budget in the hill and gully region of the Loess Plateau of China. *Catena* **2011**, *87*, 90–100. [CrossRef]

50. McCulloh, K.; Sperry, J.S.; Lachenbruch, B.; Meinzer, F.C.; Reich, P.B.; Voelker, S. Moving water well: Comparing hydraulic efficiency in twigs and trunks of coniferous, ring-porous, and diffuse-porous saplings from temperate and tropical forests. *New Phytol.* **2010**, *186*, 439–450. [CrossRef] [PubMed]

51. Kume, T.; Otsuki, K.; Du, S.; Yamanaka, N.; Wang, Y.-L.; Liu, G.-B. Spatial variation in sap flow velocity in semiarid region trees: Its impact on stand-scale transpiration estimates. *Hydrol. Process.* **2012**, *26*, 1161–1168. [CrossRef]

52. Umebayashi, T.; Utsumi, Y.; Koga, S.; Inoue, S.; Fujikawa, S.; Arakawa, K.; Matsumura, J.; Oda, K. Conducting pathways in north temperate deciduous broadleaved trees. *IAWA J.* **2008**, *29*, 247–263. [CrossRef]

53. Zimmerman, M.H.; Brown, C.L. *Trees: Structure and Function*; Springer-Verlag: New York, NY, USA, 1971.

54. Huber, B.; Schmidt, E. Weitere thermoelektrische Untersuchungen über den Transpirationsstrom der Bäume. *Tharandter Forstl. Jahrb.* **1936**, *87*, 369–412.

55. Taneda, H.; Sperry, J.S. A case-study of water transport in co-occurring ring-versus diffuse-porous trees: Contrasts in water-status, conducting capacity, cavitation and vessel refilling. *Tree Physiol.* **2008**, *28*, 1641–1651. [CrossRef] [PubMed]

56. De Oliveira Reis, F.; Campostrini, E.; de Sousa, E.F.; e Silva, M.G. Sap flow in papaya plants: Laboratory calibrations and relationships with gas exchanges under field conditions. *Sci. Hortic.* **2006**, *110*, 254–259. [CrossRef]

57. Hultine, K.R.; Nagler, P.L.; Morino, K.; Bush, S.E.; Burtch, K.G.; Dennison, P.E.; Glenn, E.P.; Ehleringer, J.R. Sap flux-scaled transpiration by tamarisk (*Tamarix* spp.) before, during and after episodic defoliation by the saltcedar leaf beetle (*Diorhabda carinulata*). *Agric. For. Meteorol.* **2010**, *150*, 1467–1475. [CrossRef]

58. Jacobsen, A.L.; Pratt, R.B.; Tobin, M.F.; Hacke, U.G.; Ewers, F.W. A global analysis of xylem vessel length in woody plants. *Am. J. Bot.* **2012**, *99*, 1583–1591. [CrossRef] [PubMed]

59. Zimmermann, M.H.; Jeje, A.A. Vessel-length distribution in stems of some American woody plant. *Can. J. Bot.* **1981**, *59*, 1882–1892.

60. Nagata, A.; Kose, K.; Terada, Y. Development of an outdoor MRI system for measuring flow in a living tree. *J. Magn. Reson.* **2016**, *265*, 129–138. [CrossRef] [PubMed]

61. Lu, P.; Urban, L.; Zhao, P. Granier's thermal dissipation probe (TDP) method for measuring sap flow in trees: Theory and practice. *Acta Bot. Sin.* **2004**, *46*, 631–646.

62. Granier, A.; Anfodillo, T.; Sabatti, M.; Cochard, H.; Dreyer, E.; Tomasi, M.; Valentini, R.; Bréda, N. Axial and radial water flow in the trunks of oak trees: A quantitative and qualitative analysis. *Tree Physiol.* **1994**, *14*, 1383–1396. [CrossRef] [PubMed]

63. Wang, Y. The hydrological influence of black locust plantations in the loess area of northwest China. *Hydrol. Process.* **1992**, *6*, 241–251.

64. Wu, Y.Z.; Huang, M.B.; Warrington, D.N. Black locust transpiration responses to soil water availability as affected by meteorological factors and soil texture. *Pedosphere* **2015**, *25*, 57–71. [CrossRef]

 *forests*

Article

Contrast Effects of Vegetation Cover Change on Evapotranspiration during a Revegetation Period in the Poyang Lake Basin, China

Ying Wang [1], Yuanbo Liu [1,*] and Jiaxin Jin [2]

[1] Key Laboratory of Watershed Geographic Sciences, Nanjing Institute of Geography and Limnology, Chinese Academy of Sciences, Nanjing 210008, China; mfacewang@niglas.ac.cn

[2] School of Earth Sciences and Engineering, Hohai University, Nanjing 211100, China; jiaxinking@hhu.edu.cn

* Correspondence: ybliu@niglas.ac.cn; Tel.: +86-025-8688-2164

Received: 5 March 2018; Accepted: 17 April 2018; Published: 19 April 2018

Abstract: It is known that evapotranspiration (ET) differs before and after vegetation change in watersheds. However, impacts of vegetation change on ET remain incompletely understood. In this paper, we investigated the process-specific, nonclimatic contribution (mainly vegetation coverage changes) to ET at grid, sub-basin, and basin scales using observation and remote sensing data. The Poyang Lake Basin was selected as the study area, which experienced a fast vegetation restoration from 1983 to 2014. Our results showed that vegetation cover change produced contrasting effects on annual ET in magnitude and direction during shifts from a less covered to a more covered stage. At the early stage (1983–1990), with vegetation cover of 30%, vegetation cover change produced negative effects on ET over the basin. At the middle stage (1990–2000), the vegetation coverage increased at a fast pace and the negative effects gradually shifted to positive. At the late stage (2000–2014), the vegetation coverage remained high (over 60%) and maintained a positive relationship with ET. In summary, the vegetation effects are collaboratively influenced by both vegetation coverage and its change rate. Our findings should be helpful for a comprehensive understanding of complicated hydrological responses to anthropogenic revegetation.

Keywords: vegetation cover change; evapotranspiration; Poyang Lake Basin; remote sensing

1. Introduction

Evapotranspiration (ET), including transpiration and evaporation, represents terrestrial water consumption [1]. It is the crucial component which couples water cycle and energy balance, and hence is a subject of focus in the field of global change. ET is dominated by climatic controls, including not only demand (e.g., air temperature, air vapor pressure and wind speed) but also supply factors (precipitation and solar radiation) [2]. More importantly, it is quite sensitive to human activities, such as vegetation cover change due to anthropogenic vegetation restoration [3]. It is well known that a vegetation cover change could alter the leaf area in a canopy, consequently altering the canopy transpiration and interception area, potentially affecting canopy transpiration and evaporation [4]. Meanwhile, a variation in vegetation cover can impact land surface albedo, affecting absorbed radiation energy for evapotranspiration [5]. Nevertheless, vegetation cover change can alter the radiation allocations in soil, and therefore influence soil evaporation [6]. All these additive and/or offsetting effects will determine the total amount of evapotranspiration in response to vegetation cover change at the basin or regional scale.

Previous studies have widely investigated the influences of vegetation restoration on ET in vegetated watersheds [7–10]. However, there remains a debate on the magnitudes and directions of the influences. Using a distributed Soil and Water Assessment Tool (SWAT) model, Fang et al. [11]

observed that changes in vegetation cover classes and areas significantly affected hydrological elements in northeastern China. In addition, the increases of vegetation cover areas all led to increases in actual ET [11]. Some other studies have also supported this view and demonstrated that an increased vegetation cover could reduce runoff because of an enhanced canopy interception and ET [12–19]. In contrast, several studies have pointed to a negative relationship between vegetation cover and annual ET at multiple spatial scales [15,20]. Although the ET difference before and after revegetation has been extensively studied with field observations, remote sensing, and hydrological model simulation, how vegetation cover variation impacts ET at different stages of revegetation remains incompletely understood.

In this paper, we attempted to shed light on the differences in the effects of vegetation cover change on ET in continuous periods of a vegetation changing process, and to detect the contribution of vegetation cover and climates to ET based on field observation and remote sensing data. Here, the Poyang Lake Basin was selected as the study area, which is a typical subtropical humid basin in the middle reaches of the Yangtze River [21]. Over the last decades, the basin has experienced a fast vegetation restoration process dominated by afforestation, where forest coverage increased from 33.1% to 63.1% [22–24]. Given the important role of water resources and ecological services in southeastern China, the hydrological responses, especially the ET response, to this fast vegetation restoration over the basin have aroused extensive attention. This study is helpful to understand feedbacks of anthropogenic revegetation to hydrological processes, and supports effective water resource management in subtropical humid basins.

2. Materials and Methods

2.1. Study Area

Poyang Lake is the biggest freshwater lake in China, and is located on the south bank of the middle reaches of the Yangtze River (24.5° N–30.1° N, 113.6° E–118.5° E) (Figure 1) [21]. Its watershed size is about 1.6225×10^5 km^2, accounting for 9% of the total Yangtze River Basin. The river systems of the Poyang Lake Basin are quite developed, containing Ganjiang River, Fuhe River, Xinjiang River, Raohe River, Xiushui River, and their multiscale branch rivers. The Poyang Lake Basin is an important component of the river systems of the Yangtze River.

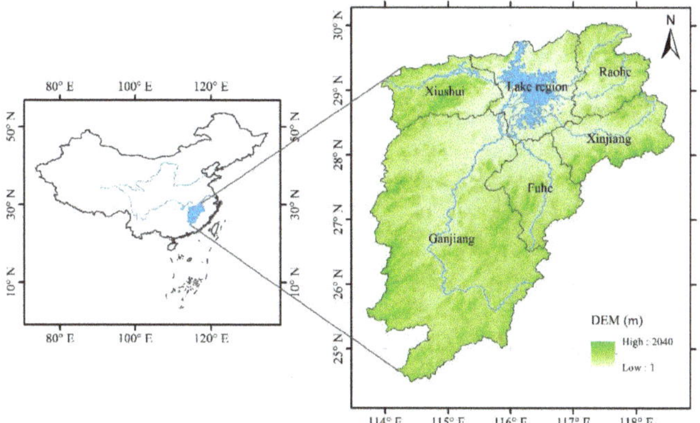

Figure 1. The geophysical location of Poyang Lake Basin, China.

The Poyang Lake Basin is a typical humid basin with a subtropical monsoon climate, of which annual temperature and precipitation range from 16.3 to 17.5 °C and from 1341 to 1943 mm, respectively. The vegetation cover in the basin experienced an outstanding dynamic process during the last decades [22–24]. The basin suffered from three severe deforestations from the 1950s to the 1970s, and the forest coverage dropped to 33.1% in 1978. Due to the reforestation in the 1990s, the forest cover gradually increased and in 2011 rose to 63.1% across the basin, implying that vegetation cover has exhibited a continuously fast increase over the past three decades, mainly resulting from anthropogenic vegetation restoration. In order to eliminate effects of the water exchange between the Poyang Lake and the Yangtze River and between runoff to evaporation, the open water body of the Poyang Lake was excluded from our present analysis.

2.2. Data Preprocessing

To calculate vegetation cover, we used the Global Inventor Modeling and Mapping Studies (GIMMS) Normalized Difference Vegetation Index (NDVI) version 3g.v1 dataset with a resolution of 0.083° covering the period of 1983–2014 at half-month intervals [25,26]. This NDVI dataset has been corrected for orbital drift effects, calibration, viewing geometry, stratospheric volcanic aerosols, and other errors unrelated to vegetation change [26,27]. The method to calculate vegetation cover is described in the following section. To eliminate white noise points from various sources, the maximum value compositing method (vegetation cover) was used to accumulate NDVI 3g dataset [28].

Two widely used satellite-based evapotranspiration datasets adopted in this study were the Advanced Very High Resolution Radiometer (AVHRR) monthly ET product [29,30] with a spatial resolution of 8 km covering the period 1983–2006, and the Moderate Resolution Imaging Spectroradiometer (MODIS) ET monthly product [31] with a resolution of 1 km covering the period 2001–2014. Both datasets were aggregated to 0.083° according to the spatial resolution of the NDVI data. Then, these two datasets were combined based on the overlapping period to obtain a long-time series of ET from 1983 to 2014.

Climatic variables potentially impacting evapotranspiration change considered in our present work were surface air temperature (temp), precipitation (prec), solar radiation (srad) and wind speed (wind) [32]. The meteorological datasets were available from China Meteorological Forcing Dataset, with a resolution of 0.1° covering the period of 1983–2014 at month intervals [31]. These meteorological data were also resampled to 0.083° based on the digital elevation model (DEM) of the Poyang Lake Basin, to match the resolution of the NDVI data.

In addition, field survey data of vegetation cover was available from the Statistics Yearbook. Monthly hydrological observation data (e.g., runoff, precipitation, etc.) were collected to validate our satellite-based vegetation cover and evapotranspiration data [33].

To reduce interannual impacts of climatic variables, we used the GIMMS NDVI data to obtain a long-time series of vegetation coverage based on a new optimized dimidiate pixel model that adopts dynamic background values and MODIS Vegetation Cover product [34].

The NDVI value of a pixel represents the combined vegetation and soil signal. According to the dimidiate pixel model, vegetation coverage (*VC*) is calculated as follows:

$$VC = \frac{NDVI - NDVI_{soil}}{NDVI_{veg} - NDVI_{soil}} \tag{1}$$

where $NDVI_{soil}$ denotes the NDVI value of a bare-soil pixel (i.e., soil background value), $NDVI_{veg}$ denotes the NDVI value of a pure-vegetation pixel (i.e., vegetation background value). To improve the precision of estimated *VC*, the NDVI background value can be optimized with observation data of *VC* as follows:

$$NDVI_{soil} = \frac{VC_{max}NDVI_{min} - VC_{min}NDVI_{max}}{VC_{max} - VC_{min}} \tag{2}$$

$$NDVI_{veg} = \frac{(1 - VC_{min})NDVI_{max} - (1 - VC_{max})NDVI_{min}}{VC_{max} - VC_{min}} \tag{3}$$

where VC_{max} and VC_{min} represent the maximum and minimum value of vegetation coverage in satellite image, respectively, $NDVI_{max}$ and $NDVI_{min}$ represent the corresponding maximum and minimum value of NDVI, respectively. Here the MODIS Vegetation Cover data were used to auxiliary determine $NDVI_{soil}$ and $NDVI_{veg}$.

2.3. Change Detection Approach

In this study, simple linear regression—the first-difference de-trending method—and multivariate linear regression were used at both grid and basin scales over the Poyang Lake Basin. The trends of vegetation cover, ET and climatic factors (temp, prec, srad and wind) were determined by the slope of the simple linear regression model. The first-difference de-trending method (i.e., the difference of values in one year to its previous year) was used to remove non-climatic influences (mainly vegetation cover change) on ET [32,35–38], thereby segregating the impacts of vegetation cover and climatic factors on ET.

In order to quantify the contribution of climatic factors, multivariate linear regression [2] was conducted with the first differences of climatic factors as the predictor variables, and the first difference in ET as the response variable,

$$\Delta ET = a_{temp}\Delta Temp + a_{prec}\Delta Prec + a_{srad}\Delta Srad + a_{wind}\Delta Wind + int \tag{4}$$

where ΔET is the de-trended ET; a_{temp}, a_{prec}, a_{srad}, and a_{wind} are the regression coefficient of the predictors; $\Delta Temp$, $\Delta Prec$, $\Delta Srad$, and $\Delta Wind$ are the de-trended climatic factors.

Assuming that the responses of dependent variables to climate trends and year-to-year climate variations are similar [35,36], climatic contribution to ET trends can be quantified by the above regression coefficients and the trends of climatic factors [35],

$$Q_C = a_{temp}Q_{Temp} + a_{prec}Q_{Prec} + a_{srad}Q_{Srad} + a_{wind}Q_{Wind} \tag{5}$$

where Q_{Temp}, Q_{Prec}, Q_{Srad}, and Q_{Wind} are the trends of climatic factors; Q_c is the climatic contribution to the trend of ET.

The surplus of the whole ET trend and the trend caused by climatic factors is identified as the trend caused by nonclimatic factor (e.g., anthropogenic vegetation cover change), and thus the nonclimatic contribution can be quantified [39],

$$Q_{nc} = Q - Q_C \tag{6}$$

where Q_{nc} is the non-climatic contribution to the trend of ET, Q is the trend of ET.

In addition, the relative climatic and nonclimatic contributions are calculated as follows,

$$RC_C = \frac{|Qc|}{|Qc| + |Qnc|} \tag{7}$$

$$RC_{nc} = \frac{|Qnc|}{|Qc| + |Qnc|} \tag{8}$$

where RC_c and RC_{nc} are the relative contributions of climatic and nonclimatic change, respectively.

To investigate the differences in the effects of vegetation cover change on ET in the continuous periods of the fast vegetation changing process, all the above analysis methods were applied over a 15-year moving window from 1983 to 2014. Within each 15-year window, both the relative nonclimatic contribution to ET variabilities and the corresponding mean and trend of vegetation cover were calculated to explore their relationships at different scales. All the spatial data preparation and statistical analysis of data were conducted with ArcGIS 10.3 (Esri, Redlands, CA, USA) and MATLAB R2017a (MathWorks, Natick, MA, USA).

3. Results

3.1. Temporal-Spatial Patterns of Vegetation Cover and ET over the Poyang Lake Basin

Figure 2 exhibits the vegetation coverage between satellite-based and field survey data over the Poyang Lake Basin during 1983–2014. Overall, the correlation coefficient (r) between the satellite-based and survey-based vegetation cover is 0.94, which indicates that the satellite-based vegetation cover could appropriately capture the vegetation changing process in the study area. Figure 2 also displays the interannual variation of ET and vegetation cover over the entire Poyang Lake Basin from 1983 to 2014. The vegetation grew nearly twice in area during the period. Vegetation cover was low in the 1980s (about 30%). After a quick revegetation from 1990 to 2000, it increased to 60% and has maintained a slow increase since 2000. The annual ET first decreased and then increased over the entire basin during the whole study period. Specifically, it decreased from 740 mm/year to 655 mm/year since 1983 and then increased since 2002. As of 2014, the annual ET rose to 708 mm/year.

Figure 3 compares the satellite-based ET with that calculated from a water balance equation using observed hydrological data over five sub-basins and the entire basin. The calculated/observed ET is the difference between precipitation (P) and runoff (R) in a long period. It indicates that the multi-year average of satellite-based ET was consistent with that of the calculated ET (=P − R) with a quite low error (6 mm/year) over the entire basin. At the sub-basin scale, the Fuhe Basin exhibited the largest error in which the observation-based ET was larger than the satellite-based one. This is due to the fact that about four billion cubic meters of water was transferred from the Fuhe Basin to the Ganjiang Basin to meet irrigation needs, leading to an overestimate in the Fuhe Basin and an underestimate in the Ganjiang Basin, compared to satellite-based ET.

Figure 4 shows the spatial distribution of the multiyear mean values of annual ET and vegetation cover during 1983–2014 over the basin. The multiyear average of annual ET was 718 mm and the annual vegetation cover was 41.9% in the Poyang Lake Basin. In general, vegetation cover in the upper reaches of each sub-basin was relatively high (over 50%). Vegetation cover gradually reduced from the inlands towards the river channels. Notably, the lake region exhibited the lowest vegetation cover in the whole study area. Annual ET exhibited a negative relationship with vegetation cover in space over the Poyang Lake Basin. That is, the ET was relatively lower in forestlands with a high vegetation cover than that in croplands and wetlands with a low vegetation cover. The ET was the highest near Poyang Lake over the study area, followed by the plain fields in the lower reaches. The ET was low in upland forest areas. Notably, ET was quite high (over 800 mm/year) in the southern Ganjiang Basin where the Masson pine forest dominated.

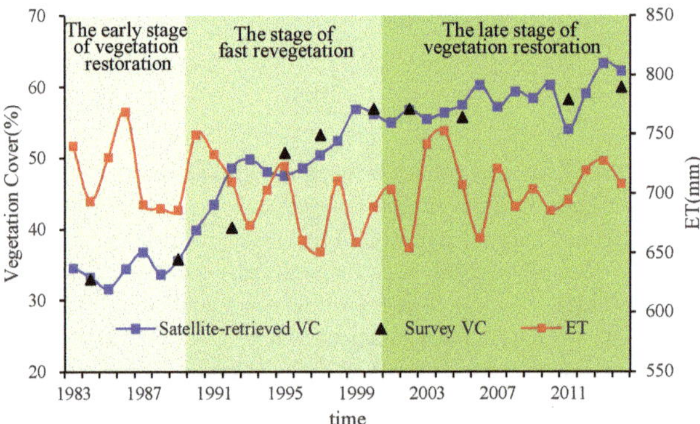

Figure 2. Interannual variation of satellite-retrieved annual evapotranspiration (ET) and vegetation cover (VC) with survey-based vegetation cover for the entire Poyang Lake Basin for the period 1983–2014.

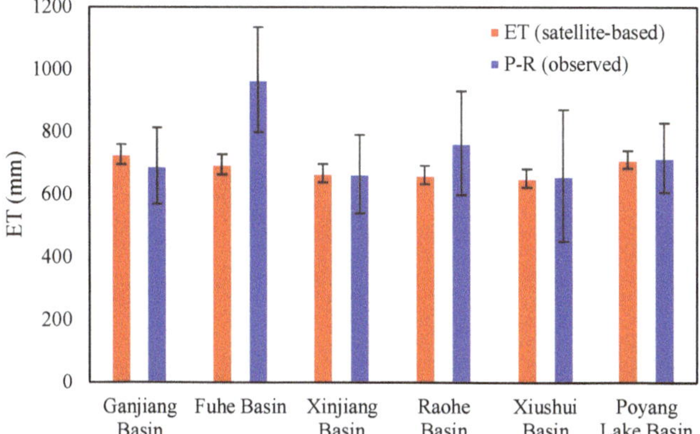

Figure 3. Comparison of satellite-retrieved multi-year mean annual ET with that calculated from a water balance equation using observed pricitpation (P) and runoff (R) over five sub-basins and the entire basin.

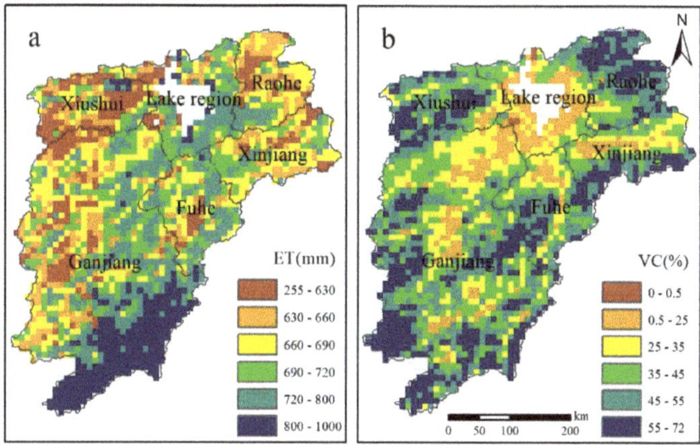

Figure 4. The spatial distribution of the mean annual ET (**a**) and vegetation coverage (VC) (**b**) during 1983–2014 in the Poyang Lake Basin.

3.2. Relative Climatic and Nonclimatic Contributions to Variabilities of ET

According to the interannual variability of vegetation cover, we divided the study period into three stages over the whole of the study area. The first stage is from 1983 to 1990, which is the early stage of vegetation restoration. The second stage is from 1990 to 2000, which is the stage of fast revegetation. The third stage is from 2000 to 2014, which is the stage of slow revegetation.

Figure 5 illustrates the relative climatic and nonclimatic contributions to ET variabilities in each grid across the whole basin. The relative non-climatic contribution (anthropogenic vegetation cover change) were negative in 1983–1990, 1990–2000, and positive in 2000–2014. Specifically, the nonclimatic contribution was −85.1% in 1983–1990, −88.6% in 1990–2000, and 56.6% in 2000–2014. By contrast, the relative climatic contribution was 14.9%, −11.4% and 43.4%, respectively. It indicates that the nonclimatic factors contributed more than the climatic factors did.

During the early stage of vegetation restoration (1983–1990), the relative nonclimatic contribution to ET change was mainly negative in the Raohe Basin, Xinjiang Basin, and Fuhe Basin, which accounts for 70% of the study area. In the second stage (1990–2000), the relative contribution was 60% for over 90% of the study area. During the third stage (2000–2014), the nonclimatic contribution became positive for nearly half the area, mostly in the Xinjiang Basin, Fuhe Basin, and Xiushui Basin. The contribution remained as a weak negative effect in the Raohe Basin and south of Ganjiang Basin.

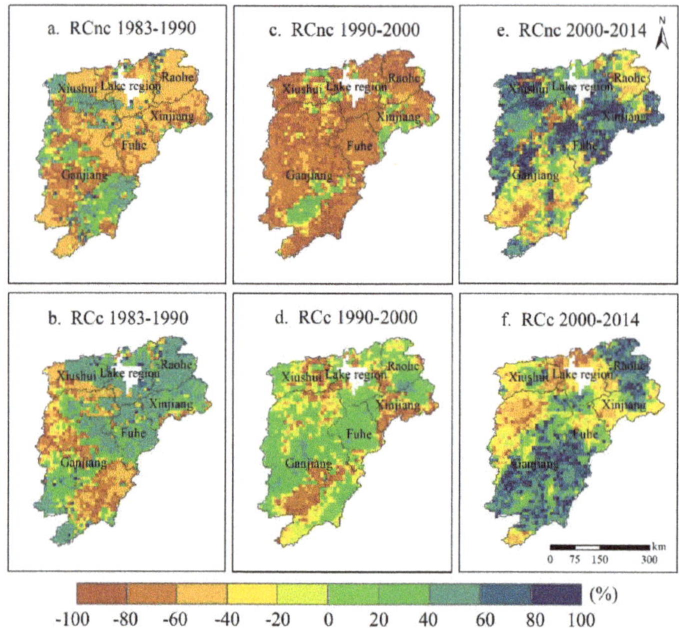

Figure 5. Relative climatic and nonclimatic contributions (RC_c and RC_{nc}) to grid-wise ET change for three stages of vegetation restoration: 1983–1990 (**a,b**), 1990–2000 (**c,d**), and 2000–2014 (**e,f**).

Figure 6 shows climatic and nonclimatic contributions to annual ET trends in absolute value in the Poyang Lake Basin and its sub-basins. At the early stage of vegetation restoration (1983–1990), the nonclimatic factors contributed to annual ET trends (Q_{nc}) by −5.1 mm/year, and the climatic factors contributed by 3.0 mm/year. Moreover, the Q_{nc} values were negative while the Q_c values were positive in all sub-basins. At the second stage of fast restoration (1990–2000), Q_c decreased to 0.4 mm/year but Q_{nc} increased to −6.6 mm/year over the entire Poyang Lake Basin. All the Q_{nc} values were negative in five sub-basins with the values higher than 5 mm/year. Q_c was lower than Q_{nc}. It was positive in the Raohe, Fuhe and Ganjiang Basins, but negative in the Xiushui and Xinjiang Basins. In the late stage (2000–2014), Q_{nc} became the lowest and was positive except for the Raohe Basin. Q_c remained positive in the sub-basins except the Xinjiang and Xiushui Basins.

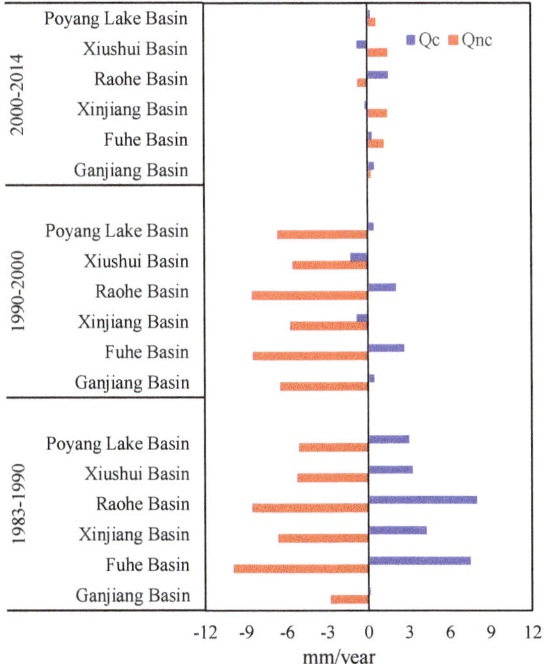

Figure 6. Climatic and nonclimatic contributions to annual ET trends (Q_c and Q_{nc}) over the entire Poyang Lake Basin and its five sub-basins.

3.3. Relationships between Non-Climatic Contribution and Vegetation Cover

The above analysis demonstrates that the nonclimatic contribution to ET varied dramatically in the study period, implying that the nonclimatic factors may have played important roles in regulating basin-scale ET. Given that the dominant nonclimatic variable was anthropogenic vegetation cover change in the present study, here we examined the relationships between the nonclimatic contribution and the corresponding vegetation cover, based on the moving window method, for further interpretation of potential effects of vegetation cover change on ET.

Figure 7 illustrates that the relative nonclimatic contribution to ET shifted from a negative phase (−85%) to a positive phase (74%) over time for the entire basin. Specifically, at the early stage of vegetation restoration, the contribution was negative. In the moving window of 1983–1998, the relative nonclimatic contribution was about −85%, while vegetation cover was the lowest (about 41%) in the period. In the subsequent moving windows, the relative nonclimatic contribution approached to the negative maximum in 1986–2001, during the region which experienced fast vegetation restoration. The change rate of vegetation cover was as fast as 0.23%/year in this period. With the development of revegetation, vegetation cover kept the increasing trend but slowed down gradually, and the change rate of vegetation cover reduced gradually in the consecutive windows. Meanwhile, the absolute nonclimatic contribution to ET dropped. Remarkably, the relative nonclimatic contribution reversed from negative to positive in the period of 1996–2011, during which vegetation coverage recovered to 55%. The positive contribution continued in the following periods, during which vegetation coverage slightly increased.

During the study period, vegetation cover increased from 41% to 57%, corresponding to its change rate from 2.1%/year to 0.5%/year. The vegetation cover change was closely related to the nonclimatic contribution with a correlation coefficient of 0.90 ($p < 0.05$) for their mean values and -0.93 ($p < 0.05$) for trend. This close correlation was also observed in each sub-basin (Figure 7). The correlation coefficient was 0.82 (-0.81) for mean (trend) across the sub-basins, which was close to the correlation coefficient calculated from all valid pixels. The correlation coefficient was the highest for mean in Xinjiang Basin ($r = 0.85$) and the lowest in Xiushui Basin ($r = 0.79$). The vegetation cover is the highest in Xiushui Basin and the lowest in Xinjiang Basin. The lower vegetation cover was, the more significant the vegetation cover influence to ET was, and vice versa. For trend, the correlation coefficient was the strongest in Xiushui Basin ($r = -0.93$) and the weakest in Xinjiang Basin ($r = -0.67$). For the sub-basin with the most increase in vegetation coverage, the trend showed a closer relationship with the effect of vegetation coverage to ET, and vice versa.

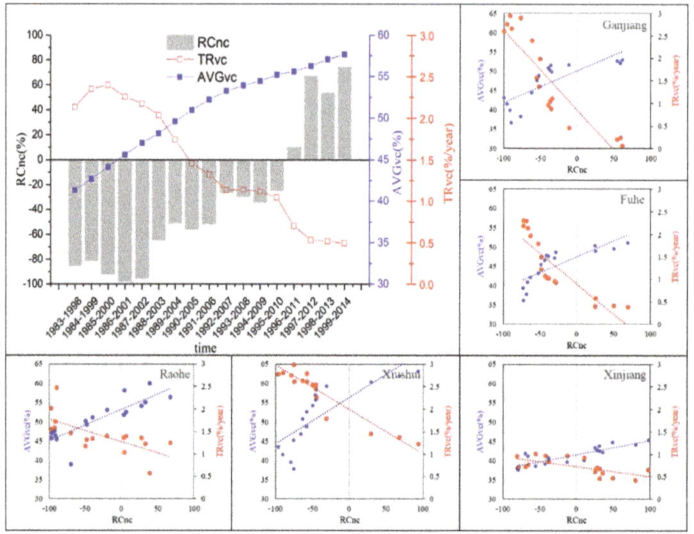

Figure 7. Relationships between vegetation cover change (VC) and nonclimatic contribution to annual ET (*RCnc*) for mean (AVGvc) and trend (TRvc) values within moving windows over the Poyang Lake Basin and its five sub-basins.

4. Discussion

4.1. Evapotranspiration in Response to Revegetation

Hydrological responses to vegetation degeneration or restoration have been widely examined. For effective basin-scale water management, not only the overall responses but also respective responses during different stages of vegetation restoration need to be investigated. This study clearly demonstrates that vegetation cover and its change rate are the main nonclimatic factors collectively affecting ET at different scales. At the early stage of vegetation restoration, when vegetation cover was relatively low, ET was from both vegetation transpiration and soil evaporation. Under this circumstance, change of vegetation cover strongly affected the composition and magnitude of transpiration, evaporation, and consequently the total ecosystem ET [40]. At the stage of fast revegetation, energy and water allocations were obviously altered in landscapes composed of vegetated and bare surfaces. Previous studies have suggested that ET was constrained by energy in the Poyang Lake Basin [24]. Two potential causes account for the ET decline with increasing vegetation cover. Firstly, solar radiation may decrease on the soil surface due to the expansion of vegetation leaf area, leading to a decrease in soil evaporation [6]. Secondly, increased interception with afforestation likely results in an increase in baseflow and runoff rather than ET, because the transpiration of new plants is lower than that of mature plants despite having the same leaf area index (LAI) [41]. A contrasting result was observed in North China where ET was constrained by water supply [40]. At the late stage of vegetation restoration, when vegetation cover was high, both vegetation and climatic factors produced relatively low effects on ET, suggesting that the forest ecosystem established a relatively stable state for ET. With a closed vegetation canopy, a slow increase of vegetation cover may have enhanced canopy transpiration but reduced contribution to surface evaporation. This is consistent with the study [41] that transpiration accounted for a high ratio of ET that marginally increased with LAI. Given the factors mentioned above, vegetation cover change on the basin-scale ET was attributed not only to coverage but also to altered local composition of vegetation and soil. Hence, the vegetation cover change exhibited a contrary effect to ET, which turned from negative to positive during the period 1996–2011 when the vegetation coverage increased to about 55%.

It should be noted that annual ET was generally lower after vegetation restoration than before, when vegetation coverage was quite low over the Poyang Lake Basin. It implies that anthropogenic revegetation may have produced a systematic shift between water cycle components [7,9,15,40]. The ET decrease accompanied with an increase of runoff, confirmed from observation data that annual runoff coefficient between runoff and precipitation increased from 0.38 in 2004 to 0.52 in 2014. The main reason may be that more precipitation was converted to runoff due to higher soil moisture, stemming from relatively constant transportation and relatively low soil evaporation for high vegetation cover. A high vegetation cover could also enhance the runoff yield due to precipitation interception of a large leaf area of canopy. A recent study has reported that increased vegetation greenness resulted in reduced water yield, which may elevate or aggravate water conflicts in droughts [19]. Our results demonstrate that vegetation restoration overall favors annual runoff yield, and its water retention plays a positive role in supplying water resources for the Poyang Lake Basin and downstream areas of the Yangtze River as well.

4.2. Uncertainties and Further Studies

There are limitations in the present report. First of all, in spite of evaluation with observation data, the remotely sensed ET contains uncertainties from satellite data quality, the retrieval algorithm, and parameter optimization [42]. The static land cover data were adopted in AVHRR and MODIS ET products used in this study [31,43]. In the algorithm logic of the products, land cover data were used to aid in identifying biome-specific physiological parameters, especially the optimized canopy conductance (g_0). Hence, the static land cover classifications may lead to some uncertainties in calculating pixel-based ET.

Based on the MODIS illustrates that land cover data in 2001, four plant types mainly occurred in the Poyang Lake basin, including evergreen broadleaf forest (EBF), mixed forest (MF), woody savanna (WSV), and cropland (CRP) which accounted for 1.64%, 40.61%, 34.43%, and 21.31% of the total study area, respectively. In light of the study of Zhang et al. [30], the g_0 varied with not only the plant type but also with the plant physiology, which could be represented by NDVI. In view of this point, the ranges of the variability in NDVI and corresponding g_0 were investigated for each plant type in our study, as shown in Figures S1 and S2. The results showed that the average g_0 was 0.0091 ± 0.0025 m S^{-1}, 0.0069 ± 0.0034 m S^{-1}, 0.0033 ± 0.0031 m S^{-1}, and 0.0033 ± 0.0021 m S^{-1} for EBF, MF, WSV and CRP, respectively. The variance analysis (ANOVA) showed that there was no significant ($p < 0.05$) difference in the g_0 amplitudes between EBF and MF, WSV and CRP, as well as MF and WSV, and MF and CRP, respectively. It indicates that the conversions between these pairs of plant types would not significantly alert the g_0 and consequently the calculation of ET. However, the g_0 for EBF exhibited a significant difference from that of WSV and CRP. To reduce this uncertainty, the distribution and conversion of the land cover types were further investigated based on the MODIS land cover data and another land cover data provided by Liu et al. [44]. The MODIS land cover data suggested that EBF was mainly located in the mountain area in the southern portion of the Poyang Lake basin (below 25° N). In spite of the difference in the classification systems between MODIS and Liu's land cover products, the latter showed that the forest land with high canopy density (>30%) in the southern mountain area rarely was converted to other land cover types (especially cropland) between 1980 and 2015. The conversion area only accounted for 0.37% of the total basin. It implies that the EBF conversion marginally impacted the variation in ET over the sub-basin or the entire basin. Hence, although the static land cover data were used, both the AVHRR and MODIS ET data had reason to be used in our present work. So far, there is a lack of continuous long-term historical land cover data which could be reliably used to derive ET. This limitation should be taken seriously in future studies.

In addition to vegetation cover change, local water resources management such as anthropogenic water storage, and diversion could also play an important role in regulating basin-scale ET [45]. It may also lead to uncertainties in interpreting our present results. Moreover, to facilitate the distinction of climatic and nonclimatic contributions to basin-scale ET, only four climatic variables were selected in first-difference model. Although the variables are commonly recognized as dominant regulators of ecosystem-scale ET, neglecting other variables may lead to a poor performance of the model, and consequently an underestimate/overestimate of climatic and nonclimatic contributions. Besides climatic variables, it has been reported that enhanced CO_2 concentration and atmospheric nitrogen deposition could also affect ET through altering plant activities and water-use efficiency [29,46]. Although there is still a debate on trade-off fertilization effects of enhanced CO_2 and nitrogen deposition on water cycles [47], it would be helpful to more accurately capture interannual variation of climatic and nonclimatic contributions to ET.

5. Conclusions

ET is a key component of both terrestrial water cycle and surface energy balance. It is impacted by both climatic and nonclimatic factors such as anthropogenic vegetation cover change. This study presented how the vegetation cover change could affect annual ET during different stages of revegetation across the Poyang Lake Basin, a typical sub-tropical humid basin that experienced a fast anthropogenic revegetation during the last decades. Our findings demonstrate that the nonclimatic contribution, in terms of magnitudes and directions, differed at the early, the middle, and the late stages of the restoration. At the early stage of low vegetation, vegetation cover change produced negative effects on ET overall. With the development of vegetation restoration, the fast increase of vegetation cover shifted the effects from negative to positive. At the late stage, when vegetation cover was high, it remained in the positive relationship with ET. This contrary response of ET to vegetation dynamic in the humid basin was quite different from that in the semi-arid/arid basin. Overall, we argue that both amount and change rate of vegetation coverage should be taken into account for a comprehensive understanding and evaluation of hydrological responses to anthropogenic vegetation change. This study provides a new insight into hydrological response to vegetation cover change during a long-term vegetation restoration process.

Supplementary Materials: The following are available online at http://www.mdpi.com/1999-4907/9/4/217/s1, Figure S1: Average potential surface conductance values (g_0) derived from tower measurements versus corresponding NDVI values from the AVHRR GIMMS dataset for evergreen broadleaf forest (EBF), mixed forest (MF), Woody savanna (WSV), and cropland (CRP) vegetation types, Figure S2: Range of potential surface conductance values (g_0) for evergreen broadleaf forest (EBF), mixed forest (MF), Woody savanna (WSV), and cropland (CRP) vegetation types in Poyang Lake Basin.

Acknowledgments: This work was supported by the State Key Program of National Natural Science Foundation of China (41430855) and Jiangsu Planned Projects for Postdoctoral Research Funds (1188000003).

Author Contributions: Yuanbo Liu and Ying Wang conceived and designed the experiments; Ying Wang and Jiaxin Jin performed the experiments and analyzed the data; Ying Wang wrote the paper.

Conflicts of Interest: The authors declare no conflict of interest.

References

1. Wang, K.; Dickinson, R.E. A review of global terrestrial evapotranspiration: Observation, modeling, climatology, and climatic variability. *Rev. Geophys.* **2012**, *50*. [CrossRef]
2. Zhang, K.; Kimball, J.S.; Nemani, R.R.; Running, S.W.; Yang, H.; Gourley, J.J.; Yu, Z. Vegetation greening and climate change promote multidecadal rises of global land evapotranspiration. *Sci. Rep.* **2015**, *5*, 75–77. [CrossRef] [PubMed]
3. Wei, X.; Li, Q.; Zhang, M.; Giles-Hansen, K.; Liu, W.; Fan, H.; Wang, Y.; Zhou, G.; Piao, S.; Liu, S. Vegetation cover—Another dominant factor in determining global water resources in forested regions. *Glob. Chang. Biol.* **2017**, *24*, 786–795. [CrossRef] [PubMed]
4. Zha, T.; Alang, B.; Garth, V.D.K.; Tandy, B.; Jharry, M.C.; Lawrenceb, F. Interannual variation of evapotranspiration from forest and grassland ecosystems in western canada in relation to drought. *Agric. Forest Meteorol.* **2010**, *150*, 1476–1484. [CrossRef]
5. Forzieri, G.; Alkama, R.; Miralles, D.G.; Cescatti, A. Satellites reveal contrasting responses of regional climate to the widespread greening of earth. *Science* **2017**, *356*, 1180–1184. [CrossRef] [PubMed]
6. Beer, C.; Ciais, P.; Reichstein, M.; Baldocchi, D.; Law, B.E.; Papale, D.; Soussana, J.F.; Ammann, C.; Buchmann, N.; Frank, D. Temporal and among-site variability of inherent water use efficiency at the ecosystem level. *Glob. Biogeochem. Cycles* **2009**, *23*, GB2018. [CrossRef]
7. Sun, G.; Zhou, G.; Zhang, Z.; Wei, X.; Mcnulty, S.G.; Vose, J.M. Potential water yield reduction due to forestation across China. *J. Hydrol.* **2006**, *328*, 548–558. [CrossRef]
8. Cao, S. Why large-scale afforestation efforts in China have failed to solve the desertification problem. *Environ. Sci. Technol.* **2008**, *42*, 1826–1831. [CrossRef] [PubMed]
9. Ellison, D.; Futter, M.N.; Bishop, K. On the forest cover–Water yield debate: From demand- to supply-side thinking. **2012**, *18*, 806–820. [CrossRef]
10. Ma, H.; Lv, Y.; Li, H. Complexity of ecological restoration in China. *Ecol. Eng.* **2013**, *52*, 75–78. [CrossRef]
11. Fang, X.; Ren, L.; Li, Q.; Zhu, Q.; Shi, P.; Zhu, Y. Hydrologic response to land use and land cover changes within the context of catchment-scale spatial information. *J. Hydrol. Eng.* **2013**, *18*, 1539–1548. [CrossRef]
12. Zhang, L.; Dawes, W.R.; Walker, G.R. Response of mean annual evapotranspiration to vegetation changes at catchment scale. *Water Resour. Res.* **2001**, *37*, 701–708. [CrossRef]
13. Huang, M.; Zhang, L.; Gallichand, J. Runoff responses to afforestation in a watershed of the Loess Plateau, China. *Hydrol. Proc.* **2003**, *17*, 2599–2609. [CrossRef]
14. Andréassian, V. Waters and forests: From historical controversy to scientific debate. *J. Hydrol.* **2004**, *291*, 1–27. [CrossRef]
15. Brown, A.E.; Zhang, L.; Mcmahon, T.A.; Western, A.W.; Vertessy, R.A. A review of paired catchment studies for determining changes in water yield resulting from alterations in vegetation. *J. Hydrol.* **2005**, *310*, 28–61. [CrossRef]
16. Jackson, R.B.; Murray, B.C. Trading water for carbon with biological carbon sequestration. *Science* **2005**, *310*, 1944–1947. [CrossRef] [PubMed]
17. Bao, Z.; Zhang, J.; Wang, G.; Fu, G.; He, R.; Yan, X.; Jin, J.; Liu, Y.; Zhang, A. Attribution for decreasing streamflow of the haihe river basin, northern China: Climate variability or human activities? *J. Hydrol.* **2012**, *460–461*, 117–129. [CrossRef]
18. Liu, W.; Wei, X.; Liu, S.; Liu, Y.; Fan, H.; Zhang, M.; Yin, J.; Zhan, M. How do climate and forest changes affect long-term streamflow dynamics? A case study in the upper reach of poyang river basin. *Ecohydrology* **2015**, *8*, 46–57. [CrossRef]
19. Tang, L.-L.; Cai, X.-B.; Gong, W.-S.; Lu, J.-Z.; Chen, X.-L.; Lei, Q.; Yu, G.-L. Increased vegetation greenness aggravates water conflicts during lasting and intensifying drought in the poyang lake watershed, China. *Forests* **2018**, *9*, 24. [CrossRef]
20. Mcvicar, T.R.; Li, L.T.; Niel, T.G.V.; Zhang, L.; Li, R.; Yang, Q.K.; Zhang, X.P.; Mu, X.M.; Wen, Z.M.; Liu, W.Z. Developing a decision support tool for China's re-vegetation program: Simulating regional impacts of afforestation on average annual streamflow in the loess plateau. *Forest Ecol. Manag.* **2007**, *251*, 65–81. [CrossRef]
21. Liu, Y.; Wu, G.; Zhao, X. Recent declines in China's largest freshwater lake: Trend or regime shift? *Environ. Res. Lett.* **2013**, *8*, 14010–14019. [CrossRef]

22. Lin, Y.; Liao, Z. Historical changes of jiangxi forest. *J. Nanchang Univ. (Sciences Edition)* **1982**, 68–74.

23. Hu, X. On the evolution of the environment of poyang lake basin in recent a century. *J. Jiangxi Normal Univ. (Nat. Sci.)* **2001**, *25*, 175–179.

24. Liu, Y. *Climatic and Hydrological Processes and Effects of Water Environment of Poyang Lake Basin*; Science Press: Beijing, China, 2012.

25. Buermann, W.; Parida, B.; Jung, M.; Macdonald, G.M.; Tucker, C.J.; Reichstein, M. Recent shift in eurasian boreal forest greening response may be associated with warmer and drier summers. *Geophys. Res. Lett.* **2014**, *41*, 1995–2002. [CrossRef]

26. Pinzon, J.E.; Tucker, C.J. A non-stationary 1981–2012 AVHRR NDVI3g time series. *Remote Sens.* **2014**, *6*, 6929–6960. [CrossRef]

27. Garonna, I.; De, J.R.; de Wit, A.J.; Mücher, C.A.; Schmid, B.; Schaepman, M.E. Strong contribution of autumn phenology to changes in satellite-derived growing season length estimates across europe (1982–2011). *Glob. Chang. Biol.* **2014**, *20*, 3457–3470. [CrossRef] [PubMed]

28. Holben, B.; Dan, K.; Fraser, R.S. Directional reflectance response in avhrr red and near-ir bands for three cover types and varying atmospheric conditions. *Remote Sens. Environ.* **1986**, *19*, 213–236. [CrossRef]

29. Norby, R.J.; Warren, J.M.; Iversen, C.M.; Medlyn, B.E.; Mcmurtrie, R.E. Co2 enhancement of forest productivity constrained by limited nitrogen availability. *Proc. Natl. Acad. Sci. USA* **2010**, *107*, 19368–19373. [CrossRef] [PubMed]

30. Zhang, K.; Kimball, J.S.; Mu, Q.; Jones, L.A.; Goetz, S.J.; Running, S.W. Satellite based analysis of northern ET trends and associated changes in the regional water balance from 1983 to 2005. *J. Hydrol.* **2009**, *379*, 92–110. [CrossRef]

31. Mu, Q.; Zhao, M.; Running, S.W. Improvements to a modis global terrestrial evapotranspiration algorithm. *Remote Sens. Environ.* **2011**, *115*, 1781–1800. [CrossRef]

32. Chen, X.; Xingguo, M.O.; Liu, S. Contributions of climate change and human activities to et and gpp trends over north China plain from 2000 to 2014. *Acta Geogr. Sin.* **2017**, *27*, 661–680. [CrossRef]

33. Liu, Y.; Wu, G. Hydroclimatological influences on recently increased droughts in China's largest freshwater lake. *Hydrol. Earth Syst. Sci.* **2016**, *20*, 93–107. [CrossRef]

34. DiMiceli, C.M.; Carroll, M.L.; Sohlberg, R.A.; Huang, C.; Hansen, M.C.; Townshend, J.R.G. Annual Global Automated Modis Vegetation Continuous Fields (MOD44B). University of Maryland: College Park, MA, USA, 2011. Available online: http://www.landcover.org/data/vcf/ (accessed on 1 September 2016).

35. Nicholls, N. Increased australian wheat yield due to recent climate trends. *Nature* **1997**, *387*, 484–485. [CrossRef]

36. Lobell, D.B.; Field, C.B. Global scale climate–crop yield relationships and the impacts of recent warming. *Environ. Res. Lett.* **2007**, *2*, 014002. [CrossRef]

37. Tao, F.; Yokozawa, M.; Liu, J.Y.; Zhang, Z. Climate-crop yield relationships at provincial scales in China and the impacts of recent climate trends. *Clim. Res.* **2008**, *38*, 83–94. [CrossRef]

38. Verón, S.R.; Abelleyra, D.D.; Lobell, D.B. Impacts of precipitation and temperature on crop yields in the pampas. *Clim. Chang.* **2015**, *130*, 235–245. [CrossRef]

39. Lobell, D.B.; Asner, G.P. Climate and Management Contributions to Recent Trends in U.S. Agricultural Yields. *Science* **2003**, *299*, 1032. [CrossRef] [PubMed]

40. Liu, Y.; Xiao, J.; Ju, W.; Xu, K.; Zhou, Y.; Zhao, Y. Recent trends in vegetation greenness in China significantly altered annual evapotranspiration and water yield. *Environ. Res. Lett.* **2016**, *11*, 094010. [CrossRef]

41. Wang, L.; Good, S.P.; Caylor, K.K. Global synthesis of vegetation control on evapotranspiration partitioning. *Geophys. Res. Lett.* **2015**, *41*, 6753–6757. [CrossRef]

42. Chen, Y.; Xia, J.; Liang, S.; Feng, J.; Fisher, J.B.; Li, X.; Li, X.; Liu, S.; Ma, Z.; Miyata, A. Comparison of satellite-based evapotranspiration models over terrestrial ecosystems in China. *Remote Sens. Environ.* **2014**, *140*, 279–293. [CrossRef]

43. Zhang, K.; Kimball, J.S.; Nemani, R.R.; Running, S.W. A continuous satellite-derived global record of land surface evapotranspiration from 1983 to 2006. *Water Resour. Res.* **2010**, *46*, 109–118. [CrossRef]

44. Liu, J.; Liu, M.; Zhuang, D.; Zhang, Z.; Deng, X. Study on spatial pattern of land-use change in China during 1995–2000. *Sci. China Ser. D Earth Sci.* **2003**, *46*, 373–384.

45. Zhou, Y.; Li, X.; Yang, K.; Zhou, J. Assessing the impacts of an ecological water diversion project on water consumption through high-resolution estimations of evapotranspiration in the downstream regions of the Heihe River Basin, China. *Agric. Forest Meteorol.* **2018**, *249*, 210–217. [CrossRef]
46. Keenan, T.F.; Hollinger, D.Y.; Bohrer, G.; Dragoni, D.; Munger, J.W.; Schmid, H.P.; Richardson, A.D. Increase in forest water-use efficiency as atmospheric carbon dioxide concentrations rise. *Nature* **2013**, *499*, 324–327. [CrossRef] [PubMed]
47. Huang, M.; Piao, S.; Zeng, Z.; Peng, S.; Ciais, P.; Cheng, L.; Mao, J.; Poulter, B.; Shi, X.; Yao, Y. Seasonal responses of terrestrial ecosystem water-use efficiency to climate change. *Glob. Chang. Biol.* **2016**, *22*, 2165–2177. [CrossRef] [PubMed]

Article

Assessing the Driving Forces in Vegetation Dynamics Using Net Primary Productivity as the Indicator: A Case Study in Jinghe River Basin in the Loess Plateau

Hao Wang [1,*], Guohua Liu [2], Zongshan Li [2], Pengtao Wang [1] and Zhuangzhuang Wang [1]

[1] Department of Geography, School of Geography and Tourism, Shaanxi Normal University, Xi'an 710119, China; yewufeige@126.com (P.W.); Geowangzz@126.com (Z.W.)

[2] State Key Laboratory of Urban and Regional Ecology, Research Center for Eco-Environmental Science, Chinese Academy of Sciences, Beijing 100085, China; ghliu@rcees.ac.cn (G.L.); zsli_st@rcees.ac.cn (Z.L.);

* Correspondence: foreva@snnu.edu.cn; Tel.: +86-029-8531-0525

Received: 5 May 2018; Accepted: 19 June 2018; Published: 21 June 2018

Abstract: An objective and effective method to distinguish the influence of climate change and human activities on vegetation dynamics has great significance in the design and implementation of ecosystem restoration projects. Based on the Moderate Resolution Imaging Spectroradiometer (MODIS) remote data and the Miami and Carnegie–Ames–Stanford Approach (CASA) model, this study simulated and used net primary productivity (NPP) as an indicator to identify vegetation dynamics and their driving forces in the Jinghe River basin from 2000 to 2014. The results showed that: (1) The vegetation in the Jinghe River basin, which accounted for 84.4% of the study area, showed an increasing trend in NPP; (2) Human activities contributed most to vegetation restoration, which accounted for 54.5% of the areas; 24.0% of the areas showed an increasing trend in the NPP that was dominated by climate factors. Degradation dominated by human activities accounted for 4.3% of the study area, and degradation dominated by climate factors resulted in 17.2%; (3) The rate of vegetation degradation in areas dominated by climate factors rose with increased slope, where the arid climate caused shortages of water resources, and the human-dominated vegetation restoration activities exacerbated the vegetation's water demand further, which surpassed the carrying capacity of regional water resources and led ultimately to vegetation degradation. We recommend that future ecological restoration programs pay more attention to maintaining the balance between ecosystem restoration and water resource demand to maximize the benefits of human activities and ensure the vegetation restoration is ecologically sustainable.

Keywords: net primary productivity; Loess Plateau; climate fluctuation; human activity; vegetation restoration; simulation modeling; CASA; MODIS; remote sensing

---

1. Introduction

Climate and human activities are the primary driving forces of changes in terrestrial ecosystems [1–3]. Commonly, regional vegetation dynamics are related closely to changes in local climate conditions and human activities [4–6]. However, it is difficult to distinguish the influence of these two driving factors when both function in the process of vegetation growth [7]. Particularly in arid and semi-arid regions such as the Loess Plateau, the water-limited environment makes the vegetation there highly sensitive to changes in temperature and precipitation [8,9], and high-intensity human activities easily may lead to degradation of the local vegetation [10–13]. To improve the ecological environment, the local government has implemented a series of ecological restoration programs, such as the Grain for Green Program (GGP), which complicates the effects of human activities on vegetation [14]. Our previous studies showed that the changes between different land use types contribute significantly to the changes in vegetation coverage, especially for the transformation between farmland and forests/grasslands, which were related closely to topographical factors; for example, the implementation of GGP requires that farmland with slopes between 15 to 25° are returned to grassland or forest, while farmland with slopes >25° should be returned to forest [15,16]. These ecological restoration programs have increased the coverage and net primary productivity (NPP) of vegetation in the Loess Plateau and improved the local ecological environment gradually [17,18]. However, recent studies have shown that restoration is reaching the plateau's sustainable water resource limits [19]. The water resources of the natural environment are unable to meet the growing demand of a large amount of vegetation planted recently [20]. Inappropriate selection of ecological restoration species increases the consumption of soil moisture and causes a low survival rate of revegetated trees and shrubs [21,22]. Therefore, studies that identify and quantify the effects of climate conditions and human activities on vegetation dynamics have great significance in the design and implementation of ecosystem restoration projects. The results of studies such as this will help in the selection of suitable sites and methods for ecological restoration that are adapted to local climate conditions or mitigate the negative effects of human activities, and achieve sustainable development of regional ecological restoration [23,24].

Previous studies designed to differentiate the effects of climate and human activities on vegetation dynamics have focused primarily on statistical analyses, such as principal component, correlation, and significance analyses. Limited by the study methods, these studies failed to tell us the spatial distribution pattern and the change trend of vegetation, which is driven by climate change or human activities [25–28]. With the development of remote sensing technology, recent studies have begun to use the Normalized Difference Vegetation Index (NDVI) remote data and residual analysis methods to distinguish the influence of climate conditions and human activities on vegetation dynamics [15,29]. The concept on which this method is based is that the NDVI and precipitation are correlated significantly, and based on the NDVI and precipitation data, a regression relation is established to simulate the NDVI expected. The difference between the expected and actual NDVI indicates the effect of human factors on vegetation dynamics [30,31]. However, research that relies solely on the relation between NDVI and precipitation fails to reflect temperature's influence on vegetation dynamics. Meanwhile, there are uncertainties about the results of the NDVI expected, which is calculated according to the precipitation–NDVI relation, and thus, the influence of climate and human activities on vegetation dynamics cannot be differentiated fully [28,32]. Therefore, it is necessary to use an objective and effective method to distinguish the influence of the two factors on vegetation [33].

NPP is the net energy that vegetation converts through photosynthesis to biomass [34]. As an important part of ecosystem function and carbon circulation, NPP often is used as an indicator of vegetation's sensitivity to climate change and human activities [33,35]. Previous studies have adopted NPP to discriminate the response of vegetation to climate change [36,37], and today, researchers have begun to use NPP to identify the effects of human activities on vegetation dynamics [4,38]. Based on models of different ecological processes and remote sensing data, the NPP expected (NPP_e) can be calculated to simulate the climate-induced production, and the actual NPP (NPP_a) to simulate the combined induced production [39,40]. The difference between the NPP_e and NPP_a indicates the effects of human factors on vegetation dynamics. Because different ecological models are used to simulate both the NPP_e and NPP_a, the results can avoid the errors and uncertainties associated with the precipitation–NDVI linear regression method effectively [41–43]. Therefore, this study adopted NPP as an indicator to assess the driving forces in vegetation dynamics.

The Jinghe River is a secondary tributary of the Yellow River that plays an important role in the ecological security of the Loess Plateau, and both the natural environment and human activities have experienced significant changes there in recent years [44,45]. However, few studies have focused on spatial quantificational analysis of the driving forces in vegetation dynamics in this region [46]. This study uses NPP as an indicator to identify the vegetation change trend and its driving forces in Jinghe River basin from 2000 to 2014. The ultimate objectives of this study were to: (1) explore the vegetation change trend in Jinghe River basin; (2) distinguish the role of climate change and human activities in vegetation dynamics; and (3) quantify the effects of these two factors and introduce topographical factors to determine their spatial distribution. This study can be considered a reproducible method for the analysis of driving factors in vegetation dynamics at the basin scale and provides a scientific basis for the development of local ecological restoration.

2. Data and Methods

2.1. Study Area

The Jinghe River basin is located in the southwest of the Loess Plateau and covers an area of 70,040 km^2 (Figure 1). The basin is in the transitional zone between the temperate semi-humid and temperate semi-arid and has a typical temperate continental climate. The temperature and precipitation in the Jinghe River basin decrease gradually from southeast to northwest. The annual average temperature and average annual precipitation in the region are approximately 10 °C and 290–560 mm, respectively. The primary vegetation types in this area are forest, shrub, and grassland (Table 1) [47]. In the past decade, high-intensity human activities in the basin have led to an increasing trend in soil erosion and decreasing trend in vegetation coverage [45]. Thus, to improve the ecological environment, the local government has implemented a series of ecological programs, such as the Grain for Green Program (GGP). However, the continuous population growth and rapidly expanding towns continue to exert considerable pressure on the natural environment [46]. Therefore, this study focused on the Jinghe River basin as the study area to analyze changes in vegetation dynamics and distinguish the effects of climate change and human activities. The results of this research are of great scientific significance in understanding the rules of regional vegetation changes, as well as summarizing and improving ecological restoration measures.

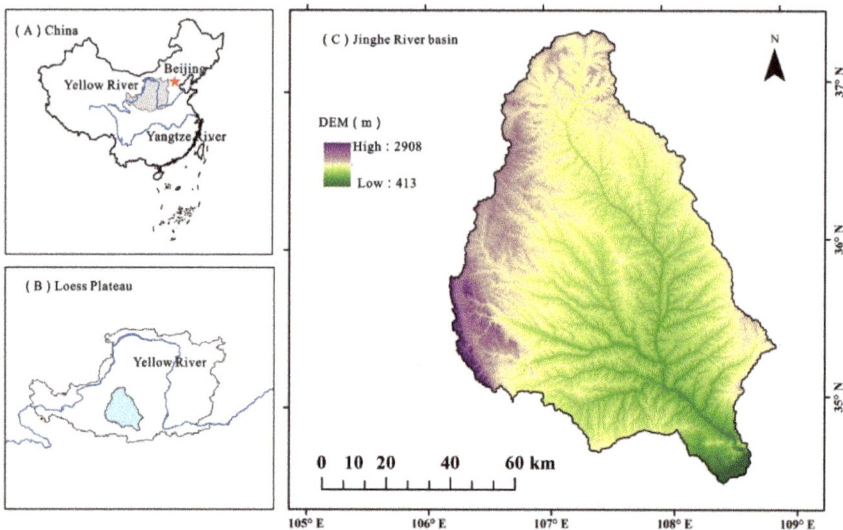

Figure 1. Location of the Jinghe River basin. DEM, Digital Elevation Model.

Table 1. The land use types in percentage terms in the Jinghe River basin (Unit: %).

Land Use Types	Forest	Shrub	Grassland	Farmland	Others
Area in percentage	28.7	10.2	32.1	27.3	1.7

2.2. Remote Sensing Data Sets

Land use map and vegetation classification map for the Jinghe River basin were obtained from the Center for Earth Observation and Digital Earth, China (http://www.ceode.cas.cn/sjyhfw/) (Table 1). Based on the Landsat remote data and the vegetation classification map, a series of 30 m resolution land use maps were created with an accuracy rate higher than 94%.

Temperature and precipitation were adopted in this study as the meteorological factors that affect vegetation dynamics, and the data were obtained from the China Meteorological Data Sharing Network (http://data.cma.cn/). The monthly temperature and precipitation data of 676 stations in China were used to calculate the spatial distribution of the meteorological factors by using ArcGIS v10.2 software (Environmental Systems Research Institute, Inc., Redlands, CA, USA) with Kriging interpolation method. The spatial resolution of the results was set to 250 m. Based on the range of the Jinghe River basin, the meteorological data for the Jinghe River basin from 2000 to 2014 were obtained using the Extract by mask function of ArcGIS. Then, the spatial meteorological data were used in the Miami model to simulate the NPP_e.

The NDVI data (2000–2014) that were used to simulate the NPP_a using the Carnegie–Ames–Stanford Approach (CASA) model were obtained from the MODIS NDVI product (MOD13Q1). This dataset can be downloaded from https://ladsweb.modaps.eosdis.nasa.gov and has a spatial resolution of 250 m and a temporal resolution of 16-day intervals. To reduce the noise attributable to bare soil and clouds, we converted all NDVI remote data to monthly data using the maximum value method, and eliminated those grid cells with a NDVI value less than 0.05 [48,49].

2.3. Net Primary Production Estimates

2.3.1. Estimation of the Expected NPP

The Miami model was used to estimate the NPP_e, which is affected only by meteorological factors [40]. This model is the first NPP estimation model used widely worldwide. The Miami model, which is based on Liebig's "Law of minimum" and the relation between vegetation NPP and annual average temperature and annual precipitation, was used to determine the values of NPP [50,51]. Because of its simple parameters and reasonable estimates of NPP, the Miami model has been used widely in NPP estimation studies in different regions of the world [52]. The formula of the model is as follows:

$$NPP_e = \min\left\{ \left(\frac{3000}{1 + \exp(1.315 - 0.119\,t)} \right), (3000[1 - \exp(-0.000664\,r)]) \right\} \tag{1}$$

where the unit of NPP_e is g C·m^{-2}·year^{-1}, t is the annual average temperature (°C), and r is the annual precipitation (mm). Based on the raster calculator function of the ArcGIS v10.2 software (Environmental Systems Research Institute, Inc., Redlands, CA, USA), the monthly spatial temperature and precipitation data obtained in Section 2.2 were converted into annual data, with a spatial resolution of 250 m. Then, the annual NPP_e was estimated based on the annual spatial meteorological data, and the spatial resolution of the results were set to 250 m.

2.3.2. Estimation of the Actual NPP

The NPP_a, which is affected both by climate and human activities factors, was estimated with the CASA model [4,53]. Based on the principle of light energy use, Monteith first proposed the concept of estimating NPP based on photosynthetically active radiation (APAR) and light energy use (ε) in 1972 [54]. Moreover, in 1993, Potter proposed the CASA model and realized the estimation of regional and global NPP using the principle of light energy use based on remote sensing data [55,56]. As it is possible to reflect the influence of climate and human factors on NPP, this is used widely in remote sensing retrieval research on NPP [17,32]. The main formula of the model is as follows:

$$NPP_a\,(x,\,t) = APAR\,(x,\,t) \times \varepsilon\,(x,\,t) \tag{2}$$

where $APAR\,(x,\,t)$ is the photosynthetically active radiation (g C·m^{-2}·month^{-1}) absorbed by vegetation in pixel x at time t, and $\varepsilon\,(x,\,t)$ is the actual light energy use (gC·MJ^{-1}) of vegetation in pixel x at time t.

$APAR\,(x,\,t)$ can be calculated as follows:

$$APAR\,(x,\,t) = SOL\,(x,\,t) \times 0.5 \times FPAR\,(x,\,t) \tag{3}$$

where $SOL\,(x,\,t)$ indicates the total solar radiation (MJ·m^{-2}) in pixel x at time t, $FPAR\,(x,\,t)$ indicates the proportion of photosynthetically active radiation vegetation absorbs, and a constant of 0.5 indicates the proportion of total solar radiation (0.4–0.7 μm) available for vegatation.

The SOL were obtained from the China Meteorological Data Sharing Network (http://data.cma.cn/). The monthly SOL data of meteorological stations were used to calculate the spatial distribution of the SOL by using ArcGIS v10.2 software (Environmental Systems Research Institute, Inc., Redlands, CA, USA) with Kriging interpolation method. The spatial resolution of the results was set to 250 m.

FPAR can be expressed as follows:

$$FPAR = \frac{(NDVI(x,\,t) - NDVI_{i,\text{min}})(FPAR_{\text{max}} - FPAR_{\text{min}})}{NDVI_{i,\text{max}} - NDVI_{i,\text{min}}} + FPAR_{\text{min}} \tag{4}$$

where $NDVI\,(x,\,t)$ indicates the NDVI value in pixel x at time t, $NDVI_{i,\text{max}}$ and $NDVI_{i,\text{min}}$ are the maximum and minimum NDVI value of the vegetation type i. $FPAR_{\text{max}}$ and $FPAR_{\text{min}}$ are constants of 0.95 and 0.001, respectively.

$\varepsilon\,(x,\,t)$ can be calculated as follows:

$$\varepsilon\,(x,\,t) = T_{\varepsilon 1}\,(x,\,t) \times T_{\varepsilon 2}\,(x,\,t) \times W_\varepsilon\,(x,\,t) \times \varepsilon_{max} \tag{5}$$

where $T_{\varepsilon 1}\,(x,\,t)$ and $T_{\varepsilon 2}\,(x,\,t)$ are the temperature stress coefficients at low and high temperatures, $W_\varepsilon\,(x,\,t)$ is the water stress coefficient, and ε_{max} is the maximum light energy conversion rate under ideal conditions, which is 0.389 g C·MJ^{-1}.

$T_{\varepsilon 1}\,(x,\,t)$ and $T_{\varepsilon 2}\,(x,\,t)$ can be presented as follows:

$$T_{\varepsilon 1}(x,\,t) = 0.8 + 0.02 \cdot T_{opt}(x) - 0.0005 \cdot \left[T_{opt}(x)\right]^2 \tag{6}$$

$$T_{\varepsilon 2}(x,\,t) = 1.184/\left\{1 + \exp\left[0.2 \cdot \left(T_{opt}(x) - 10 - T(x,\,t)\right)\right]\right\} \\ \cdot 1/\left\{1 + \exp\left[0.3 \cdot \left(-T_{opt}(x) - 10 + T(x,\,t)\right)\right]\right\} \tag{7}$$

where $T_{opt}\,(x)$ is the optimum temperature for vegetation growth, which is the average monthly temperature (°C) when the NDVI value in pixel x reaches the maximum within one year. T is the annual average temperature (°C).

$W_\varepsilon\,(x,\,t)$ can be calculated as follows:

$$W_\varepsilon(x,\,t) = 0.5 + 0.5 \cdot EET(x,\,t)/EPT(x,\,t) \tag{8}$$

where EET is the actual evapotranspiration (mm), EPT is the potential evapotranspiration (mm), which are both obtained from the meteorological data in Section 2.2.

The time and spatial resolution of all the parameters of the CASA model for estimating NPP are set to monthly and 250 m, respectively. The monthly NPP was calculated and then summed to the annual NPP used in this study.

2.3.3. Estimation of the NPP_h and Condition Analysis

The difference between the NPP_a and the NPP_e is the human-induced NPP (NPP_h), which is affected only by human activities. The formula can be expressed as follows:

$$NPP_h = NPP_a - NPP_e \tag{9}$$

To measure the change trend in the NPP, the ordinary least-squares regression formula was used, which is as follows:

$$Slope = \frac{\sum_{i=1}^{n} x_i y_i - \frac{1}{n}\left(\sum_{i=1}^{n} x_i\right)\left(\sum_{i=1}^{n} y_i\right)}{\sum_{i=1}^{n} x_i^2 - \frac{1}{n}\left(\sum_{i=1}^{n} x_i\right)^2} \tag{10}$$

where x_i is 1 to n for years 2000 to 2014, and y_i is the NPP in year x_i. Areas with a positive slope indicate that both the NPP and vegetation dynamics in these areas showed an increasing trend, while conversely, areas with a negative value indicate a decreasing trend [48,57].

Consequently, five types of possible conditions that lead to vegetation dynamics change can be defined by the slopes of the NPP_a (S_a), NPP_e (S_e), and the NPP_h (S_h) (Table 2). Combined with the effects of climate change and human activities on vegetation dynamics, Condition 1 is the vegetation with no change (NC), Condition 2 is the restoration of vegetation dominated by meteorological conditions (RDC), Condition 3 is the restoration of vegetation dominated by human activities (RDH), Condition 4 is the degradation of vegetation dominated by meteorological conditions (DDC), and Condition 5 is the degradation of vegetation dominated by human activities (DDH) [32,58,59].

Table 2. Conditions to assess the effects of climate change and human activities on vegetation dynamics.

Number	Method	Cause of Vegetation Dynamics Change
Condition 1	$S_a = 0$	the vegetation had no change (NC)
Condition 2	$S_a > 0$ and $S_e > S_h$	the restoration of vegetation dominated by climate factors (RDC)
Condition 3	$S_a > 0$ and $S_e < S_h$	the restoration of vegetation dominated by human factors (RDH)
Condition 4	$S_a < 0$ and $S_e > S_h$	the degradation of vegetation dominated by climate factors (DDC)
Condition 5	$S_a < 0$ and $S_e < S_h$	the degradation of vegetation dominated by human factors (DDH)

2.4. Correlation Coefficient and Significance Test

Correlation analysis can be used to indicate the relevance and change trend of research factors [15,60], therefore, this study used the Pearson's correlation coefficient formula to calculate the significance of the NPP change trend. The calculation formula is as follows:

$$r = \frac{n \sum_{i=1}^{n} x_i y_i - \sum_{i=1}^{n} x_i \cdot \sum_{i=1}^{n} y_i}{\sqrt{n \sum_{i=1}^{n} x_i^2 - \left(\sum_{i=1}^{n} x_i\right)^2} \cdot \sqrt{n \sum_{i=1}^{n} y_i^2 - \left(\sum_{i=1}^{n} y_i\right)^2}} \tag{11}$$

where x_i is 1 to n for years 2000 to 2014 ($n = 15$), and y_i is the NPP_a in year x_i, r is the Pearson's correlation coefficient for each pixel. When $r > 0$, the pixel experienced an increasing trend of NPP, while conversely, when $r < 0$, the pixel experienced a decreasing trend of NPP. When $0.514 < r < 1$ or $-1 < r < -0.514$, the pixel experienced a significant increasing or decreasing trend of NPP at the $p < 0.05$ confidence intervals.

2.5. Validating NPP

The measured aboveground NPP data included 45 sites (five plots per site) of different vegetation types. The details of the sampling time and methods can be found in [19,60]. NPP simulated by the CASA model was compared with the measured NPP (Figure 2). The result indicated that the simulated NPP showed a good correlation with the measured NPP ($R^2 = 0.8$, $p < 0.001$).

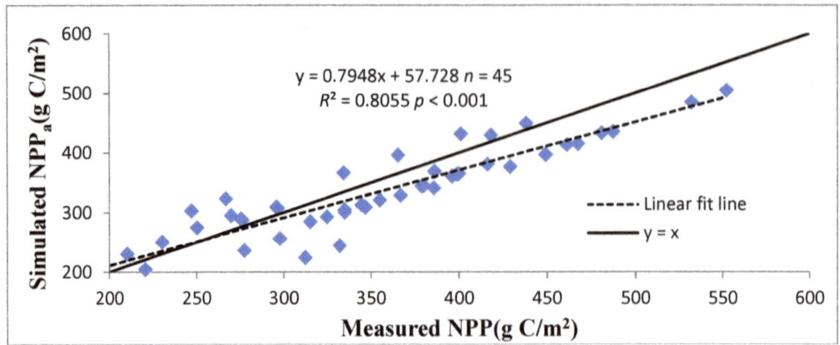

Figure 2. Comparison between the CASA model simulated NPP and the measured NPP in the Jinghe River basin. CASA model, Carnegie–Ames–Stanford Approach model; NPP, net primary productivity.

3. Results

3.1. Spatio-Temporal Trends of NPP

The annual average NPP in the Jinghe River basin from 2000 to 2014 was calculated and is shown in Figure 3. Generally, the NPP in the study area showed an increasing trend, with an increase rate of 9.438 g C·m^{-2}·year^{-1}. The highest value of annual average NPP in the 15 years was in 2014, while the lowest was in 2000. The change process can be divided into two parts: from 2000 to 2006, the NPP increased relatively moderately, then increased rapidly with fluctuations from 2007 to 2014.

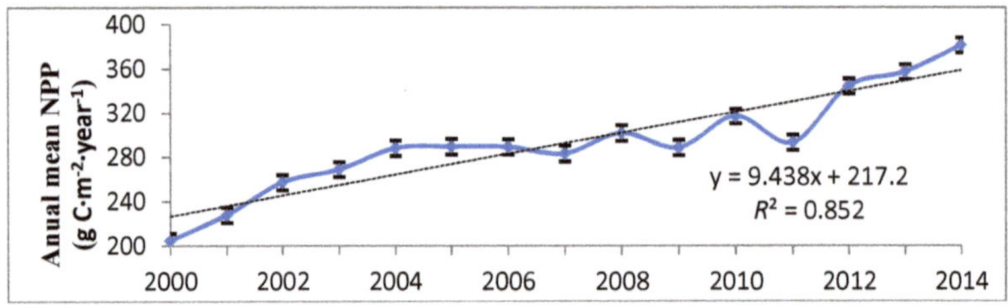

Figure 3. Interannual variations in the annual average NPP in the Jinghe River basin from 2000 to 2014.

The spatial distribution of the annual mean NPP and the change trend in the NPP are shown in Figure 4. The annual mean NPP of the Jinghe River basin showed a decreasing trend from the southeast to the northwest, which largely is consistent with the regional water and heat distribution (Figure 4a). There was a relatively clear dividing line around Weiyuan, Zhenyuan, and Heshui counties, in which the annual average NPP value was lower north of the dividing line, with the lowest value 52.6 g C· m^{-2}, and was higher in the south, with the highest value 677.33 g C· m^{-2}, which is more than ten times the low value and indicates obvious spatial changes.

Figure 4b shows the change trend in the NPP of the Jinghe River basin from 2000 to 2014. The results indicated that less than 0.1% of the study area showed no change trend, while 84.4% of the area showed an increasing trend. Specifically, 34.3% of the total area showed a significant increasing trend ($p < 0.05$), which was located primarily in the middle of the basin where the terrain is gentler and human activities are more frequent. Meanwhile, areas with decreasing trends in NPP accounted for 15.5% of the Jinghe River Basin area, 3.0% of which showed a significant decreasing trend ($p < 0.05$). These areas are concentrated primarily in the Ziwuling Mountain and Liupan Mountain areas on the east and west sides of the basin, respectively. The vegetations in these areas are forests and shrubs, which had a high average value of NPP (Figure 4a). In addition, compared with the areas in which the NPP increased significantly, the terrain in these areas is relatively steep and human activities are limited.

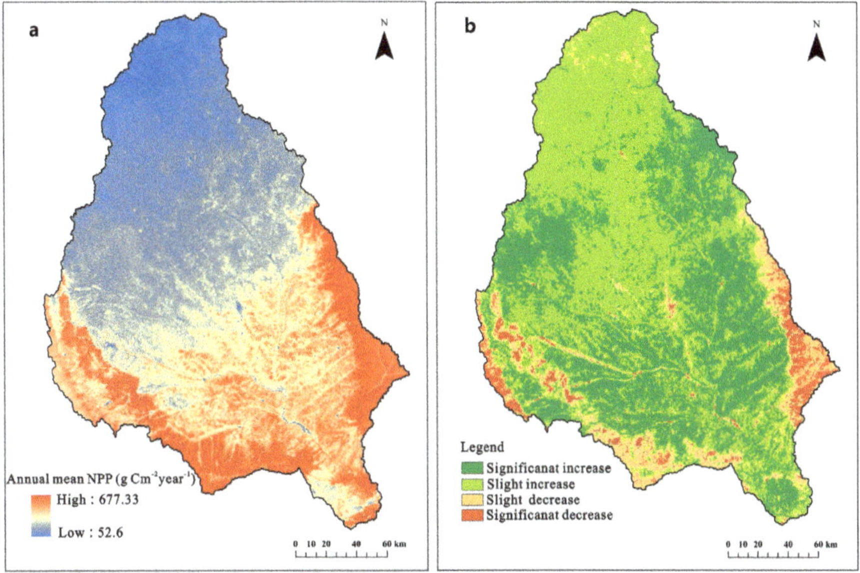

Figure 4. Spatial pattern of the mean NPP (**a**) and the change trend in the NPP (**b**) in the Jinghe River basin from 2000 to 2014.

3.2. Driving Forces in Vegetation Dynamics

The spatial pattern and area statistics of the NPP change caused by different driving factors in the Jinghe River basin from 2000 to 2014 were analyzed and are shown in Figures 5a and 6. The results indicated that human activities contributed most to the vegetation restoration in the 54.5% of the areas in which the NPP changed, which were located largely in the middle and south of the study area. Meanwhile, 24.0% of the areas in which NPP changed showed an increasing trend in the NPP that was dominated by climate factors and was located primarily in the north of the study area. Climate factors and human activities also caused vegetation degradation. Degradation dominated by human activities accounted for 4.3% of the areas in which NPP changed and were concentrated primarily in the middle of the study area. Climate factors produced 17.2% of the vegetation degradation in the areas in which NPP changed and were concentrated largely in the Ziwuling Mountain and Liupan Mountain regions.

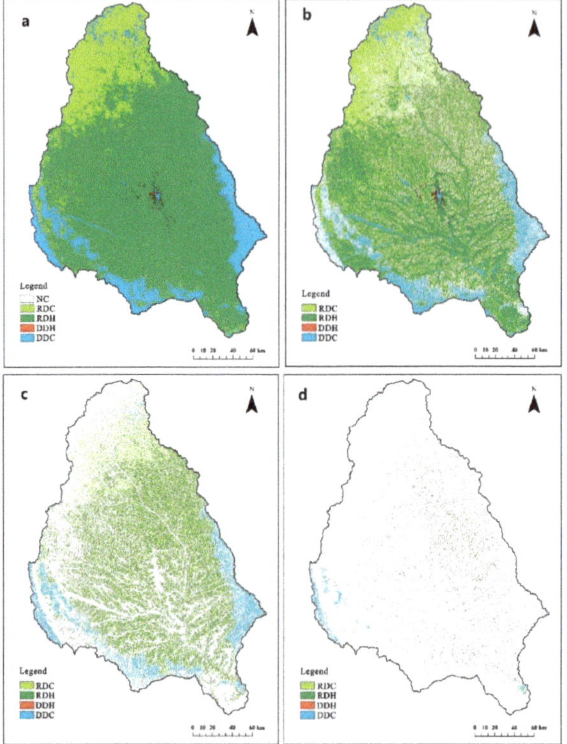

Figure 5. Spatial pattern of different conditions of the NPP change of the entire basin (**a**); areas with a slope less than 15° (**b**); areas with a slope between 15 and 25° (**c**); and areas with a slope greater than 25° (**d**) in the Jinghe River basin from 2000 to 2014. NC is the vegetation with no change, RDC is the restoration of vegetation dominated by meteorological conditions, RDH is the restoration of vegetation dominated by human activities, DDC is the degradation of vegetation dominated by meteorological conditions, DDH is the degradation of vegetation dominated by human activities.

As human activities including urban expansion and ecological restoration were closely related to topographical factors, this study adopted the requirements of the GGP to introduce topographical factors to achieve a better understanding of the spatial patterns in the NPP change trend and its driving forces. Slope gradients were divided into three levels according to the GGP requirements, slopes <15°, those between 15 and 25°, and slopes >25°, respectively (Figure 5b–d). The area statistics results in Figure 6 show that the positive effect of climate factors on the NPP declined continuously as the slope increased. A total of 26.6% of areas with slopes <15° demonstrated a restoration trend in vegetation dominated by climate factors (RDC). However, in areas with slopes >25°, the rate decreased to only 11.9%. Furthermore, the rate of vegetation degradation in areas dominated by climate factors (DDC) increased from 16.4% to 19.6% with increasing slope. Conversely, the positive effect of human activities on the NPP continued to increase as slope increased; 52.6% of areas with slopes <15° indicated a restoration trend of vegetation dominated by human activities (RDH), and the rate increased to 68.4% in areas with slopes >25°. At the same time, the rate of vegetation degradation in areas dominated by human activities (DDH) decreased from 4.4 to 0.1% with increased slope. Thus, the factors that drove vegetation dynamics changed clearly depending on the terrain.

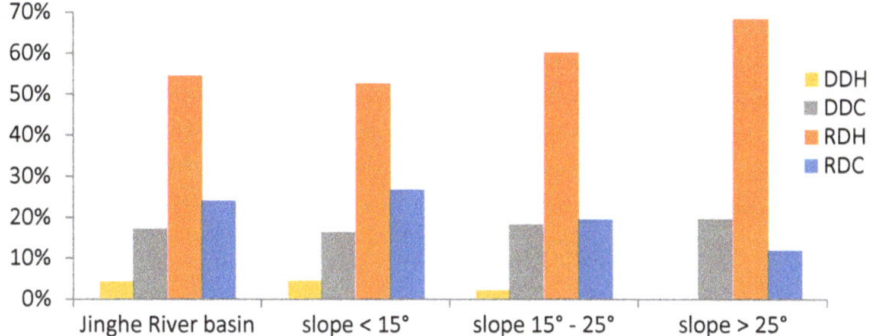

Figure 6. Area statistics of the driving factors in NPP change in the Jinghe River basin and different slopes. RDC is the restoration of vegetation dominated by meteorological conditions, RDH is the restoration of vegetation dominated by human activities, DDC is the degradation of vegetation dominated by meteorological conditions, DDH is the degradation of vegetation dominated by human activities.

4. Discussion

4.1. Methodology

Assessing the spatial patterns of the influence of climate change and human activities on vegetation dynamics accurately is of great significance in the management and restoration of regional ecological environments. However, distinguishing the effect of human activities on vegetation dynamics from those of climate factors traditionally has been difficult [40]. Several studies have adopted the NPP, which is an efficient and accurate indicator of vegetation growth status, to distinguish vegetation change dominated by human factors from that dominated by climate by comparing the difference between the expected and actual NPP [61,62]. Both the Miami and CASA models that estimate the expected and actual NPP have been used successfully in several studies at the global and regional scale [63–65]. The results of this study showed that the actual NPP in the Jinghe River basin increased from 2000 to 2014, which is consistent with previous studies and supports the feasibility of applying NPP models in this region [17,46]. Therefore, this study adopted NPP as an indicator to assess the relative roles of climate factors and human activities in vegetation change.

Although the expected and actual NPP distinguished the effects of climate factors and human activities on vegetation dynamics successfully, this method may have its own limitations. In the process of estimating the NPP expected, the Miami model includes only temperature and precipitation as the two climate factors that simulate an ideal environment of vegetation growth. Similarly, we assessed the actual NPP and the relative roles of climate and human factors in vegetation dynamics based on the NPP variation, and established conditions based on the hypothesis that vegetation dynamics is only affected by climate and human activities. However, vegetation productivity and its simulation results may be affected by several factors, such as wind, soil organic matter, vegetation types, herbivore activities, and the accuracy of the remote sensing data used in the NPP estimate models [42,66–68] Future studies should incorporate additional driving factors based on the characteristics of the study area. Meanwhile, because of the errors inherent in the remote sensing data and the NPP simulation methods itself, there can be some errors in the NPP simulation results and the differences of NPP_e and NPP_a. However, according to previous studies, these errors may exist in the assessment of the slope, vegetation communities, and other small-scale studies. For regional and global scales, the methodology introduced in this study can be considered as a feasible method of evaluating the spatial distribution of the relative roles of climate and human activities [28,40,42].

4.2. Driving Forces

Previous studies have shown that both the NPP and vegetation in the Loess Plateau have increased significantly because of human activities, such as reducing grazing pressure and returning farmland to forests [69–71]. The results of this study confirmed that vegetation in the Jinghe River basin experienced similar change trends, with 85.5% of the vegetation in the study area showing an increasing trend. Among them, 54.5% of the increased vegetation was dominated by human activities. This rate is similar to Li's research, which indicated that human activities account for 55% of vegetation changes from 2000 to 2015 in the Loess Plateau [72]. Meanwhile, the rate of vegetation increase dominated by human activities rose with increased slope, from 52.6% in areas with slopes less than 15° to 68.4% in those with slopes greater than 25°. These areas were located primarily in valleys in the middle of the basin, where the land use changed more dramatically during the past decade [73]. These results are consistent with the implementation of a series of ecological projects, including the GGP. Under the guidance of government policies, farmland in valleys with steep slopes has been converted to grassland and forest, which enhances vegetation and soil carbon fixation effectively [72,74]. However, the study also confirmed that 4.3% of the NPP showed a decreasing trend in the Jinghe River basin that was dominated by negative human activities. These areas were concentrated largely in the middle of the study area, in which Qingyang city is located and has the largest population density in the Jinghe River basin. Because of its rapid population growth, the pace of urbanization has accelerated significantly and has led to drastic changes in the local environment around the city that have decreased the vegetation cover and carbon fixation [60,75].

Changes in climate factors are another important force that affects the vegetation dynamics, and the vegetation changes in the Jinghe River basin that climate forcing dominated showed clear spatial characteristics. The results of this study indicated that vegetation restoration dominated by climate factors in the study area is distributed primarily in the northern part of the basin (Figure 5a). Based on the zonal statistics results of the spatial annual temperature and precipitation data in the Jinghe River basin from 2000 to 2014, the annual mean temperature in the study area is between 7.8 to 12.5 °C and increased at a change rate of 0.2 °C/10 year; while the annual precipitation is between 334.2 to 620.8 mm and decreased at a change rate of 24 mm/10 year over the past 15 years, respectively (Figures 7 and 8). These results indicate that the climate in the study area is becoming warmer and drier, which leads to drought, as Li et al. and Zhao et al. reported [76,77]. In a water-limited area, the spatial distribution of precipitation determines the vegetation distribution and growth [9]. Zhang et al. pointed out that, compared with other areas that suffered drought, precipitation in the northern part of the Jinghe River basin is relatively sufficient, which is likely to be the reason that vegetation restoration dominated by climate factors is concentrated in that area [78].

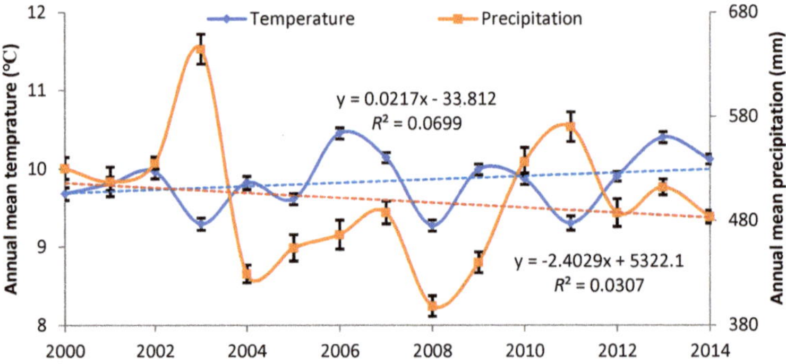

Figure 7. Interannual variations in the annual mean temperature and annual precipitation in the Jinghe River basin from 2000 to 2014.

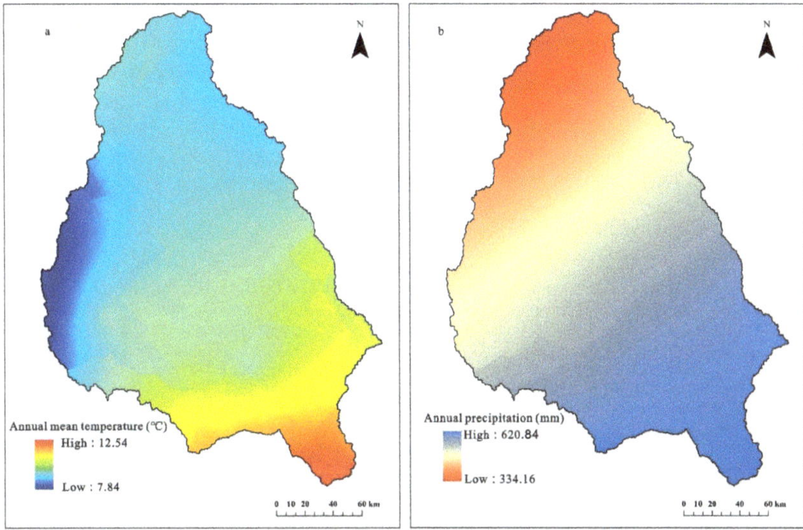

Figure 8. Spatial pattern of the annual mean temperature (**a**) and the annual precipitation (**b**) in the Jinghe River basin from 2000 to 2014.

The vegetation degradation dominated by climate factors was located generally in the Ziwuling Mountain and Liupan Mountain regions on the east and west sides of the basin, respectively. Because of the drought climate, the vegetation in the mountain areas showed significant vegetation degradation dominated by climatic factors. Although ecological restoration measures, such as returning farmland to forests and protecting vegetation, also must be implemented in these areas, the terrain there restricts follow-up human management activities such as irrigation. Therefore, the water demand of the vegetation in the mountain areas relies primarily on natural precipitation and soil moisture [19,79]. However, vegetation planted recently has increased the local water demand and accelerated the consumption of regional water resources, which eventually had led to degradation of the vegetation [80,81].

5. Conclusions

This study assessed the driving forces in vegetation dynamics in the Jinghe River basin from 2000 to 2014 using NPP as the indicator. The results showed that the vegetation increased in the study area, and human activities played an active role in the vegetation restoration, especially in valleys in the middle of the basin, where the rate of vegetation change in the areas dominated by human activities rose continuously with the increase in slope. This result is consistent with the implementation of ecological projects such as GGP. The degradation of vegetation caused by human activities was located primarily in populous areas and was related closely to urban expansion. The vegetation restoration that was dominated by climate factors was concentrated largely in the northern part of the basin, where the precipitation was relatively sufficient. However, the vegetation degradation dominated by climate factors generally was located in the Ziwuling Mountain and Liupan Mountain regions on the east and west sides of the basin, where the vegetation degradation rate in areas attributable to climate factors rose with increases in slope. In these regions, the arid climate caused a shortage of water resources, and the human dominated vegetation restoration activities exacerbated the water demand of vegetation further and surpassed the carrying capacity of the regional water resources, which led ultimately to vegetation degradation.

The methodology of comparing the expected and actual NPP to distinguish the effect of climate factors and human activities on vegetation dynamics in this study demonstrated a relatively higher accuracy and can be applied at different regional scales. Further, as unsustainable vegetation restoration measures may cause regional imbalances in water supply, and lead eventually to vegetation degradation, we recommend that future ecological restoration programs pay more attention to maintaining the balance between ecosystem restoration and water resource demands to maximize the benefits of human activities and ensure that the vegetation restoration is ecologically sustainable.

Author Contributions: H.W. analyzed the related data and wrote the manuscript; G.L. and Z.L. designed the framework of the research and revised the manuscript; and P.W. and Z.W. provided the climate and NDVI data.

Acknowledgments: This research was funded by the Science and Technology Service Network Initiative Project of Chinese Academy of Sciences (No. KFJ-STS-ZDTP-036), the National Key R&D Program of China (2017YFC0504701), Fundamental Research Funds for the Central Universities (No. GK201703053), China Postdoctoral Science Foundation (No. 2017M623114).

Conflicts of Interest: The authors declare no conflict of interest.

References

1. Field, C.B. Global change. Sharing the garden. *Science* **2001**, *294*, 2490–2491. [CrossRef] [PubMed]
2. Root, T.L.; Price, J.T.; Hall, K.R.; Schneider, S.H.; Rosenzweig, C.; Pounds, J.A. Fingerprints of global warming on wild animals and plants. *Nature* **2003**, *421*, 47–60. [CrossRef] [PubMed]
3. Li, A.; Wu, J.G.; Huang, J.H. Distinguishing between human-induced and climate-driven vegetation changes: A critical application of RESTREND in Inner Mongolia. *Landsc. Ecol.* **2012**, *27*, 969–982. [CrossRef]
4. Zhang, Y.; Zhang, C.; Wang, Z.; Chen, Y.; Gang, C.; An, R.; Li, J. Vegetation dynamics and its driving forces from climate change and human activities in the Three-River Source Region, China from 1982 to 2012. *Sci. Total Environ.* **2016**, *563–564*, 210–220. [CrossRef] [PubMed]
5. Ma, W.; Wang, X.; Zhou, N.; Jiao, L. Relative importance of climate factors and human activities in impacting vegetation dynamics during 2000–2015 in the Otindag Sandy Land, northern China. *J. Arid Land* **2017**, *9*, 558–567. [CrossRef]
6. Huang, K.; Zhang, Y.; Zhu, J.; Liu, Y.; Zu, J.; Zhang, J. The Influences of Climate Change and Human Activities on Vegetation Dynamics in the Qinghai-Tibet Plateau. *Remote Sens.* **2016**, *8*, 876. [CrossRef]
7. Wessels, K.J.; Prince, S.D.; Malherbe, J.; Small, J.; Frost, P.E.; VanZyl, D. Can human-induced land degradation be distinguishedfrom the effects of rainfall variability? A case study in South Africa. *J. Arid Environ.* **2007**, *68*, 271–297. [CrossRef]
8. Yao, J.Q.; Yang, Q.; Chen, Y.N. Climate change in arid areas of Northwest China in past 50 years and its effects on the local ecological environment. *Chin. J. Ecol.* **2013**, *32*, 1283–1291.
9. Wang, H.; Liu, G.H.; Li, Z.S.; Ye, X.; Wang, M.; Gong, L. Driving force and changing trends of vegetation phenology in the Loess Plateau of China from 2000 to 2010. *J. Mt. Sci.* **2016**, *13*, 844–856. [CrossRef]
10. An, R.; Wang, H.L.; Feng, X.Z.; Wu, H.; Wang, Z.; Wang, Y.; Shen, X.J.; Lu, C.H.; Quaye-Ballard, J.A.; Chen, Y.H.; et al. Monitoring rangeland degradation using a novel local NPP scaling based scheme over the "Three-River Headwaters" region, hinterland of the Qinghai-Tibetan Plateau. *Quat. Int.* **2017**, *444*, 97–114. [CrossRef]
11. Zhang, B.; Wu, P.; Zhao, X.; Wang, Y.; Gao, X. Changes in vegetation condition in areas with different gradients (1980–2010) on the Loess Plateau, China. *Environ. Earth Sci.* **2012**, *68*, 2427–2438. [CrossRef]
12. Zhang, J.; Wang, T.; Ge, J. Assessing Vegetation Cover Dynamics Induced by Policy-Driven Ecological Restoration and Implication to Soil Erosion in Southern China. *PLoS ONE* **2015**, *10*, e0131352. [CrossRef] [PubMed]
13. Yang, H.F.; Yao, L.; Wang, Y.D.; Li, J.L. Relative contribution of climate change and human activities to vegetation degradation and restoration in North Xinjiang, China. *Rangel. J.* **2017**, *39*, 289–302. [CrossRef]
14. Cao, S.; Chen, L.; Shankman, D.; Wang, C.; Wang, X.; Zhang, H. Excessive reliance on afforestation in China's arid and semi-arid regions: Lessons in ecological restoration. *Earth-Sci. Rev.* **2011**, *104*, 240–245. [CrossRef]

15. Wang, H.; Liu, G.; Li, Z.; Ye, X.; Fu, B.; Lü, Y. Analysis of the Driving Forces in Vegetation Variation in the Grain for Green Program Region, China. *Sustainability* **2017**, *9*, 1853. [CrossRef]

16. Wang, H.; Liu, G.; Li, Z.; Ye, X.; Fu, B.; Lv, Y. Impacts of Drought and Human Activity on Vegetation Growth in the Grain for Green Program Region, China. *Chin. Geogr. Sci.* **2018**, *28*, 470–481. [CrossRef]

17. Feng, X.; Fu, B.; Lu, N.; Zeng, Y.; Wu, B. How ecological restoration alters ecosystem services: An analysis of carbon sequestration in China's Loess Plateau. *Sci. Rep.* **2013**, *3*, 2846. [CrossRef] [PubMed]

18. Deng, L.; Shangguan, Z.P.; Sweeney, S. "Grain for Green" driven land use change and carbon sequestration on the Loess Plateau, China. *Sci. Rep.* **2014**, *4*, 7039. [CrossRef] [PubMed]

19. Feng, X.M.; Fu, B.J.; Piao, S.L.; Wang, S.; Ciais, P.; Zeng, Z.Z.; Lv, Y.H.; Zeng, Y.; Li, Y.; Jiang, X.H.; et al. Revegetation in China's Loess Plateau is approaching sustainable water resource limits. *Nat. Clim. Chang.* **2016**, *6*, 1019–1022. [CrossRef]

20. Wang, Y.; Cao, S. Carbon sequestration may have negative impacts on ecosystem health. *Environ. Sci. Technol.* **2011**, *45*, 1759–1760. [CrossRef] [PubMed]

21. Cao, S.; Chen, L.; Yu, X. Impact of China's Grain for Green Project on the landscape of vulnerable arid and semi-arid agricultural regions: A case study in northern Shaanxi Province. *J. Appl. Ecol.* **2009**, *46*, 536–543. [CrossRef]

22. Wang, X.M.; Zhang, C.X.; Hasi, E.; Dong, Z.B. Has the Three Norths Forest Shelterbelt Program solved the desertification and dust storm problems in arid and semiarid China? *J. Arid Environ.* **2010**, *74*, 13–22. [CrossRef]

23. Aldous, A.; Fitzsimons, J.; Richter, B.; Bach, L. Droughts, floods and freshwater ecosystems: Evaluating climate change impacts and developing adaptation strategies. *Mar. Freshw. Res.* **2011**, *62*, 223–231. [CrossRef]

24. Lawler, J.J. Climate change adaptation strategies for resource managementand conservation planning. *Ann. N. Y. Acad. Sci.* **2009**, *1162*, 79–98. [CrossRef] [PubMed]

25. Chang, X.L.; Lu, C.X.; Gao, Y.B. Impacts of human economic activities on wind and sand environment in Kerqin sandy land. *Resour. Sci.* **2003**, *25*, 78–83. (In Chinese)

26. Ma, Y.H.; Fan, S.Y.; Zhou, L.H.; Dong, Z.H.; Zhang, K.C.; Feng, J.M. The temporal change of driving factors during the course of land desertification in arid region of North China: The case of Minqin County. *Environ. Geol.* **2007**, *51*, 999–1008. [CrossRef]

27. Zhang, Y.S.; Wang, L.X.; Zhang, H.Q.; Li, X.Y. Influence of environmental factor changes on desertification process in Shule River. *Resour. Sci.* **2003**, *25*, 60–65. (In Chinese)

28. Gang, C.; Zhou, W.; Chen, Y.; Wang, Z.; Sun, Z.; Li, J.; Qi, J.; Odeh, I. Quantitative assessment of the contributions of climate change and human activities on global grassland degradation. *Environ. Earth Sci.* **2014**, *72*, 4273–4282. [CrossRef]

29. Evans, J.; Geerken, R. Discrimination between climate and human-induced dryland degradation. *J. Arid Environ.* **2004**, *57*, 535–554. [CrossRef]

30. Prince, S.D.; Wessels, K.J.; Tucker, C.J.; Nicholson, S.E. Desertification in the Sahel: A reinterpretation of a reinterpretation. *Glob. Chang. Biol.* **2007**, *13*, 1308–1313. [CrossRef]

31. Wessels, K.J.; Prince, S.D.; Frost, P.E.; Zyl, D. Assessing the effects of human-induced land degradation in the former homelands of northern South Africa with a 1 km AVHRR NDVI time-series. *Remote Sens. Environ.* **2004**, *91*, 47–67. [CrossRef]

32. Chen, B.; Zhang, X.; Tao, J.; Wu, J.; Wang, J.; Shi, P.; Zhang, Y.; Yu, C. The impact of climate change and anthropogenic activities on alpine grassland over the Qinghai-Tibet Plateau. *Agric. For. Meteorol.* **2014**, *189–190*, 11–18. [CrossRef]

33. Wessels, K.J.; Prince, S.D.; Reshef, I. Mapping land degradation by comparison of vegetation production to spatially derived estimates of potential production. *J. Arid Environ.* **2008**, *72*, 1940–1949. [CrossRef]

34. Yeganeh, H.; Khajedein, S.J.; Amiri, F.; Shariff, A.R.B.M. Monitoring rangeland ground cover vegetation using multitemporal MODIS data. *Arabian J. Geosci.* **2014**, *7*, 287–298. [CrossRef]

35. Zheng, Y.R.; Xie, Z.X.; Robert, C.; Jiang, L.H.; Shimizu, H. Did climate drive ecosystem change and induce desertification in Otindag sandy land, China over the past 40 years? *J. Arid Environ.* **2006**, *64*, 523–541. [CrossRef]

36. Wang, H.; Liu, G.; Li, Z.; Ye, X.; Wang, M.; Gong, L. Impacts of climate change on net primary productivity in arid and semiarid regions of China. *Chin. Geogr. Sci.* **2015**, *26*, 35–47. [CrossRef]
37. Nemani, R.R.; Keeling, C.D.; Hashimoto, H.; Jolly, W.M.; Piper, S.C.; Tucker, C.J.; Myneni, R.B.; Running, S.W. Climate driven increases in global terrestrial net primary production from 1982 to 1999. *Science* **2003**, *300*, 1560–1563. [CrossRef] [PubMed]
38. Li, S.; Yan, J.; Liu, X.; Wan, J. Response of vegetation restoration to climate change and human activities in Shaanxi-Gansu-Ningxia Region. *J. Geogr. Sci.* **2013**, *23*, 98–112. [CrossRef]
39. Haberl, H.; Krausmann, F.; Erb, K.; Schulz, N.B. Human Appropriation of Net Primary Production. *Science* **2002**, *296*, 1968–1969. [CrossRef] [PubMed]
40. Zhou, W.; Gang, C.; Zhou, F.; Li, J.; Dong, X.; Zhao, C. Quantitative assessment of the individual contribution of climate and human factors to desertification in northwest China using net primary productivity as an indicator. *Ecol. Indic.* **2015**, *48*, 560–569. [CrossRef]
41. Haberl, H.; Erb, K.; Krausmann, F.; Gaube, V.; Bondeau, A.; Plutzar, C.; Gingrich, S.; Lucht, W.; Fischer-Kowalski, M. Quantifying and Mapping the Human Appropriation of Net Primary Production in Earth's Terrestrial Ecosystems. *Proc. Natl. Acad. Sci. USA* **2007**, *104*, 12942–12947. [CrossRef] [PubMed]
42. Yang, Y.; Wang, Z.; Li, J.; Gang, C.; Zhang, Y.; Zhang, Y.; Odeh, I.; Qi, J. Comparative assessment of grassland degradation dynamics in response to climate variation and human activities in China, Mongolia, Pakistan and Uzbekistan from 2000 to 2013. *J. Arid Environ.* **2016**, *135*, 164–172. [CrossRef]
43. Xu, D.; Kang, X.; Liu, Z.; Zhuang, D.; Pan, J. Assessing the relative role of climate change and human activities in sandy desertification of Ordos region, China. *Sci. China* **2009**, *52*, 855–868. (In Chinese) [CrossRef]
44. Xie, F.; Qiu, G.; Yin, J.; Xiong, Y.J.; Wang, P. Comparison of Land Use/Land Cover Change in Three Sections of the Jinghe River Basin between the 1970s and 2006. *J. Nat. Resour.* **2009**, *24*, 1354–1365. (In Chinese)
45. Yue, D.X.; Du, J.; Liu, J.Y.; Gou, J.J.; Zhang, J.J.; Ma, J.H. Spatio-temporal analysis of ecological carrying capacity in Jinghe Watershed based on Remote Sensing and Transfer Matrix. *Acta Ecol. Sin.* **2011**, *31*, 2550–2558. (In Chinese)
46. Qi, Q.; Wang, T.; Kou, X.J.; Ge, J.P. Temporal and spatial changes of vegetation cover and the relationship with precipitation in Jinghe watershed of china. *J. Plant Ecol.* **2009**, *33*, 246–253. (In Chinese)
47. Zhang, Y.S.; Chen, X.; Gao, M.; Zhang, Z.C.; Cheng, Q.B. Detecting Temporal Variations of Temperature Characteristics in Jinghe Watershed. *J. Water Resour. Res.* **2017**, *6*, 33–41. (In Chinese) [CrossRef]
48. Slayback, D.A.; Pinzon, J.E.; Los, S.O.; Tucker, C.J. Northern hemisphere photosynthetic trends 1982–99. *Glob. Chang. Biol.* **2003**, *9*, 1–15. [CrossRef]
49. Wang, X.; Piao, S.L.; Ciais, P.; Li, J.; Friedlingstein, P.; Koven, C.; Chen, A. Spring temperature change and its implication in the change of vegetation growth in North America from 1982 to 2006. *Proc. Natl. Acad. Sci. USA* **2011**, *108*, 1240–1245. [CrossRef] [PubMed]
50. Lieth, H. Primary production: Terrestrial ecosystems. *Hum. Ecol.* **1973**, *1*, 303–332. [CrossRef]
51. Lin, H.; Feng, Q.; Liang, T.; Ren, J. Modelling global-scale potential grassland changes in spatio-temporal patterns to global climate change. *Int. J. Sustain. Dev. World Ecol.* **2012**, *20*, 83–96. [CrossRef]
52. Adams, B.; White, A.; Lenton, T.M. An analysis of some diverse approaches to modelling terrestrial net primary productivity. *Ecol. Model.* **2004**, *177*, 353–391. [CrossRef]
53. Mu, S.; Zhou, S.; Chen, Y.; Li, J.; Ju, W.; Odeh, I.O.A. Assessing the impact of restoration-induced land conversion and management alternatives on net primary productivity in Inner Mongolian grassland, China. *Glob. Planet. Chang.* **2013**, *108*, 29–41. [CrossRef]
54. Monteith, J.L. Solar Radiation and Porductivity in Tropical Exosystems. *J. Appl. Ecol.* **1972**, *9*, 747–766. [CrossRef]
55. Potter, C.S.; Randerson, J.T.; Field, C.B.; Matson, P.A.; Vitousek, P.M.; Mooney, H.A.; Klooster, S.A. Terrestrial ecosystem production: A process model based on global satellite and surface data. *Glob. Biogeochem. Cycles* **1993**, *7*, 811–841. [CrossRef]
56. Wen, Y.; Liu, X.; Du, G. Nonuniform Time-Lag Effects of Asymmetric Warming on Net Primary Productivity across Global Terrestrial Biomes. *Earth Interact.* **2018**, *22*, 1–26. [CrossRef]
57. Shi, Y.; Shen, Y.; Kang, E.; Li, D.; Ding, Y.; Zhang, G.; Hu, R. Recent and Future Climate Change in Northwest China. *Clim. Chang.* **2006**, *80*, 379–393. [CrossRef]

58. Zhang, C.X.; Wang, X.; Li, J.C.; Hua, T. Roles of climate changes and human interventions in land degradation: A case study by net primary productivity analysis in China's Shiyanghe Basin. *Environ. Earth Sci.* **2011**, *64*, 2183–2193. [CrossRef]

59. Zhou, W.; Li, J.; Mu, S.J.; Gang, C.C.; Sun, Z.G. Effects of ecological restoration-induced land-use change and improved management on grassland net primary productivity in the Shiyanghe River Basin, north-west China. *Grass Forage Sci.* **2013**, *10*, 1111. [CrossRef]

60. Lu, Y.; Fu, B.; Feng, X.; Zeng, Y.; Liu, Y.; Chang, R.; Sun, G.; Wu, B. A policy-driven large scale ecological restoration: Quantifying ecosystem services changes in the Loess Plateau of China. *PLoS ONE* **2012**, *7*, e31782.

61. Wu, S.H.; Zhou, S.L.; Chen, D.X.; Wei, Z.Q.; Dai, L.; Li, X.G. Determining the contributions of urbanisation and climate change to NPP variations over the last decade in the Yangtze River Delta, China. *Sci. Total Environ.* **2014**, *472*, 397–406. [CrossRef] [PubMed]

62. Mu, S.J.; Chen, Y.Z.; Li, J.L.; Ju, W.M.; Odeh, I.O.A.; Zou, X.L. Grassland dynamics in response to climate change and human activities in Inner Mongolia, China between 1985 and 2009. *Rangel. J.* **2013**, *35*, 315–329. [CrossRef]

63. Zhu, W.Q.; Pan, Y.Z.; Zhang, J.S. Estimation of net primary productivity of Chinese terrestrial vegetation based on remote sensing. *J. Plant Ecol.* **2007**, *31*, 413–424. (In Chinese)

64. Yu, D.Y.; Shi, P.J.; Shao, H.B.; Zhu, W.Q.; Pan, Y.Z. Modelling net primary productivity of terrestrial ecosystems in East Asia based on an improved CASA ecosystem model. *Int. J. Remote Sens.* **2009**, *30*, 4851–4866. [CrossRef]

65. Zaks, D.P.; Ramankutty, N.; Barford, C.C.; Foley, J.A. From Miami to Madison: Investigating the relationship between climate and terrestrial net primary production. *Glob. Biogeochem. Cycles* **2007**, *21*, GB3004. [CrossRef]

66. Cao, X.; Gu, Z.H.; Chen, J.; Liu, J.; Shi, P.J. Analysis of human-induced steppe degradation based on remote sensing in Xilin Gole, Inner Mongolia, China. *J. Plant Ecol.* **2006**, *30*, 268–277. (In Chinese)

67. Yi, L.; Ren, Z.Y.; Zhang, C.; Liu, W. Vegetation cover, climate and human activities on the loess plateau. *Resour. Sci.* **2014**, *36*, 166–174. (In Chinese)

68. Wang, Q.; Zhang, B.; Dai, S.P.; Zhang, F.F.; Zhao, Y.F.; Yin, H.X.; He, X.Q. Analysis of the vegetation cover chang and its relationship with factors in the Three-North Shelter Forest Program. *China Environ. Sci.* **2012**, *32*, 1302–1308. (In Chinese)

69. Su, C.; Fu, B. Evolution of ecosystem services in the Chinese Loess Plateau under climatic and land use changes. *Glob. Planet. Chang.* **2013**, *101*, 119–128. [CrossRef]

70. Wang, Y.; Fu, B.; Lü, Y.; Chen, L. Effects of vegetation restoration on soil organic carbon sequestration at multiple scales in semi-arid Loess Plateau, China. *Catena* **2011**, *85*, 58–66. [CrossRef]

71. Li, S.; Liang, W.; Fu, B.; Lu, Y.; Fu, S.; Wang, S.; Su, H. Vegetation changes in recent large-scale ecological restoration projects and subsequent impact on water resources in China's Loess Plateau. *Sci. Total Environ.* **2016**, *569–570*, 1032–1039. [CrossRef] [PubMed]

72. Li, J.; Peng, S.; Li, Z. Detecting and attributing vegetation changes on China's Loess Plateau. *Agric. For. Meteorol.* **2017**, *247*, 260–270. [CrossRef]

73. Li, J.; Li, Z.; Lü, Z. Analysis of spatiotemporal variations in land use on the Loess Plateau of China during 1986–2010. *Environ. Earth Sci.* **2016**, *75*, 997. [CrossRef]

74. Deng, L.; Liu, G.B.; Shangguan, Z.P. Land-use conversion and changing soil carbon stocks in China's 'Grain-for-Green' Program: A synthesis. *Glob. Chang. Biol.* **2014**, *20*, 3544–3556. [CrossRef] [PubMed]

75. Luck, G.W.; Smallbone, L.T.; O'Brien, R. Socio-Economics and Vegetation Change in Urban Ecosystems: Patterns in Space and Time. *Ecosystems* **2009**, *12*, 604–620. [CrossRef]

76. Li, Z.; Wang, J.; Liu, W.Z. Climate Changes in Jinghe Watershed and Its Relationship with ENSO. *Prog. Geogr.* **2010**, *29*, 833–839. (In Chinese)

77. Zhao, C.; Li, Z.; Liu, W.Z. Downscaling GCMs to Project the Potential Changes of Precipitation in Jinghe Basin. *Res. Soil Water Conserv.* **2014**, *21*, 23–28. (In Chinese)

78. Zhang, H.B.; Gu, L.; Xin, C.; Yu, Q.J. Investigation on the Spatial-temporal Variation of Drought Characteristics in Jinghe River Basin. *J. North China Univ. Water Resour. Electr. Power* **2016**, *37*, 1–10. (In Chinese)

79. Lal, R. Carbon Sequestration in Dryland Ecosystems. *Environ. Manag.* **2004**, *33*, 528–544. [CrossRef] [PubMed]

80. Wang, H.S.; Huang, M.; Zhang, L. Impacts of re-vegetation on water cycle in a small watershed of the Loess Plateau. *J. Nat. Resour.* **2004**, *19*, 344–350.
81. Tian, F.; Feng, X.; Zhang, L.; Fu, B.; Wang, S.; Lv, Y.; Wang, P. Effects of revegetation on soil moisture under different precipitation gradients in the Loess Plateau, China. *Hydrol. Res.* **2017**, *48*, 1378–1390. [CrossRef]

Article

Different Influences of Vegetation Greening on Regional Water-Energy Balance under Different Climatic Conditions

Dan Zhang [1], Xiaomang Liu [2,*] and Peng Bai [2]

[1] Key Laboratory of Watershed Geographic Sciences, Nanjing Institute of Geography and Limnology, Chinese Academy of Sciences, Nanjing 210008, China; dzhang@niglas.ac.cn

[2] Institute of Geographic Sciences and Natural Resources Research, Chinese Academy of Sciences, Chaoyang District, Beijing 100011, China; Baip@igsnrr.ac.cn

* Correspondence: hydroliu@163.com; Tel.:+86-10-6488-9083

Received: 4 June 2018; Accepted: 5 July 2018; Published: 9 July 2018

Abstract: Vegetation serves as a key element in the land-atmospheric system, and changes in vegetation can impact the regional water-energy balance via several biophysical processes. This study proposes a new water-energy balance index that estimates the available-water-to-available-energy ratio (WER) by improving upon the Budyko framework, which evaluates climate variation and vegetation change. Moreover, the impact of vegetation greening on WER is quantified in 34 catchments under different climatic conditions. The results show that the normalized difference vegetation index (NDVI) increased at all the catchments, which indicates that there was a vegetation greening trend in the study area. There are negative relationships between the NDVI and runoff at both water-limited and energy-limited catchments, which demonstrates that both types of catchments became drier due to vegetation greening. Four numerical experiments were executed to quantify the contribution of vegetation greening and climate variations to WER changes. The results show that the calculated WER trends by numerical tests fit well with the observed WER trends (R^2 = 0.96). Vegetation greening has positive influences on WER changes under energy-limited conditions, which indicates that residual energy decreases faster than water availability, resulting in less energy for sensible heat, i.e., a cooling effect. Nevertheless, vegetation greening has negative influences on WER under water-limited conditions, which indicates that water availability decreases faster than residual energy, resulting in more energy for sensible heat. Notably, the WER decrease in water-limited catchments is dominated by potential evapotranspiration and NDVI variation, whereas the WER change in energy-limited catchments is dominated by climate variation. This study provides a comprehensive understanding of the relationships among water, energy and vegetation greening under different climatic conditions, which is important for land-atmosphere-vegetation modeling and designing strategies for ecological conservation and local water resource management.

Keywords: vegetation greening; water-energy balance; quantification; different climatic conditions

---

1. Introduction

Water and energy are essential elements of regional hydrology and are closely connected via evapotranspiration [1–3]. Vegetation is a crucial component of the land-atmospheric system, and changes in vegetation alter the regional water and energy balance through several biophysical processes [4–6]. Large-scale vegetation variation would lead to changes in climate dynamics, atmospheric-land surface interaction and global hydrological processes. According to reports from around the world, vegetation greening occurs in response to the effects of climate change and the impact of policies on ecological conservation and restoration in recent decades [7,8], including increased growth of the savannas in Australia, Africa and South America due to increased rainfall, the reforestation of abandoned farmlands in Russia and tree planting projects in China [9–11]. Investigating the impact of vegetation greening on the regional water-energy balance is an important issue in Earth system science with impacts in the present and in the future [12,13].

The regional water-energy balance is determined by the water supply (precipitation) and the energy availability (evaporative demand, PET) and is modified by land-surface characteristics, such as vegetation, soil and topography [14,15]. The Budyko framework, which is the primary approach used to investigate the regional water-energy balance, considers both climatic conditions and the underlying characteristics using a land-surface parameter w [16,17], which is related to vegetation, soil type, soil infiltration capacity, topography, etc. Recently, Liu et al. [18] proposed a new water-balance index (the available-water-to-available-energy ratio, WER) to reflect the regional energy balance and hydrological processes based on the Budyko framework and further investigated the impact of land-surface changes on WER. However, the impact of vegetation change on WER remains unknown because the study by Liu et al. [18] did not provide data on the relationships between vegetation greening and WER.

Several previous studies have noted the importance of considering vegetation dynamics when using the Budyko framework to evaluate the water-energy balance [19–21]. For example, Donohue et al. [22] concluded that the accuracy of runoff estimation was improved by incorporating dynamic vegetation into the Budyko framework when it was applied at finer spatial timescales. In addition, some researchers attempted to establish relationships between the variations in vegetation and the land-surface parameter w [23–25]. Li et al. [23] found that w was linearly correlated with vegetation coverage at long timescales in large river basins. Zhang et al. [25] presented an exponential relationship between the change in w and the vegetation variation using the Budyko framework. In general, these studies found positive relationships between changes in w and variations in vegetation under all climatic conditions; in other words, vegetation greening leads to an increase in evapotranspiration (i.e., a decrease in runoff), and the inverse is also true. This phenomenon is known as the "trade-off relationship" between vegetation and water [26]. The essence of this "trade-off relationship" is that precipitation and energy are redistributed via evapotranspiration because of vegetation changes.

Vegetation in China has increased considerably during the past 30 years because of the large-scale soil and water conservation projects implemented by the Chinese government [27,28]. The coverage of forest increased from 11% in the 1980s to 22% in 2010, especially in the Yangtze River and Yellow River basins [29,30]. Although some studies have investigated the influence of vegetation greening on evapotranspiration and runoff in China, no consensus regarding this subject has yet been achieved [31,32]. This study aims to propose a comprehensive approach to quantify the contribution of vegetation greening to the water-energy balance index (WER). Furthermore, the new approach is used to study 34 catchments under different climatic conditions in China. The results are expected to provide a comprehensive understanding of the interactions among vegetation, energy and water, which would be helpful for ecological restoration, forest protection and water resources management. The objectives of this study are (1) to investigate the relationships among WER changes, vegetation greening and climate variation under different climatic conditions, and (2) to quantify the contribution of vegetation greening and climate variation to WER changes.

2. Study Area

In this study, 34 catchments located in the middle reaches of the Yangtze River and Yellow River in China are selected to investigate the impact of vegetation greening on WER under different climatic conditions (Figure 1). The drainage areas range from 656 to 30,661 km^2, and the aridity index (AI) values range from 0.5 to 3.2. Catchments in the Yangtze River (nos. 1–16) belong to an energy-limited condition (PET < precipitation), and catchments in the Yellow River (nos. 17–34) belong to a water-limited condition (PET > precipitation) [33]. The details are shown in Table 1. The dominant vegetation of the energy-limited catchments is forest, and the dominant vegetation of the water-limited catchments is grass [34,35].

Table 1. Information of the studied catchments under different climatic conditions.

Catchments Characteristics	Energy-Limited	Water-Limited
number	16	18
catchment area (km^2)	656–15,307	1121–30,661
air temperature (°C)	14.8–19.7	0.9–11.3
annual rainfall (mm)	1279–1852	388–683
aridity index	0.5–0.8	1.7–3.2

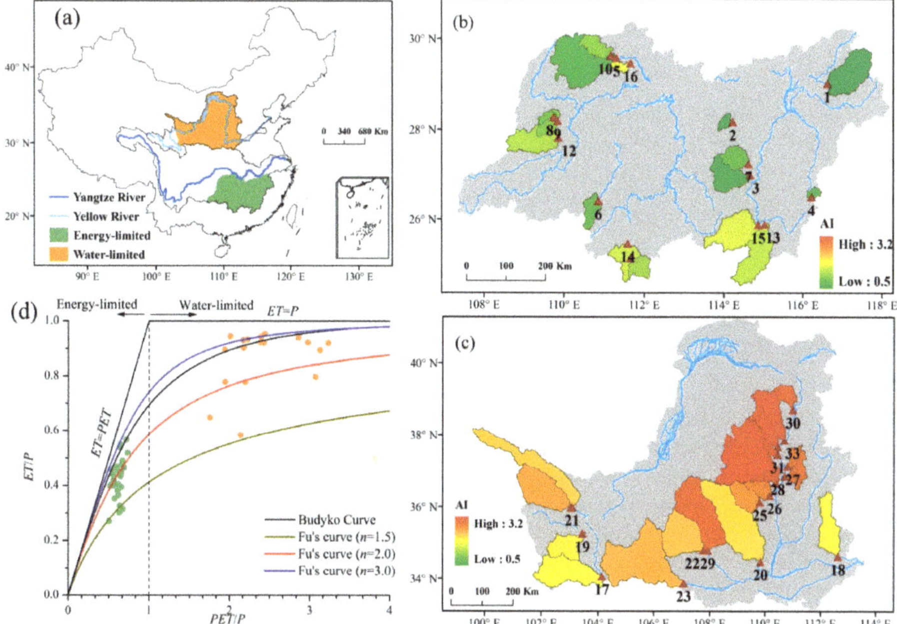

Figure 1. (**a**) Sketch map of the study area, (**b**) catchments under energy-limited condition, (**c**) catchments under water-limited condition, and (**d**) the long-term hydroclimatological characteristics of the 34 selected catchments based on the Budyko hypothesis. AI is the aridity index, P is precipitation, PET is potential evapotranspiration and ET is actual evaporation. Triangle in (**b**,**c**) represents the location of hydrological station.

3. Materials and Methods

3.1. Materials

Monthly runoff datasets from 1982 to 2013 under energy-limited conditions were collected from the Hydrological Yearbook of the Yangtze River in China. These energy-limited catchments are in the headwaters of the tributaries of the Yangtze River, and human influences, such as dams and reservoirs, on runoff are limited. Runoff under water-limited conditions in the Yellow River is influenced by intensive human activities, such as irrigation and water impoundment by dams. As a result, monthly naturalized runoff time series were provided by the Hydrological Bureau of the Yellow River. Monthly precipitation data from 1954 rainfall stations were obtained from the National Climate Center of the China Meteorological Administration. The maximum and minimum air temperatures, sunshine duration, relative humidity, and wind speed from 753 meteorological stations are used to calculate the PET based on the Penman formula, which has been shown to be the most appropriate form when considering a changing climate [36–38]. The AI is calculated as the ratio of the mean annual PET to the mean annual precipitation averaged from 1982–2013 and therefore reflects the average conditions of the regional climate. The climate variables and PET for each catchment are weighted by the Thiessen polygon method.

Monthly Advanced Very High Resolution (AVHRR) NDVI (normalized difference vegetation index)datasets processed by the Global Inventory Modeling and Mapping Studies are employed to explore the trends in vegetation during the past 30 years and are provided by the Global Land Cover Facility of the University of Maryland [39,40]. These data have a good accuracy and have been used widely in hydrological and ecological research [40]. The spatial resolution is 8 km × 8 km. The NDVI data are averaged over all grid cells whose centers are in the corresponding catchment.

3.2. Water-Energy Balance Index Considering Vegetation Change

The water-energy balance index proposed by Liu et al. [18] is defined as the ratio of water availability to energy availability, which is written as:

$$\text{WER} = \frac{P - ET}{PET - ET},\tag{1}$$

Where P is precipitation, PET is potential evapotranspiration and ET is actual evaporation. WER reflects the regional dryness and wetness condition comprehensively, which is the balanced state of the catchment's energy and water. ET is estimated by Fu's equation based on the Budyko framework [17]:

$$ET = P \cdot (1 + \text{AI} - (1 + \text{AI}^w)^{1/w})\tag{2}$$

Where AI is the aridity index, which is the ratio of precipitation to potential evapotranspiration. The parameter w is related to the land-surface characteristics and can be estimated using long-term hydroclimatic data. Previous studies have demonstrated that the Budyko-based Fu's equation is appropriate for ET estimation at different timescales [31,41]. For a reasonable application of Fu's equation and to better use the available data, a 60-month (5-year timescale) moving average is adopted for water balance analysis.

Substituting Equation (2) into Equation (1), WER can be simplified as:

$$\text{WER} = \frac{(1 + \text{AI}^w)^{1/w} - \text{AI}}{(1 + \text{AI}^w)^{1/w} - 1}\tag{3}$$

Previous studies have indicated that w is related to vegetation, soil type, soil infiltration capacity, topography, and other factors. When all the affecting factors for w are considered, w can be estimated. Li et al. [23] found a linear relationship between w and NDVI over a wide range of river basins, which is expressed as:

$$w = a \cdot NDVI + b\tag{4}$$

Where a and b are the regression coefficients. Substituting Equation (4) into Equation (3), WER can be simplified as:

$$\text{WER} = \frac{(1 + \text{AI}^{(a \cdot NDVI + b)})^{1/(a \cdot NDVI + b)} - \text{AI}}{(1 + \text{AI}^{(a \cdot NDVI + b)})^{1/(a \cdot NDVI + b)} - 1}\tag{5}$$

Equation (5) shows the nonlinear correlations among WER, vegetation and climate variables (precipitation and PET). There are three advantages of WER. First, it considers both the regional climatic condition and the land-surface features and is therefore closer to the actual condition than the standardized precipitation index, standardized precipitation evapotranspiration index or Palmer drought severity index. Second, it provides a nonlinear relationship between WER changes and NDVI changes, as well as AI, which can be used to quantify WER changes. Third, the index has a simple form and can be easily combined with remote sensing products, making it appropriate for use in data-scarce catchments.

3.3. Sensitivity Analysis

The sensitivity of WER to climate variation and vegetation change is evaluated by the following equation [42,43]:

$$\varepsilon_{x_i} = \frac{\partial WER}{\partial x_i} \times \frac{x_i}{WER} \tag{6}$$

Where $\partial WER/\partial x_i$ is the partial derivative of WER to the influencing factor x_i (precipitation, PET and NDVI). A positive sensitivity coefficient indicates WER will increase as x_i increases, whereas a negative sensitivity coefficient indicates WER will decrease as x_i increases. A sensitivity coefficient of 0.1 indicates that a 10% increase of x_i will lead to an increase in WER by 1%. The larger the absolute value of ε_{xi} the larger the influence of change in x_i on WER.

3.4. Contribution Method

In this study, a numerical experiment approach is employed to quantify the impacts of vegetation changes and climate variation on WER. Four numerical experiments are designed, including one control experiment and three sensitivity experiments (Table 2). The WER values calculated by the control experiment represent the combined influence of climate and vegetation changes, and the WER values calculated by each sensitivity experiment represent the influence of the two factors other than the unchanged factor. Thus, the impact of each factor on WER changes can be estimated as:

$$C\_x_i = T_{WER\_ctr} - T_{WER\_x_i} \tag{7}$$

Where x_i represents precipitation, PET or NDVI, C_x_i is the contribution of the corresponding factor to WER changes, T_{WER_ctr} is the slope of WER_ctr based on linear regression, and T_{WER_xi} is the slope of WER based on the ith sensitivity experiment.

Table 2. Numerical experiment design for WER contributions.

Experiment	Description
Control test:WER_ctr	Precipitation, PET and NDVI from 1982–2013
Sensitivity test:WER_Prcp	Precipitation maintained at the initial year, the others same as the control test
Sensitivity test:WER_PET	PET maintained at the initial year, the others same as the control test
Sensitivity test:WER_NDVI	NDVI maintained at the initial year, the others same as the control test

NDVI: normalized difference vegetation index; PET: potential evapotranspiration.

4. Results

4.1. Relationship between Vegetation and the Land-Surface Parameter w

Figure 2 shows the relationships between w and NDVI for the selected catchments. The coefficient of determination R^2 ranges from 0.11 to 0.68, and all catchment values are significant at the level of 0.01 by the t-test, which indicates that changes in NDVI explain 11–68% of the changes in w (Table 3). Parameter a is positive in all 34 catchments, indicating that evapotranspiration increases with vegetation greening. The average value of a in energy-limited catchments is 6.3, whereas the average value of a in water-limited catchments is 11.4. This difference indicates that the impact of vegetation greening on w (e.g., evapotranspiration) is larger under a water-limited condition than under an energy-limited condition. Parameter *b* ranges from −9.22 to 0.16 under an energy-limited condition, whereas it ranges from −4.55 to 1.58 under a water-limited condition.

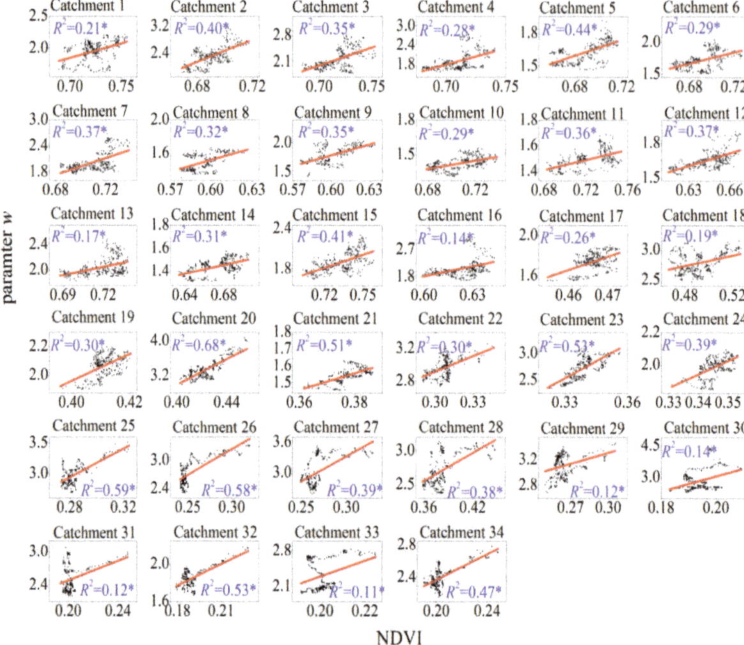

Figure 2. Relationships between NDVI and the land-surface parameter *w* in the 34 catchments. The red line is the fitting curve of scatters, and * indicates that the relationship is significant at the level of 0.01 by the *t*-test. NDVI: normalized difference vegetation index.

Table 3. Relationship between NDVI and the land surface parameter *w*.

Catchment No.	AI	R^2	*a*	*b*	Catchment No.	AI	R^2	*a*	*b*
1	0.52	0.21 *	4.5	−1.3	18	1.94	0.19 *	4.39	0.61
2	0.54	0.40 *	16.71	−9.22	19	1.95	0.30 *	9.38	−1.79
3	0.55	0.35 *	9.44	−4.59	20	2.01	0.68 *	14.9	−2.98
4	0.56	0.28 *	10.09	−5.26	21	2.14	0.51 *	4.91	−0.31
5	0.57	0.44 *	3.6	−0.85	22	2.17	0.30 *	6.41	0.99
6	0.61	0.29 *	4.39	−1.32	23	2.18	0.53 *	21.44	−4.54
7	0.61	0.37 *	9.47	−4.7	24	2.2	0.39 *	8.29	−0.88
8	0.61	0.32 *	4.66	−1.28	25	2.23	0.59 *	11.7	−0.36
9	0.62	0.35 *	6.99	−2.42	26	2.39	0.58 *	11.02	−0.06
10	0.64	0.29 *	1.79	0.16	27	2.4	0.39 *	9.91	0.35
11	0.64	0.36 *	1.96	0.08	28	2.43	0.38 *	6.86	0.08
12	0.66	0.37 *	3.63	−0.68	29	2.44	0.12 *	5.76	1.58
13	0.67	0.17 *	4.36	−1.09	30	2.86	0.14 *	36.12	−4.18
14	0.68	0.31 *	1.97	0.11	31	2.98	0.12 *	8.81	0.72
15	0.69	0.41 *	6.54	−2.89	32	3.07	0.53 *	7.66	0.39
16	0.73	0.14 *	11.28	−4.95	33	3.13	0.11 *	15.72	−0.85
17	1.76	0.26 *	13.49	−4.55	34	3.23	0.47 *	7.92	0.79

Note: * indicates the relationship is significant at the level of 0.01 by the *t*-test. AI: aridity index.

4.2. Changes in WER, Climate Variables and Vegetation

Figure 3 show the changes in WER in the 34 catchments during the past three decades. WER decreased significantly in 21 catchments ($p < 0.01$) (Table 4), whereas WER increased significantly ($p < 0.01$) in 12 catchments, and all the increasing trends were present in energy-limited catchments. Precipitation increased significantly ($p < 0.01$) 8 catchments and decreased significantly ($p < 0.01$) in 9 catchments. PET increased significantly ($p < 0.01$) in 28 catchments and decreased significantly ($p < 0.01$) in 4 catchments. Overall, runoff decreased significantly ($p < 0.01$) in 31 catchments and only increased significantly ($p < 0.01$) in 1 catchment, in which precipitation increased significantly ($p < 0.01$). The NDVI values in all 34 catchments increased significantly ($p < 0.01$), which means that the vegetation recovered during the last 30 years in the selected catchments. The general conclusion that vegetation greening leads to a decrease in runoff was observed in almost all catchments [25].

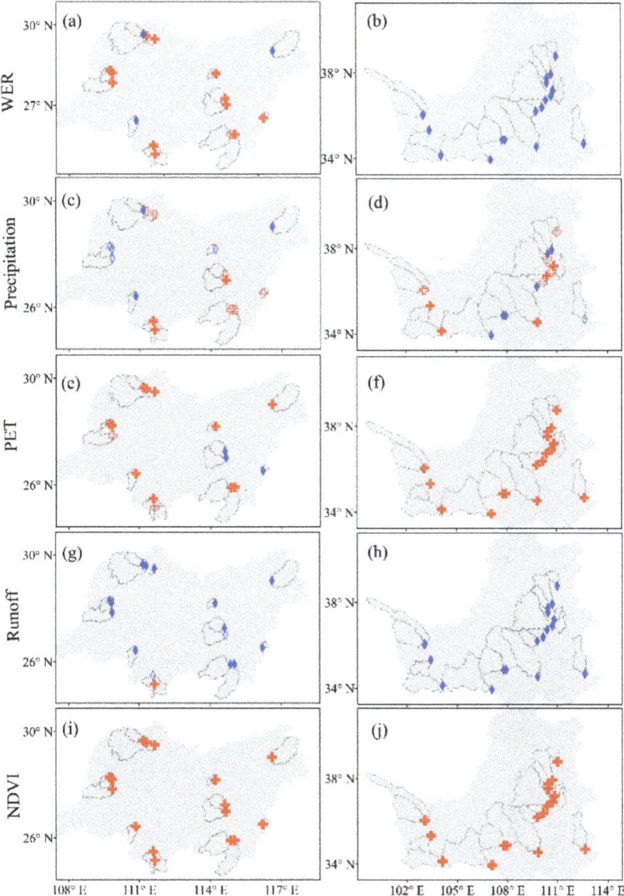

Figure 3. Changes in WER, climate variables and NDVI in the 34 selected catchments. Red indicates increase trends, and blue indicates decrease trends. Solid indicates that the trend is significant ($p < 0.01$), and hollow indicates that the trend is insignificant ($p > 0.01$). The left column, from (**a–i**), represents the energy-limited condition, and the right column, from (**b–j**), represents the water-limited condition. WER: available-water-to-available-energy ratio.

Table 4. Numbers of catchments in four categories of trends for WER, climate variables and NDVI.

Group	Trend	WER	Precipitation	PET	Runoff	NDVI
Energy-limited	Increase ($p < 0.01$)	12	3	11	1	16
	Increase	1	6	2	0	0
	Decrease	0	4	0	2	0
	Decrease ($p < 0.01$)	3	3	3	13	0
Water-limited	Increase ($p < 0.01$)	0	5	17	0	18
	Increase	0	6	0	0	0
	Decrease	0	1	0	0	0
	Decrease ($p < 0.01$)	18	6	1	18	0

4.3. Sensitivity of WER to Climate Variables and Vegetation

Figure 4 shows the sensitivities of WER to precipitation, PET and NDVI in the 34 catchments. Under the energy-limited condition, the sensitivity of WER to precipitation (ε_{prcp}) ranges from 0.26 to 28.2, which means a 10% increase in precipitation would result in WER decreasing by 2.6% to 282%. However, ε_{prcp} ranges from 0.73 to 1.60 under the water-limited condition, which would result in a much smaller WER change than that under the energy-limited condition. Because the sensitivity of WER to PET (ε_{PET}) is the inverse of the sensitivity of WER to precipitation, its spatial distribution is similar to that of precipitation. The sensitivity of WER to NDVI (ε_{NDVI}) under the energy-limited condition ranges from 0.21 to 11.2, which means a 10% increase in NDVI would result in WER increasing by 2.1 to 112%. However, ε_{NDVI} under the water-limited condition ranges from -0.23 to -1.09, which means a 10% increase in NDVI would result in WER decreasing by 2.3% to 10.9%. Moreover, the absolute value of the sensitivity of WER to NDVI under the water-limited condition is much smaller than that under the energy-limited condition, indicating that WER changes in energy-limited catchments are more sensitive to vegetation greening.

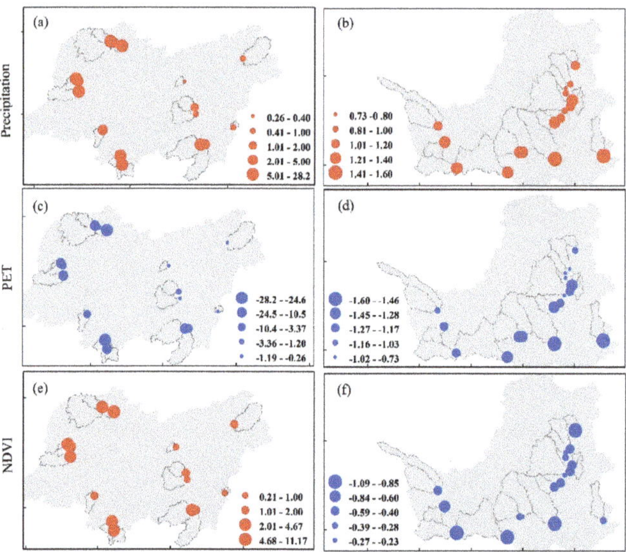

Figure 4. Spatial distribution of the sensitivity of WER to precipitation, PET and NDVI. The left column, from (**a**–**e**), represents the energy-limited condition, and the right column, from (**b**–**f**), represents the water-limited condition.

Interestingly, there are positive relationships between sensitivity coefficients and AI (Figure 5). ε_{prcp} increases with AI under an energy-limited condition ($r = 0.74$, $p < 0.01$), whereas ε_{prcp} decreases with AI under a water-limited condition ($r = -0.76$, $p < 0.01$). The relationship between ε_{PET} and AI is the inverse of that between ε_{prcp} and AI. There are negative relationships between ε_{NDVI} and AI under a water-limited condition ($r = -0.16$, $p > 0.01$), whereas there are positive relationships between them under an energy-limited condition ($r = 0.69$, $p < 0.01$). Please note that there is a large absolute value of the sensitivities when AI is approaching 1 because WER is the ratio of water availability to energy availability. When AI is approaching 1, evapotranspiration is close to precipitation and PET. A small change in water availability and energy availability will lead to a substantial change in WER.

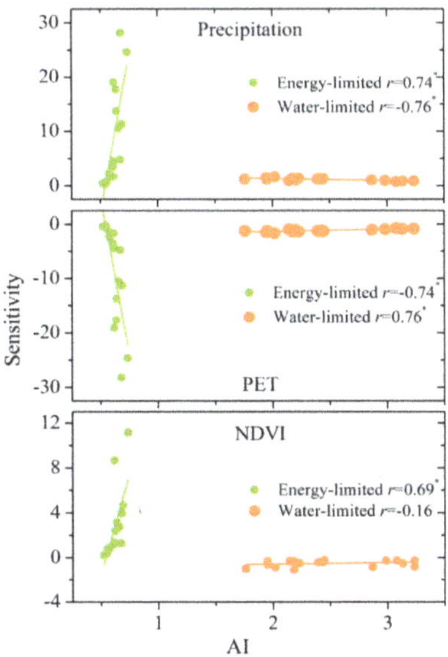

Figure 5. Correlations between the sensitivity of WER to precipitation, PET, NDVI and AI. * indicates that the relationship is significant at the level of 0.01 by the *t*-test.

4.4. Quantification of Climate Variation and Vegetation Greening to WER

The contributions of precipitation, PET and NDVI changes to the WER variation were quantified via numerical experiments. As shown in Figure 6a, the calculated WER trends based on the control test fit well with the detected WER trends based on linear regression for the 34 catchments ($R^2 = 0.96$). Figure 6b shows the relationship between WER changes based on the control test and the cumulative WER trends based on the three sensitivity tests. Please note that the two trends are very close to the 1:1 line ($R^2 = 0.99$). This result confirms the effectiveness of the proposed approach based on the numerical experiments.

Under the energy-limited condition, the contributions of precipitation, PET and NDVI to WER changes are $-57.9\sim80.4\%$, $-62.1\sim49.7\%$ and $11.8\sim90.8\%$, respectively (Figure 7). The WER changes are dominantly influenced by precipitation, PET and NDVI in five, five and six of the 34 catchments, respectively. Climate variation is the major factor controlling WER changes. Under the water-limited condition, the contributions of precipitation, PET and NDVI to WER changes are $-41.4\sim27.8\%$, $-69.9\sim63.2\%$ and $-71.7\sim-5.7\%$, respectively. Increased PET and NDVI are the dominant factors responsible for the WER decreases in ten and eight of the 34 catchments, respectively.

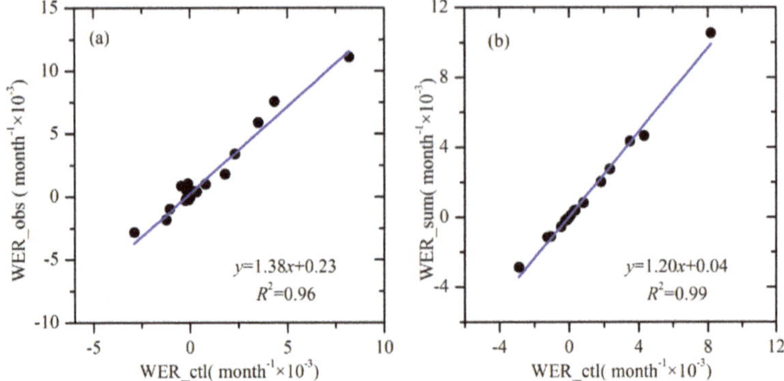

Figure 6. Comparisons between (**a**) trends in WER based on the control test and observations, and (**b**) trends in WER based on the control test and the cumulative contributions of the three sensitivity tests.

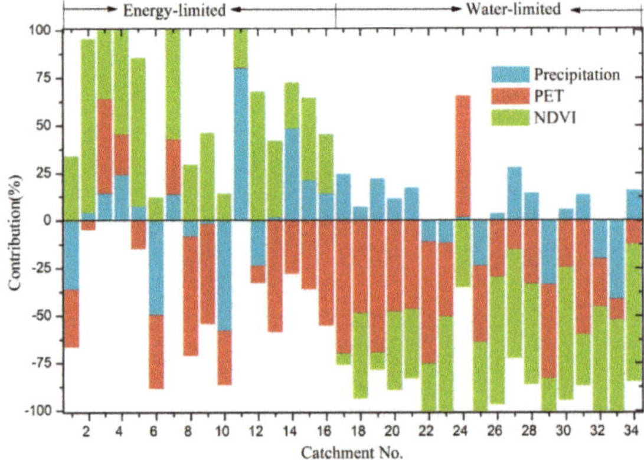

Figure 7. Contributions (%) of precipitation, PET and NDVI variations to WER changes in the 34 selected catchments.

Notably, although NDVI increased in all the catchments, the relationship between NDVI and WER varied under different climatic conditions. Positive relationships between NDVI and w existed under both energy-limited and water-limited conditions. However, an increased NDVI resulted in a WER increase in energy-limited catchments, whereas the increase in NDVI resulted in a WER decrease in water-limited catchments. This pattern exists because, although increasing NDVI results in increasing evapotranspiration under both water-limited and energy-limited conditions, the sensitivity of WER to increasing evapotranspiration is different. WER is more sensitive to water availability changes under a water-limited condition but is more sensitive to energy availability changes under an energy-limited condition.

5. Discussion

5.1. Why Do the Influences of Vegetation Greening on WER Vary?

This study employed the new water-energy balance index WER to reflect regional dryness/wetness conditions, which is similar in form to AI. However, WER is more suitable to reflect the regional water-energy balance than AI. Under climatic conditions with the same AI value, regional water and energy availability vary because of different land-surface characteristics, such as vegetation, soil, and topography [44–47]. The isolines among WER, parameter w and AI are helpful for understanding WER changes and land-surface changes under different climatic conditions. Figure 8 shows the ln(WER) changes along with the land-surface parameter w variations under different climatic conditions. WER increases as w increases under the energy-limited condition, while WER decreases as w increases under the water-limited condition. When AI approaches 1, the isolines of WER become dense, which indicates that WER would be sensitive to AI changes. WER directly reflects the regional water and energy balance conditions by considering both the climate and land-surface features.

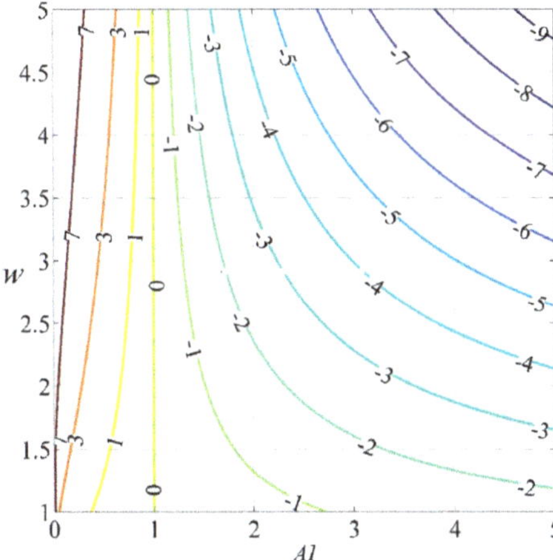

Figure 8. Isoline among WER, parameter w and AI under different climatic conditions.

Because of ecological protection and restoration in recent decades, the coverage of vegetation is increasing [28,30]. The quantitative contribution of vegetation greening on water-energy balance under different climatic conditions has not been assessed; such assessment is necessary for evaluating the effectiveness of ecological management on water resources in China. Vegetation greening impacts the water and energy balance via the linkage of evapotranspiration. This study found that WER decreases in water-limited catchments with vegetation greening. Generally, vegetation greening enhances the capacity of soil water storage. Water stored in the soil is finally consumed through actual evaporation, given a sufficient energy supply [6]. The amount of water available changes more quickly than the amount of energy available in these catchments, which results in more energy for sensible heat, i.e., a warming effect. However, WER increases because of vegetation greening under an energy-limited condition. Evapotranspiration increases with vegetation greening, which is similar to what occurs under the water-limited condition. Nevertheless, the amount of energy available changes more quickly than the amount of water available, resulting in less energy for sensible heat, which known as a cooling effect. This reflects a synergistic reaction mechanism between water availability and energy availability. However, the dominant factor for the WER change is different under the two types of climatic conditions. The primary tree species involved in forest recovery are coniferous (*Pinus massoniana*) in the middle reaches of the Yangtze River Basin. Many studies have indicated that this low-quality and low-efficiency tree type, planted because of insufficient management, has a limited ability to control water loss and soil erosion [48]. Therefore, the impacts of vegetation greening on WER in the Yangtze River are not as obvious as those in the Yellow River.

However, vegetation cannot continue to green persistently under certain climatic conditions. For example, Feng et al. [29] indicated that vegetation recovery has already reached its maximum in the middle reaches of the Yellow River, which features an arid climate. In populous and developing regions, vegetation greening can enhance drought conditions and exacerbate the scarcity of water resources. Determining how to balance vegetation recovery and water resources requires further investigation.

5.2. Uncertainties

Uncertainties exist in this study. First, this study quantifies the impact of vegetation greening on a new water-energy balance index that considers both the climatic condition and the vegetation, which are assumed to be independent. However, these factors are not independent of each other. For example, regional greening of vegetation can affect regional precipitation and PET. Moreover, changes in regional precipitation and PET will influence the distribution of vegetation. Thus, complex interactions exist between the vegetation the and climatic condition [49,50]. Second, the land-surface parameter w in the control test is estimated via NDVI, which only reflects the impact of vegetation greening on WER. This process thus introduces a difference between WER estimated by the control test and WER based on observations (Figure 6a). This difference between the two trends is caused by the influence of the other factors on WER [41,51] Third, the total water storage change can be neglected for long-term timescales, such as 10 years or longer. To determine the optimal use of the available data for detecting the relationship between vegetation and the hydrological cycle, a 60-month (five-year) moving window is applied to the hydrometeorological data.

6. Conclusions

In this study, the changes evaluated by the new water-energy balance index (WER) were quantified in terms of vegetation greening and climate variation in 34 catchments under different climatic conditions. Negative relationships were detected between vegetation greening and WER in water-limited catchments, whereas positive relationships were detected between those factors in energy-limited catchments. The results of numerical experiments showed that climate variations (precipitation and PET) were the dominant factors controlling WER changes under the energy-limited condition, whereas increased PET and NDVI were the dominant factors responsible for WER decreases under the water-limited condition.

In general, vegetation greening had negative influences on WER under the water-limited condition, indicating that water availability decreases faster than residual energy, which would result in more energy for sensible heat. However, vegetation greening had positive influences on WER changes under the energy-limited condition, which indicated that the amount of residual energy decreases faster than the amount of water availability, which would result in less energy for sensible heat. This study provides a comprehensive understanding of the relationship between water-energy balance and vegetation greening under different climatic conditions and highlights the complex relationships among water, energy and vegetation, especially under an energy-limited condition. Our results are helpful for land-atmosphere-vegetation modeling, hydrological prediction under changing environments, and designing strategies to account for ecological environmental construction and local water resource management.

Author Contributions: D.Z. performed the design of experiments and the data processing as well as wrote the manuscript. X.m.L. provided constructive suggestions towards the whole structure and edited the manuscript. P.B. processed the NDVI dataset and provided useful comments on the results.

Funding: This research was funded by the Natural Science Foundation of China (41771039, 41330529 and 41571023) and Youth Innovation Promotion Association, CAS (2018067).

Acknowledgments: The authors appreciate the Hydrological Bureau of the Yellow River by providing the runoff data used in this study. The meteorological data is provided by the National Climate Center of the China Meteorological Administration. We also appreciate the Editor, Associate Editor, and the two anonymous reviewers for their efforts and constructive comments on the manuscript.

Conflicts of Interest: The authors declare no conflict of interest.

References

1. Ivanov, V.Y.; Bras, R.L.; Vivoni, E.R. Vegetation-hydrology dynamics in complex terrain of semiarid areas: 2. Energy-water controls of vegetation spatiotemporal dynamics and topographic niches of favorability. *Water Resour. Res.* **2008**, *44*, 380–384. [CrossRef]
2. Liu, X.; Liu, C.; Brutsaert, W. Investigation of a generalized nonlinear form of the complementary principle for evaporation estimation. *J. Geophys. Res. Atmos.* **2018**, *123*, 3933–3942. [CrossRef]
3. Zhang, D.; Hong, H.; Zhang, Q.; Nie, R. Effects of climatic variation on pan-evaporation in the Poyang Lake Basin, China. *Clim. Res.* **2014**, *61*, 29–40. [CrossRef]
4. Arneth, A. Climate science: Uncertain future for vegetation cover. *Nature* **2015**, *524*, 44–45. [CrossRef] [PubMed]
5. Richardson, A.D.; Keenan, T.F.; Migliavacca, M.; Ryu, Y.; Sonnentag, O.; Toomey, M. Climate change, phenology, and phenological control of vegetation feedbacks to the climate system. *Agric. For. Meteorol.* **2013**, *169*, 156–173. [CrossRef]
6. Zhang, M.; Liu, N.; Harper, R.; Li, Q.; Liu, K.; Wei, X.; Ning, D.; Hou, Y.; Liu, S. A global review on hydrological responses to forest change across multiple spatial scales: Importance of scale, climate, forest type and hydrological regime. *J. Hydrol.* **2016**, *546*, 44–59. [CrossRef]
7. Hansen, M.; Potapov, P.; Margono, B.; Stehman, S.; Turubanova, S.; Tyukavina, A. High-resolution global maps of 21st-century forest cover change. *Science* **2013**, *342*, 850–853. [CrossRef] [PubMed]
8. Zhu, Z.; Piao, S.; Myneni, R.B.; Huang, M.; Zeng, Z.; Canadell, J.G.; Ciais, P.; Sitch, S.; Friedlingstein, P.; Arneth, A. Greening of the Earth and its drivers. *Nat. Clim. Chang.* **2016**, *6*, 791–795. [CrossRef]
9. Liu, Y.Y.; de Jeu, R.A.; McCabe, M.F.; Evans, J.P.; van Dijk, A.I. Global long-term passive microwave satellite-based retrievals of vegetation optical depth. *Geophys. Res. Lett.* **2011**, *38*. [CrossRef]
10. Liu, Y.Y.; Dijk, A.I.; McCabe, M.F.; Evans, J.P.; Jeu, R.A. Global vegetation biomass change (1988–2008) and attribution to environmental and human drivers. *Glob. Ecol. Biogeogr.* **2013**, *22*, 692–705. [CrossRef]
11. Piao, S.; Wang, X.; Ciais, P.; Zhu, B.; Wang, T.; Liu, J. Changes in satellite-derived vegetation growth trend in temperate and boreal Eurasia from 1982 to 2006. *Glob. Chang. Biol.* **2011**, *17*, 3228–3239. [CrossRef]
12. McDowell, N.G.; Coops, N.C.; Beck, P.S.; Chambers, J.Q.; Gangodagamage, C.; Hicke, J.A.; Huang, C.Y.; Kennedy, R.; Krofcheck, D.J.; Litvak, M.; et al. Global satellite monitoring of climate-induced vegetation disturbances. *Trends Plant Sci.* **2015**, *20*, 114–123. [CrossRef] [PubMed]
13. Piao, S.; Nan, H.; Huntingford, C.; Ciais, P.; Friedlingstein, P.; Sitch, S.; Peng, S.; Ahlström, A.; Canadell, J.G.; Cong, N. Evidence for a weakening relationship between interannual temperature variability and northern vegetation activity. *Nat. Commun.* **2014**, *5*. [CrossRef] [PubMed]
14. Gerten, D.; Schaphoff, S.; Haberlandt, U.; Lucht, W.; Sitch, S. Terrestrial vegetation and water balance—Hydrological evaluation of a dynamic global vegetation model. *J. Hydrol.* **2004**, *286*, 249–270. [CrossRef]
15. Sivapalan, M.; Yaeger, M.A.; Harman, C.J.; Xu, X.; Troch, P.A. Functional model of water balance variability at the catchment scale: 1. Evidence of hydrologic similarity and space-time symmetry. *Water Resour. Res.* **2011**, *47*, 2144–2150. [CrossRef]
16. Budyko, M.I. *Climate and Life*; Academic Press: London, UK, 1974.
17. Zhang, L.; Dawes, W.R.; Walker, G.R. Response of mean annual evapotranspiration to vegetation changes at catchment scale. *Water Resour. Res.* **2001**, *37*, 701–708. [CrossRef]
18. Liu, M.; Xu, X.; Xu, C.; Sun, A.Y.; Wang, K.; Scanlon, B.R.; Zhang, L. A new drought index that considers the joint effects of climate and land surface change. *Water Resour. Res.* **2017**, *53*. [CrossRef]
19. Azzali, S.; Menenti, M. Mapping vegetation-soil-climate complexes in southern Africa using temporal Fourier analysis of NOAA-AVHRR NDVI data. *Int. J. Remote Sens.* **2000**, *21*, 973–996. [CrossRef]
20. Donohue, R.J.; Roderick, M.L.; Mcvicar, T.R. On the importance of including vegetation dynamics in Budyko's hydrological model. *Hydrol. Earth Syst. Sci.* **2006**, *3*, 983–995. [CrossRef]
21. Gentine, P.; D'Odorico, P.; Lintner, B.R.; Sivandran, G.; Salvucci, G. Interdependence of climate, soil, and vegetation as constrained by the Budyko curve. *Geophys. Res. Lett.* **2012**, *39*. [CrossRef]
22. Donohue, R.J.; Roderick, M.L.; Mcvicar, T.R. Can dynamic vegetation information improve the accuracy of Budyko's hydrological model? *J. Hydrol.* **2010**, *390*, 23–34. [CrossRef]

23. Li, D.; Pan, M.; Cong, Z.; Zhang, L.; Wood, E. Vegetation control on water and energy balance within the Budyko framework. *Water Resour. Res.* **2013**, *49*, 969–976. [CrossRef]

24. Yang, D.; Sun, F.; Liu, Z.; Cong, Z.; Ni, G.; Lei, Z. Analyzing spatial and temporal variability of annual water-energy balance in nonhumid regions of China using the Budyko hypothesis. *Water Resour. Res.* **2007**, *43*, 436–451. [CrossRef]

25. Zhang, S.; Yang, H.; Yang, D.; Jayawardena, A.W. Quantifying the effect of vegetation change on the regional water balance within the Budyko framework. *Geophys. Res. Lett.* **2016**, *43*. [CrossRef]

26. Jackson, R.B.; Jobbágy, E.G.; Avissar, R.; Roy, S.B.; Barrett, D.J.; Cook, C.W.; Farley, K.A.; Maitre, D.C.L.; Mccarl, B.A.; Murray, B.C. Trading Water for Carbon with Biological Carbon Sequestration. *Science* **2005**, *310*, 1944–1947. [CrossRef] [PubMed]

27. Lü, Y.; Zhang, L.; Feng, X.; Zeng, Y.; Fu, B.; Yao, X.; Li, J.; Wu, B. Recent ecological transitions in China: Greening, browning, and influential factors. *Sci. Rep.* **2015**, *5*. [CrossRef] [PubMed]

28. Wei, X.; Sun, G.; Liu, S.; Jiang, H.; Zhou, G.; Dai, L. The forest-streamflow relationship in China: A 40-year retrospect. *J. Am. Water Resour. Assoc.* **2008**, *44*, 1076–1085. [CrossRef]

29. Feng, X.; Fu, B.; Piao, S.; Wang, S.; Ciais, P.; Zeng, Z.; Lü, Y.; Zeng, Y.; Li, Y.; Jiang, X. Revegetation in China[rsquor]s Loess Plateau is approaching sustainable water resource limits. *Nat. Clim. Chang.* **2016**, *6*, 1019–1022. [CrossRef]

30. Piao, S.; Yin, G.; Tan, J.; Cheng, L.; Huang, M.; Li, Y.; Liu, R.; Mao, J.; Myneni, R.B.; Peng, S. Detection and attribution of vegetation greening trend in China over the last 30 years. *Glob. Chang. Biol.* **2015**, *21*, 1601–1609. [CrossRef] [PubMed]

31. Huang, S.; Huang, Q.; Leng, G.; Zhao, M.; Meng, E. Variations in annual water-energy balance and their correlations with vegetation and soil moisture dynamics: A case study in the Wei River Basin, China. *J. Hydrol.* **2017**, *546*, 515–525. [CrossRef]

32. Ning, T.; Li, Z.; Liu, W. Vegetation dynamics and climate seasonality jointly control the interannual catchment water balance in the Loess Plateau under the Budyko framework. *Hydrol. Earth Syst. Sci.* **2017**, *21*, 1–25. [CrossRef]

33. Zhang, D.; Liu, X.; Zhang, Q.; Liang, K.; Liu, C. Investigation of factors affecting intra-annual variability of evapotranspiration and streamflow under different climate conditions. *J. Hydrol.* **2016**, *543*, 759–769. [CrossRef]

34. Foley, J.A.; DeFries, R.; Asner, G.P.; Barford, C.; Bonan, G.; Carpenter, S.R.; Chapin, F.S.; Coe, M.T.; Daily, G.C.; Gibbs, H.K. Global consequences of land use. *Science* **2005**, *309*, 570–574. [CrossRef] [PubMed]

35. Liu, J.; Liu, M.; Zhuang, D.; Zhang, Z.; Deng, X. Study on spatial pattern of land-use change in China during 1995–2000. *Sci. China Ser. D Earth Sci.* **2003**, *46*, 373–384.

36. McVicar, T.R.; Roderick, M.L.; Donohue, R.J.; Li, L.T.; Van Niel, T.G.; Thomas, A.; Grieser, J.; Jhajharia, D.; Himri, Y.; Mahowald, N.M. Global review and synthesis of trends in observed terrestrial near-surface wind speeds: Implications for evaporation. *J. Hydrol.* **2012**, *416*, 182–205. [CrossRef]

37. Penman, H.L. *Natural Evaporation from Open Water, Bare Soil and Grass*; The Royal Society: London, UK, 1948; pp. 120–145.

38. Shuttleworth, W.J. Evaporation. In *Handbook of Hydrology*; Maidment, D.R., Ed.; McGRaw Hill: New York, NY, USA, 1993.

39. Beck, H.E.; McVicar, T.R.; van Dijk, A.I.; Schellekens, J.; de Jeu, R.A.; Bruijnzeel, L.A. Global evaluation of four AVHRR–NDVI data sets: Intercomparison and assessment against Landsat imagery. *Remote Sens. Environ.* **2011**, *115*, 2547–2563. [CrossRef]

40. Pinzon, J.E.; Tucker, C.J. A non-stationary 1981–2012 AVHRR NDVI3g time series. *Remote Sens.* **2014**, *6*, 6929–6960. [CrossRef]

41. Carmona, A.; Poveda, G.; Sivapalan, M.; Vallejo-Bernal, S.; Bustamante, E. A scaling approach to Budyko's framework and the complementary relationship of evapotranspiration in humid environments: Case study of the Amazon River basin. *Hydrol. Earth Syst. Sci.* **2016**, *20*, 589–603. [CrossRef]

42. Arora, V.K. The use of the aridity index to assess climate change effect on annual runoff. *J. Hydrol.* **2002**, *265*, 164–177. [CrossRef]

43. Liu, X.; Zhang, D. Trend analysis of reference evapotranspiration in Northwest China: The roles of changing wind speed and surface air temperature. *Hydrol. Process.* **2013**, *27*, 3941–3948. [CrossRef]

44. Ukkola, A.; Keenan, T.; Kelley, D.I.; Prentice, I. Vegetation plays an important role in mediating future water resources. *Environ. Res. Lett.* **2016**, *11*. [CrossRef]

45. Bai, P.; Liu, X.; Liu, C. Improving hydrological simulations by incorporating GRACE data for model calibration. *J. Hydrol.* **2018**, *557*, 291–304. [CrossRef]

46. Miao, C.; Ashouri, H.; Hsu, K.L.; Sorooshian, S.; Duan, Q. Evaluation of the PERSIANN-CDR Daily Rainfall Estimates in Capturing the Behavior of Extreme Precipitation Events over China. *J. Hydrometeorol.* **2015**, *16*, 1387–1396. [CrossRef]

47. Miao, C.; Kong, D.; Wu, J.; Duan, Q. Functional degradation of the water–sediment regulation scheme in the lower Yellow River: Spatial and temporal analyses. *Sci. Total Environ.* **2016**, *551*, 16–22. [CrossRef] [PubMed]

48. Yan, F.; Wang, Q. Characterization and causes of low-quality and low-efficiency forest in Ganzhou of Jiangxi Province. *Prot. For. Sci. Technol.* **2016**. [CrossRef]

49. Behling, H. Late Quaternary vegetation, climate and fire history of the Araucaria forest and campos region from Serra Campos Gerais, ParanáState (South Brazil). *Rev. Palaeobot. Palynol.* **1997**, *97*, 109–121. [CrossRef]

50. Lehmann, C.E.; Anderson, T.M.; Sankaran, M.; Higgins, S.I.; Archibald, S.; Hoffmann, W.A.; Hanan, N.P.; Williams, R.J.; Fensham, R.J.; Felfili, J. Savanna vegetation-fire-climate relationships differ among continents. *Science* **2014**, *343*, 548–552. [CrossRef] [PubMed]

51. Hundecha, Y.; Bárdossy, A. Modeling of the effect of land use changes on the runoff generation of a river basin through parameter regionalization of a watershed model. *J. Hydrol.* **2004**, *292*, 281–295. [CrossRef]

 *forests*

Article

Understory Plants Regulate Soil Respiration through Changes in Soil Enzyme Activity and Microbial C, N, and P Stoichiometry Following Afforestation

Fazhu Zhao [1,2,3], **Jieying Wang** [1,3], **Lu Zhang** [2], **Chengjie Ren** [4,*], **Xinhui Han** [4], **Gaihe Yang** [4], **Russell Doughty** [5] and **Jian Deng** [6]

1   Shaanxi Key Laboratory of Earth Surface System and Environmental Carrying Capacity, Northwest University, Xi'an 710127, China; zhaofazhu@nwu.edu.cn (F.Z.); wjy9276@126.com (J.W.)
2   Key Laboratory of Degraded and Unused Land Consolidation Engineering the Ministry of Land and Resources of China, Xi'an 710075, China; luluqiaofeng@126.com
3   College of Urban and Environmental Sciences, Northwest University, Xi'an 710127, China
4   College of Agronomy, Northwest A&F University, Yangling 712100, China; hanxinhui@nwsuaf.edu.cn (X.H.); ygh@nwsuaf.edu.cn (G.Y.)
5   Department of Microbiology and Plant Biology, Center for Spatial Analysis, University of Oklahoma, Norman, OK 73019, USA; russell.doughty@ou.edu
6   College of Life Sciences, Yan'an University, Yan'an 716000, China; deng050702@126.com
*   Correspondence: rencj1991@nwsuaf.edu.cn; Tel.: +86-021-8830-8417

Received: 18 June 2018; Accepted: 18 July 2018; Published: 20 July 2018

Abstract: Soil respiration (SR) is an important process in the carbon cycle. However, the means by which changes in understory plant community traits affect this ecosystem process is still poorly understood. In this study, plant species surveys were conducted and soil samples were collected from forests dominated by black locust (*Robinia pseudoacacia* L.), with a chronosequence of 15, 25, and 40 years (RP15, RP25, and RP40, respectively), and farmland (FL). Understory plant coverage, evenness, diversity, and richness were determined. We investigated soil microbial biomass carbon (MBC), nitrogen (MBN), phosphorus (MBP), and stoichiometry (MBC:MBN, MBC:MBP, and MBN:MBP). Soil enzyme assays (catalase, saccharase, urease, and alkaline phosphatase), heterotrophic respiration (HR), and autotrophic respiration (AR) were measured. The results showed that plant coverage, plant richness index (R), evenness, and Shannon-Wiener diversity were higher in RP25 and RP40 than in RP15. SR, HR, and AR were significantly higher in the forested sites than in farmland, especially for SR, which was on average 360.7%, 249.6%, and 248.2% higher in RP40, RP25, and RP15, respectively. Meanwhile, catalase, saccharase, urease, and alkaline phosphatase activities and soil microbial C, N, P, and its stoichiometry were also higher after afforestation. Moreover, significant Pearson linear correlations between understory plants (coverage, evenness, diversity, and richness) and SR, HR, and AR were observed, with the strongest correlation observed between plant coverage and SR. This correlation largely depended on soil enzymes (i.e., catalase, saccharase, urease, and alkaline phosphatase), and soil microbial biomass C, N, and P contents and its stoichiometry, particularly urease activity and the MBC:MBP ratio. Therefore, we conclude that plant communities are drivers of soil respiration, and that changes in soil respiration are associated with shifts in soil enzyme activities and nutrient stoichiometry.

Keywords: understory plants; soil enzymes; soil microbial biomass; soil respiration; afforestation ecosystem

1. Introduction

Understory plants play an important role in soil carbon cycling and the future carbon (C) balance of terrestrial ecosystems under climate change [1,2]. The C flux through soil respiration (SR) is a vital component of the global C cycle; it represents approximately 10% of the atmospheric C pool, and is 10 times greater than that from fossil fuel combustion [3]. Consequently, even slight changes in understory plant community composition and traits could affect SR through shifts in productivity [4], changing litter inputs and altering the soil microclimate [5,6]. Many studies have reported that plants can control the balance between plant C inputs and losses [7–10]. However, the means by which changes in plant community traits affect this ecosystem process, especially in afforested ecosystems, is poorly understood. For example, different results are usually reported the effects of plant diversity on SR. Both positive [11] and non-significant relationships [12] between plant diversity and SR have been documented. More importantly, Dias [13] found no significant effect of plant diversity on SR, while Liu [14] observed that plant diversity was the most important driver of SR. Therefore, it is imperative to further investigate plant communities as drivers of SR, and the possible pathways, mechanisms, and significance of SR for global climate change.

Plants affect SR in many ways, one of which is soil enzyme activities [15,16]. Enzyme activity is the most basic driving factor of SR. More than 50% of SR is produced by the enzyme-related decomposition of litter and soil organic matter (SOM) [17,18]. On the other hand, considering the metabolic activities of microorganisms during C deposition into soil, enzymes transform plant residues, decompose plant-derived C, and thus affect SR [19]. These relationships have been revealed through field experiments [15], models [20], and meta-analyses [21,22]. For example, Chen [15] showed a positive correlation between glycosidase activity and SR for most types of vegetation. Ren [9] also documented that enzyme activities, especially oxidative C-degrading enzyme activities, were significantly correlated with SR due to plant litter inputs. However, a lack of clarity regarding afforested ecosystems still remains, because the distribution of both recalcitrant and labile C varies depending on the plant community composition [23]. Furthermore, understanding the effects of the aggradation of afforested ecosystems on belowground C cycles is important for quantifying and predicting the dynamics of terrestrial C, especially under the current scenarios of global climate change. Thus, investigating the role of soil enzyme activities during the aggradation of afforested ecosystems in SR will help to address this knowledge gap.

Plant communities and litter can affect soil respiration rates through nutrient availability [24]. Additionally, ecological stoichiometry is usually regarded as an indicator of microbial nutrient requirements and nutrient availability, especially for C, nitrogen (N), and phosphorus (P) content in microbial biomass [25]. Plant communities also affect soil respiration by plant-trait driven shifts in microbial biomass C, N, and P contents and stoichiometric ratio. Although results from previous studies suggest that ecological stoichiometry, especially the C:N:P ratios of organisms and substrates, could be used as a tool to acquire knowledge to the cycling of these elements [26,27], the question remains as to how plant communities influence SR through changes in the stoichiometry of soil microbial biomass during the aggradation of afforested ecosystems. We raise this question because changes in the plant community composition during aggradation generally produce more litter with higher N content, which is more easily degraded by soil microbes [28,29]. In turn, microorganisms consume nutrients in excess amounts and store them in the form of glycogen or polyphosphates, because afforestation leads to changes in their biomass C:N:P ratio [27,30]. Consequently, understanding how shifts in plant community composition affect SR due to changes in the C:N:P ratio of microbial biomass is key to predicting the dynamics of terrestrial C under future climate change. Thus, to advance our understanding of plant-soil interactions during the aggradation of afforested ecosystems, more information about the regulation of terrestrial C dynamics by plant community through shifts in microbial biomass C, N, and P stoichiometry is needed.

The Loess Plateau, which covers approximately 640,000 km^2 in China, has experienced severe soil erosion and decreased vegetation cover [31]. The abandonment of farmland with slopes >15°

to allow for afforestation is an important management practice to prevent soil erosion and restore ecosystems native to this area [32]. In recent years, numerous studies have been conducted to examine the effects of afforestation on soil physicochemical properties, microbial dynamics, and soil enzyme activities [33–35]. However, information on the relationships among understory plant communities, SR, soil enzymes, and microbial biomass nutrient stoichiometry is scarce. Therefore, we investigated plant community composition, SR, soil enzyme activities, and microbial biomass C, N, and P stoichiometry at three forest stands (aged 15, 25, and 40 years) with *Robinia pseudoacacia* L. (RP) succession after afforestation of former farmland (FL) in the Loess Plateau. We hypothesized that SR changed with plant community traits as aggradation progressed, and that this change in soil respiration was stimulated by the soil C, N, and P stoichiometry of microbial biomass and soil enzymes. The objectives of this study were to (i) evaluate the changes in plant community traits after afforestation, (ii) characterize the changes in soil enzymes and microbial biomass C, N, and P stoichiometry after afforestation, and (iii) demonstrate the relationships between plant community traits, soil enzyme activities, and soil microbial biomass C, N, and P stoichiometry after aggradation of the afforested farmland.

2. Material and Method

2.1. Study Area

The study was conducted at Wuliwan Watershed, Ansai County, Shaanxi Province, northern China (36°46′42″–36°46′28″ N, 109°13′46″–109°16′03″ E) (Figure 1). This area is a fragile, semiarid ecosystem, and has one of the largest global loess areas [35]. The average monthly temperature ranges from −6.2 °C in January to 37.2 °C in July, with a mean annual temperature of 8.8 °C and mean annual precipitation of 505 mm [36]. In this region, the growing season for deciduous species occurs from April to October [37]. The soil is highly erodible, and classified as loessial soil (Calcaric Cambisols, WRB classification, 2014) (Table 1). The dominant tree species in this area is *R. pseudoacacia* L., which was replanted on farmland. The main crop species is *Setaria italica* (L.) P. Beauvois (millet). Water resources for crop growth are dependent entirely on rainfall; irrigation is not practiced during the growing season. The Wuliwan catchment is an experimental site of the Chinese Academy of Science (CAS), and vegetation restoration has been implemented due to serious soil degradation since the 1970s. After 30 years of afforestation, the area of forest has increased significantly from 5% to 40% [38]. Prior to afforestation, all land-use types were essentially farmland, which had been subjected to similar farming practices for more than 20 years with millet and soybean rotations [39]. The understory *Stipa bungeana* Trin. community is the most extensive species in afforested sites. *Stipa grandis* P.A.Smirn. and *Pinus bungeana* Zucc. are the dominant grass species, while *Thymus mongolicus* Ronn. and *Artemisia sacrorum* Ledeb. are the dominant forb species (Table 1).

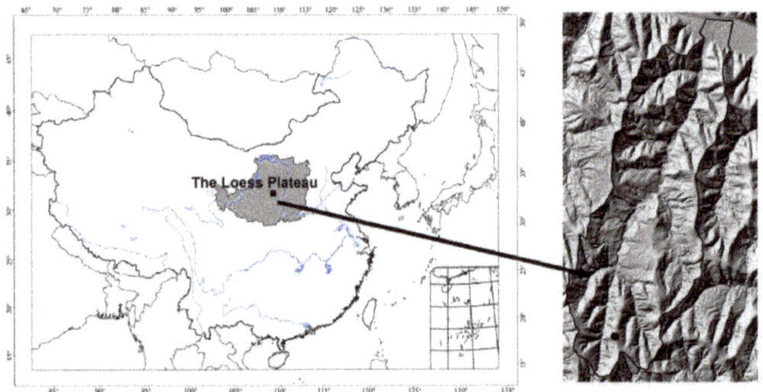

Figure 1. Location of the Loess Plateau and the study site.

Table 1. Geographical information for the four *R. pseudoacacia* L. sites.

Sites	Farmland	*Robinia pseudoacacia* **Linn.** (RP15 year)	*Robinia pseudoacacia* **Linn.** (RP25 year)	*Robinia pseudoacacia* **Linn.** (RP40 year)
Elevation (m)	1205	1303	1298	1293
Clay (%)	8.12 ± 0.21 A	8.55 ± 0.14 A	9.54 ± 0.13 A	10.11 ± 0.12 A
Silt (%)	60.45 ± 0.22 A	62.53 ± 0.19 A	60.36 ± 0.19 A	64.35 ± 0.21 A
Sand (%)	31.43 ± 0.10 A	28.92 ± 0.09 A	30.10 ± 0.18 A	25.54 ± 0.11 A
Dominate species		*Artemisia capillaries* Thunb., *Heteropappus altaicus* (Willd.) Novopokr., *Stipa bungeana* Trin., *Oxytropis bicolor* Bunge., *Cleistogenes squarrosa* (Trin.) Keng	*Artemisia capillaries* Thunb., *Heteropappus altaicus* (Willd.) Novopokr., *Stipa bungeana* Trin., *Salsola collina* Pall., *Oxytropis bicolor* Bunge., *Cleistogenes squarrosa* (Trin.) Keng	*Artemisia capillaries* Thunb., *Heteropappus altaicus* (Willd.) Novopokr., *Artemisia sacrorum* Ledeb., *Stipa bungeana* Trin., *Oxytropis bicolor* Bunge., *Cleistogenes squarrosa* (Trin.) Keng

Capital letters indicate significant difference among different land use types ($p < 0.05$); the error bars.

2.2. Experimental Design

Experiments were carried out in June and October, 2014. Based on land use history, three afforested lands, *R. pseudoacacia* L. (RP40), *R. pseudoacacia* L. (RP25), and *R. pseudoacacia* L. (RP15), as well as farmland (FL), were selected. In each different aged stand, three plots with similar slope, gradient, and altitude [38,39] were established. In total, 12 plots (four land use types × three replicate plots) were setup in the study area. In addition, six quadrats (0.5 m × 0.5 m) (three trenched and three untrenched quadrats) were randomly established within the replicate plots, and the trenches (0.5 m wide and 0.8 m deep) were excavated in October 2013. After covering the trenches with a 2 mm thick plastic sheet, we refilled them with soil.

2.3. Soil Respiration Measurement and Soil Sampling

Polyvinyl chloride collars (PVC; 16 cm in diameter × 12 cm in height) were used to measure soil respiration, as described in our previous study [22]. In the experiment sites, six PVC collars were installed to a depth of 10 cm. Three PVC collars in trenched quadrats were used to measure the soil heterotrophic respiration (HR). The other three PVC collars in untrenched quadrats were used to measure SR. In June and August, 2014, AR and HR were measured on a single rain-free day between 9:00 and 11:00 a.m., using the portable soil CO_2 flux system (GXH-3010E1, LI-COR Inc., Lincoln, NE, USA) (μmol CO_2 m^{-2} s^{-1}), and obtained from consecutive 2- or 3-day measurements to represent the average monthly soil respiration. Finally, the three respiration rate observations were averaged to obtain the results for a given plot for both AR and HR.

After removing the litter layer, soil samples were collected at 0–10 cm soil depth using a soil auger (diameter 5 cm) from ten points within an "S" shape in each subplot, and then homogenized to provide one final soil sample per subplot. Overall, 12 samples (four stand age types × three plots) were collected. The samples were sieved through a 2 mm screen, and roots and other debris were removed [39,40]. A fraction of each soil sample was air dried and stored at room temperature prior to analysis of its properties, including water content (SWC) and pH. The other portion of each soil sample was immediately transported to the laboratory (on ice, and then stored at −80 °C) for microbial biomass carbon (MBC), nitrogen (MBN), phosphorus (MBP), and enzymatic assay analyses.

2.4. Analysis of Soil Properties and Enzymes Activities

SWC was determined by oven drying to a constant mass at 105 °C. BD was taken by undisturbed soil and calculated from the gravimetric weight of the cores (using 100 cm^3 cores with a height of 5 cm) before and after oven drying at 105 °C for 24 h from the individual core volume, while soil pH was measured using a pH meter after shaking the soil water (1:5 w/v) suspension for 30 min [22,39]. MBC, MBN, and MBP were estimated from fresh soil samples using a chloroform fumigation-extraction method [39]. Soil catalase, saccharase, urease, and alkaline phosphatase activities were determined as described in our previous study [40].

2.5. Plant Species Identification and Species Diversity Index

Plant species identification was done in situ in June and October of 2014, which was described in our previous study [41]. Five 1 m × 1 m quadrats were established in each replicate plot in June and October, respectively (3 stand age × 3 replicates × 5 quadrats, a total of 45 quadrats at each site in one season). Vegetation surveys of herbaceous plants in the plantation understory were done by tallying stem quantity and plant height for each species. Plant coverage of herb layers was visually estimated using a metal frame of 1 m × 1 m with 100 equally distributed grids above the subplot, and then understory coverage was calculated as the average percentage of ground surface area covered by the shadow of the foliage in each quadrat [41].

Species richness is the number of species in each quadrat [41]. The Richness index (R) was calculated as the total number of species in each community (*S*), Shannon-Wiener diversity index (H), and Evenness index (E) of the afforested and abandoned land plant communities were calculated using the following equations:

$$H = \sum_{i=1}^{S} (P_i ln P_i) \tag{1}$$

$$E = \frac{H}{ln S} \tag{2}$$

where *S* = total number of species in each community, H = Shannon-Wiener diversity index, P_i = density proportion of species "*i*", *ln* = natural log.

2.6. Statistical Analyses

All statistical analyses were carried out using SPSS for Windows (version 17.0, SPSS Inc., Chicago, IL, USA). Analysis of variance (ANOVA) and Duncan's Multiple Range Test (DMRT) at 5% level of significance were used to compare the differences in plant community coverage, height, and plant density; R, H, and E indexes; MBC:MBN, MBC:MBP, and MBN:MBP ratios; and soil catalase, saccharase, urease, and alkaline phosphatase activity among different sites. Spearman's rank correlation coefficients were used to investigate the relationships among the plant and soil characteristics at each site. In addition, we used the Beerkan Estimation of Soil Transfer (BEST PRIMER-E, Plymouth, UK) model building procedure, which utilizes all possible combinations of factors to determine which combination of factors (Coverage, Evenness, Diversity, Richness, catalase, alkaline phosphatase, urease, saccharase, MBC:MBN, MBC:MBP, MBN:MBP) account for the greatest proportion of SR (SR, AR and HR). The factors additions were evaluated stepwise and were based on sufficient improvement in the model's *R* value.

3. Results

3.1. Changes in Soil Properties and Plant Community Traits after Afforestation

We found that after afforestation, soil bulk density (SBD) and Water Holding Capacity (WHC) increased significantly. SBD was higher at RP40 than RP25, RP15, and FL by 3.33%, 5.98%, and 8.77%. WHC was higher at RP40 than RP25, RP15, and FL by 43.79%, 54.48%, and 136.90%. For the increase of humus, pH value was decreased, and the soil was gradually acidic (Table 2). Understory plants showed remarkable variability during aggradation in our study (Figure 2). We found that after afforestation and during aggradation, plant coverage and plant R, E, and H indices increased. These parameters were higher at RP40 than at RP15 both in June and in October. Compared to RP15, plant coverage and plant R, E, and H indices were higher at RP25 by 26.69%, 20.00%, 14.72%, and 8.28% in June and by 27.15%, 89.45%, 12.47%, and 16.72% in October, respectively.

Table 2. Soil properties after afforestation.

Sites	SBD (g·cm^{-3}) [a]	pH	WHC (%) [b]
Farmland	1.14 ± 0.02 A	9.38 ± 0.01 A	10.27 ± 0.74 C
R. pseudoacacia (RP15 year)	1.17 ± 0.01 A	8.67 ± 0.11 A	15.75 ± 0.98 B
R. pseudoacacia (RP25 year)	1.20 ± 0.01 A	8.65 ± 0.01 A	16.92 ± 0.79 B
R. pseudoacacia (RP40 year)	1.24 ± 0.01 A	8.48 ± 0.02 A	24.33 ± 1.21 A

[a] SBD is soil bulk density; [b] WHC is Water Holding Capacity Note: ±SE, Capital letter represents significant difference among sites.

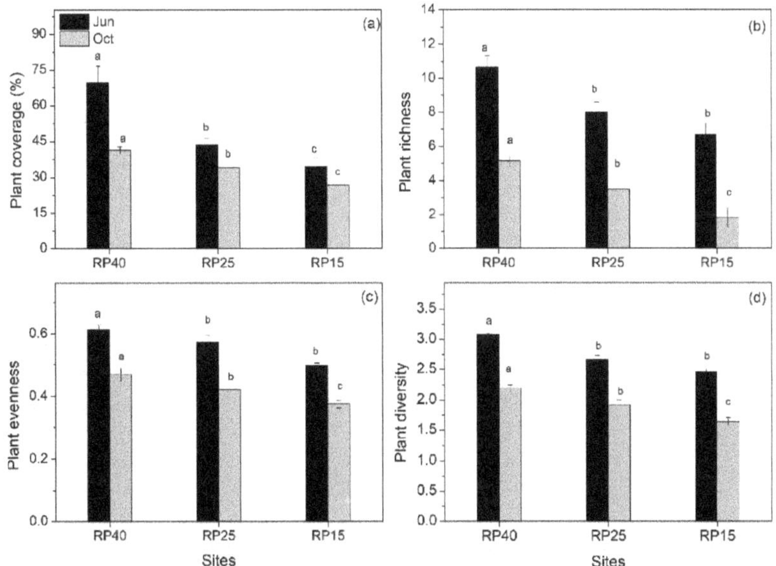

Figure 2. Plant community characteristics ((**a**) plant coverage, (**b**) plant richness, (**c**) plant evenness, (**d**) plant diversity) after afforestation. Different letters denote significant ($p < 0.05$) differences among sites in same month; error bar represents standard error. Note: 40, 25, and 15 years of *Robinia pseudoacacia* L. indicated as RP40, RP25, and RP15.

3.2. Changes in Soil Respiration and Its Components after Afforestation

Significant differences were found in SR, HR, and AR (Figure 3). SR at RP40, RP25, and RP15 was significantly higher than that in FL in June and in October. HR increased as aggradation progressed at afforested sites, with HR at RP40 being 23.68% and 40.06% higher than that at RP25 and RP15 in June, and 14.46% and 90.56% higher in October, respectively. Meanwhile, AR increased following afforestation and yielded average values of 0.54 µmol CO$_2$ m^{-2} s^{-1} in June and 0.73 µmol CO$_2$ m^{-2} s^{-1} in October.

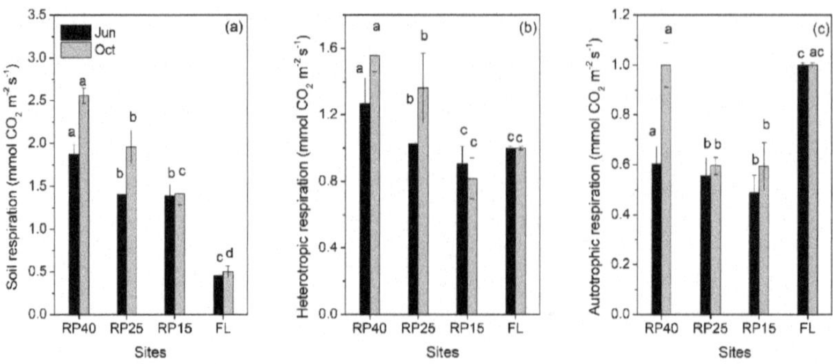

Figure 3. Soil respiration (**a**) and its components ((**b**) soil heterotrophic respiration, (**c**) soil autotrophic respiration) after afforestation. Different letters denote significant ($p < 0.05$) differences among sites in same month; error bar represents standard error. Note: 40, 25, and 15 years of *Robinia pseudoacacia* L. indicated as RP40, RP25, and RP15.

3.3. Changes in Soil Enzyme Activities after Afforestation

Changes in soil enzyme activities are shown in Figure 4. Catalase, saccharase, urease, and alkaline phosphatase contents increased following afforestation. For example, catalase, alkaline phosphatase, urease, and saccharase activities at RP40 were higher than at RP25 in June and October, respectively. Catalase, alkaline phosphatase, urease, and saccharase activities at RP25 were higher than those at RP15 in June and October, respectively. Compared with FL, the average increases of catalase, alkaline phosphatase, urease, and saccharase activities in RP sites were higher in June and October.

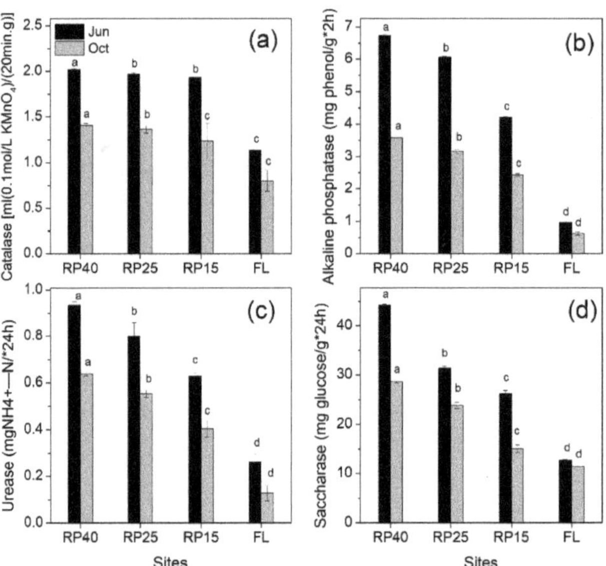

Figure 4. Soil Catalase enzyme activities (**a**), Alkaline phosphatase enzyme activities (**b**), Ureas enzyme activities (**c**) and Saccharase enzyme activities (**d**) after afforestation. Different letters denote significant ($p < 0.05$) differences among sites in same month; error bar represents standard error. Note: 40, 25, and 15 years of *Robinia pseudoacacia* L. indicated as RP40, RP25, and RP15.

3.4. Changes in Soil Microbial Biomass C, N, and P Contents and Its Stoichiometry after Afforestation

Soil microbial biomass C, N, and P contents and its stoichiometry responded differently during aggradation at the afforested sites (Figure 5). The results showed that MBC, MBN, and MBP contents increased significantly at RP40, compared with FL, by 581.7%, 231.5%, and 204.9% in June, and 347.1%, 215.2%, and 113.8% in October, respectively. Further, MBC, MBN, and MBP contents at RP40 were higher than at RP25 and RP15 by approximately 57.62%, 30.90%, and 36.99% in June, and 45.44%, 32.02%, and 18.79% in October, respectively. In addition, MBC:MBN, MBC:MBP, and MBN:MBP ratios were also significantly higher after afforestation. Compared with FL, MBC:MBN, MBC:MBP, and MBN:MBP ratios at RP40 were higher by 105.7%, 127.8%, and 10.69% in June, and 41.96%, 105.4%, and 44.16% in October, respectively. However, among the afforested sites, there were no significant trends observed in these ratios.

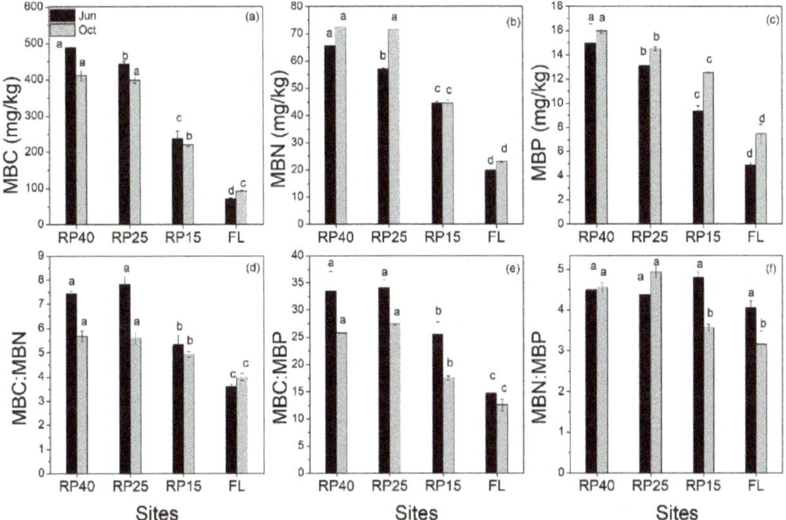

Figure 5. Soil microbial C (**a**), N (**b**), P (**c**) and soil MBC: MBN (**d**), soil MBC: MBP (**e**) and soil MBN: MBP (**f**) after afforestation. Different letters denote significant ($p < 0.05$) differences among sites in same month and error bar represents standard error. Note: 40, 25, and 15 years of *Robinia pseudoacacia* L. indicated as RP40, RP25, and RP15.

3.5. Relationships between Plant and SR Linked to Microbial Biomass C, N, and P Contents and Its Stoichiometry and Soil Enzyme Activities after Afforestation

Spearman's rank correlation coefficients also showed significant relationships among microbial biomasses (C, N, and P), their stoichiometries, soil enzyme activities, and soil respiration components (Table 3). Linear regression results were observed between most of these parameters ($p < 0.05$) (Figure 6), especially for plant coverage (Table 4). The results showed that changes in SR and its components were significantly correlated with catalase, saccharase, urease, alkaline phosphatase, and microbial biomass C, N, and P contents, and MBC:MBN, MBC:MBP, and MBN:MBP ratios ($p < 0.05$). In addition, after performing a "best" model building procedure, we found that urease, MBP, and MBC:MBP ratio were the best predictive factors influencing SR (Table 5).

Table 3. Spearman's rank correlation coefficients between the microbial biomass (C, N, and P) and its stoichiometry (MBC:MBN, MBC:MBP and MBN:MBP) and the soil enzyme actives (catalase, saccharase, urease, and alkaline phosphatase), as well as soil respiration components.

	Soil Enzyme Actives			
	catalase	Alkaline phosphatase	urease	saccharase
SR	−0.653 **	−0.649 **	−0.677 **	−0.757 **
HR	−0.520 *	−0.518 *	−0.410	−0.558 *
AR	−0.595 **	−0.649 **	−0.774 **	−0.717 **

	MBC	MBN	MBP	MBC:MBN	MBC:MBP	MBN:MBP
SR	0.42	0.582 *	0.676 **	0.613 **	−0.695 **	0.447
HR	0.553 *	0.364	0.517 *	0.658 **	−0.699 **	0.481 *
AR	0.24	0.586 *	0.568 *	0.620 **	−0.717 **	0.478 *

Soil respiration (SR); Soil heterotrophic respiration (HR); Soil autotrophic respiration (AR); Microbial biomass carbon (MBC); nitrogen (MBN); phosphorus (MBP); Microbial biomass carbon and nitrogen ratio (MBC:MBN); Microbial biomass carbon and phosphorus ratio (MBC:MBP); Microbial biomass nitrogen and phosphorus ratio (MBN:MBP) ** Correlation is significant at the 0.01 level. * Correlation is significant at the 0.05 level.

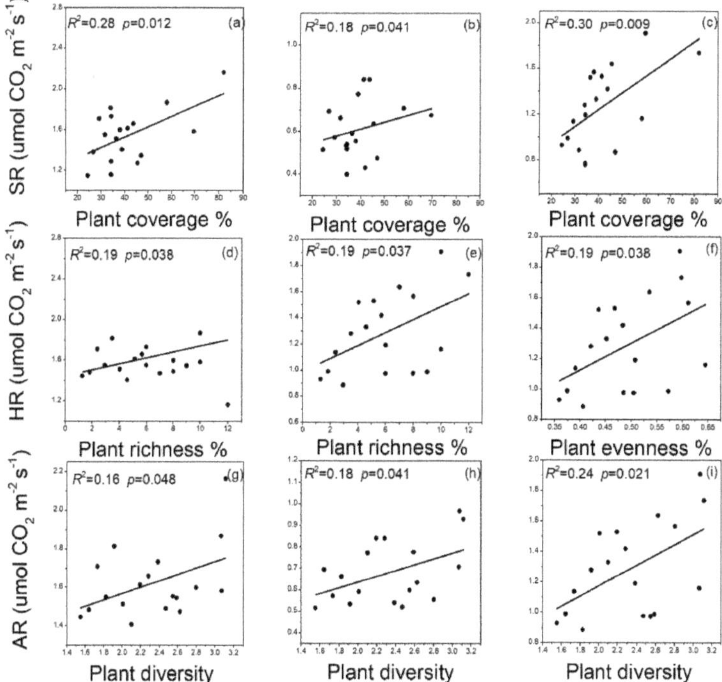

Figure 6. Linear Regression between SR and plant coverage in RP40 (**a**), RP20 (**b**), and in RP15 (**c**); between HR and plant richness in RP40 (**d**), RP20 (**e**), and in RP15 (**f**); between AR and plant diversity in RP40 (**g**), RP20 (**h**), and in RP15 (**i**); Soil respiration (SR); Soil heterotrophic respiration (HR); Soil autotrophic respiration (AR).

Table 4. Results from the BEST model for each number of predictor variables.

Number Variables	R	Predictor Variables **
1	0.142	Coverage
2	0.116	Coverage, Evenness
3	0.093	Coverage, Evenness, Diversity
4	0.037	Coverage, Evenness, Diversity, Richness

Number of permutations: 999 (random sample). ** Significance level of sample statistic: 0.1%.

Table 5. Results from the BEST model for each number of predictor variables.

Number Variables	Soil Enzyme Actives		Soil Microbial C, N, P		Soil Microbial Stoichiometry	
	R	Predictor Variables	R	Predictor Variables	R	Predictor Variables
1	0.131	urease	0.112	MBP	0.136	MBC:MBP
2	0.126	alkaline phosphatase, urease	0.158	MBP	0.193	MBC:MBP, MBN:MBP
3	0.118	catalase, alkaline phos-phatase, urease	0.138	MBP, MBN	0.132	MBC:MBP, MBN:MBP
4	0.036	catalase, alkaline phosphatase, urease, saccharase				

Number of permutations: 999 (random sample). Microbial biomass carbon (MBC); nitrogen (MBN); phosphorus (MBP); Microbial biomass carbon and nitrogen ratio (MBC:MBN); Microbial biomass carbon and phosphorus ratio (MBC:MBP); Microbial biomass nitrogen and phosphorus ratio (MBN:MBP).

4. Discussion

In line with our expectations, plant diversity and coverage significantly influenced SR across our study sites. The most important factors influencing SR were changes in C inputs into soil [7]. These changes in C inputs have been observed by species diversity experiments in grasslands, which documented that more diverse grasslands were more productive [42], and that C inputs further stimulated SR. However, increased plant diversity also showed increased use efficiency [1]. This reduction in N concentration in the soil organic matter may negatively affect SR, including AR and HR [10,13]. On the other hand, understory plant characteristics can have important and diverse effects on soil properties, microclimates, and SR [2]. In particular, the effects of plants on soil temperature and water content are important, because they are drivers of microbial activity and SR [43,44]. For example, a previous study showed that plant canopies often reduce ground level radiation and soil evaporation rates, resulting in lower soil temperature and greater soil moisture content [45]. Therefore, a higher availability of plant-derived resources has a strong and positive affect on SR following afforestation.

The effects of plant diversity on SR may be more pronounced in the presence of soil enzymes [46], which is consistent with our result that species richness mediate soil respiration, partly through changes in soil enzyme activities. Several previous studies also supported our result that soil enzyme activity might significantly affect SR in terrestrial ecosystems [47,48]. For example, Chen [15] reported that changes in microbial enzymatic activities might be the basically drivers of SR, because more than 50% of SR is came from enzyme-related decomposition of litter and SOM. Ren [9] observed that C-degrading extracellular enzyme activities were significantly correlated with SR. A possible explanation was that the litter inputs during afforestation not only provide the building blocks for enzymatic production, because enzymes are basically N-rich molecules, but also increase microbial C demands due to stoichiometry of microbial biomass nutrients [49]. Consequently, increases in microbial C demands were expected to be alleviated by promoting the activities of C-degrading enzymes, influencing SR [50].

More importantly, soil microbial biomass C, N, and P content and its stoichiometry increased significantly with increasing plant diversity in both months of our study. These findings were consistent with previous studies, which showed that the positive effects of plant diversity on soil microbial biomass and functions led to changes in SR [25,51]. In particular, HR came from the microbial decomposition of root exudates in the rhizosphere. Increased microbial biomass C, N, and P content and its stoichiometry could augment labile soil carbon [52,53], alter microbial communities [9,10], or increase soil moisture [6,54], all of which can impact SR. However, our results contrasted with those of Cong [4] who reported that no significant difference was observed in SR between plant community traits and species richness. This was probably a consequence of the low soil fertility and dry conditions at the time of sampling in their experiment [4]. Therefore, plant species diversity may exceed plant production as a driver of the shifts in soil microbial biomass and stoichiometry in the long term by providing more diverse plant-derived resources, and thus, potentially influence SR. In addition, the species composition of plant communities influences microbial biomass C, N, and P content and stoichiometry, which can lead to changes in microbial bioenergetics and qCO_2. Consequently, changes in these factors influence SR, especially microbial respiration. This conclusion was supported by a previous study that reported the qCO_2 increase with soil and litter carbon to nutrient ratios and underlying stoichiometric controls [55], which may be because microbes in soils with lower microbial C:N ratios have higher growth efficiency and lower release of C through respiration. Conversely, microbes in soils with higher microbial C:N ratios have more C available to be converted into biomass [56–59]. Therefore, changes in plant communities influences the soil microbial C, N, and P stoichiometry, and are important drivers of the trends in microbial bioenergetics and respiration rates of soil microorganisms per unit microbial biomass.

In summary, understory vegetation affects SR through many mechanisms. However, the most important factor is the quantity and quality of organic inputs into the soil from plants. Our results indicate that plants regulate C dynamics through changes in soil enzyme activities and microbial biomass C, N, and P content and stoichiometry. This study has highlighted the role of plants within the plant-soil system, and the fact that plants are necessary for better understanding and simulating the patterns of SR across terrestrial ecosystems.

5. Conclusions

Our study showed that changes in understory plant community traits (coverage, evenness, diversity, and richness), soil respiration and its components, the activities of four enzymes (catalase, saccharase, urease, and alkaline phosphatase), and soil microbial biomass C, N, and P contents and its stoichiometry were significantly driven by afforestation and the aggradation of afforested sites. Plant community traits mediated SR through changes in soil enzyme activities and soil microbial biomass C, N, and P contents and its stoichiometry. Therefore, our results highlight the importance of understory plant community traits in regulating belowground carbon dynamics, and suggested that plant traits, especially plant coverage, altered the components of soil respiration by affecting soil enzyme activities and soil microbial biomass C, N, and P contents and its stoichiometry.

Author Contributions: L.Z., C.R., X.H., G.Y. conceived and designed the experiment. J.W. and F.Z. performed data analysis and wrote the manuscript. F.Z., J.W., L.Z., C.R. and J.D. conducted the sampling, pre-treatment and experiment work. R.D. revised the English grammar.

Funding: This work was supported by The National Natural Science Foundation of China (Grants: 41601578) and by the Key Laboratory of Degraded and Unused Land Consolidation Engineering, Ministry of Land and Resources (Grants: SXDJ2018-02).

Acknowledgments: We thank Zi ting Wang help us plot figure.

Conflicts of Interest: The authors declare no competing financial interest.

References

1. Fornara, D.A.; Tilman, D. Plant functional composition influences rates of soil carbon and nitrogen accumulation. *J. Ecol.* **2008**, *96*, 314–322. [CrossRef]
2. Metcalfe, D.B.; Fisher, R.A.; Wardle, D.A. Plant communities as drivers of soil respiration: Pathways, mechanisms, and significance for global change. *Biogeosciences* **2011**, *8*, 2047–2061. [CrossRef]

3. Schlesinger, W.H.; Andrews, J.A. Soil respiration and the global carbon cycle. *Biogeochemistry* **2000**, *48*, 7–20. [CrossRef]

4. Cong, W.F.; Ruijven, J.V.; Werf, W.V.D.; Deyn, G.B.D.; Mommer, L.; Berendse, F.; Hoffland, E. Plant species richness leaves a legacy of enhanced root litter-induced decomposition in soil. *Soil. Biol. Biochem.* **2015**, *80*, 341–348. [CrossRef]

5. Fang, H.; Cheng, S.; Wang, Y.; Yu, G.; Xu, M.; Dang, X.; Li, L.; Wang, L. Changes in soil heterotrophic respiration, carbon availability, and microbial function in seven forests along a climate gradient. *Ecol. Res.* **2014**, *29*, 1077–1086. [CrossRef]

6. Jewell, M.D.; Shipley, B.; Low-Décarie, E.; Tobner, C.M.; Paquette, A.; Messier, C.; Reich, P.B. Partitioning the effect of composition and diversity of tree communities on leaf litter decomposition and soil respiration. *Oikos* **2017**, *126*, 959–971. [CrossRef]

7. Janssens, I.A.; Lankreijer, H.; Matteucci, G.; Kowalski, A.S.; Buchmann, N.; Epron, D.; Pilegaard, K.; Kutsch, W.; Longdoz, B.; Grünwald, T. Productivity overshadows temperature in determining soil and ecosystem respiration across European forests. *Glob. Chang. Biol.* **2001**, *7*, 269–278. [CrossRef]

8. Esch, E.H.; Lipson, D.; Cleland, E.E. Direct and indirect effects of shifting rainfall on soil microbial respiration and enzyme activity in a semi-arid system. *Plant Soil* **2016**, *411*, 333–346. [CrossRef]

9. Ren, C.; Zhao, F.; Shi, Z.; Chen, J.; Han, X.; Yang, G.; Feng, Y.; Ren, G. Differential responses of soil microbial biomass and carbon-degrading enzyme activities to altered precipitation. *Soil Biol. Biochem.* **2017**, *115*, 1–10. [CrossRef]

10. Xiao, H.; Li, Z.; Dong, Y.; Chang, X.; Deng, L.; Huang, J.; Nie, X.; Liu, C.; Liu, L.; Wang, D.; et al. Changes in microbial communities and respiration following the revegetation of eroded soil. *Agric. Ecosyst. Environ.* **2017**, *246*, 30–37. [CrossRef]

11. Craine, J.M.; Wedin, D.A.; Reich, P.B. The response of soil CO_2 flux to changes in atmospheric CO_2, nitrogen supply and plant diversity. *Glob. Chang. Biol.* **2001**, *7*, 947–953. [CrossRef]

12. Johnson, D.; Phoenix, G.K.; Grime, J.P. Plant community composition, not diversity, regulates soil respiration in grasslands. *Biol. Lett.* **2008**, *4*, 345–348. [CrossRef] [PubMed]

13. Dias, A.T.C.; Ruijven, J.V.; Berendse, F. Plant species richness regulates soil respiration through changes in productivity. *Oecologia* **2010**, *163*, 805–813. [CrossRef] [PubMed]

14. Liu, M.; Xia, H.; Fu, S.; Eisenhauer, N. Tree diversity regulates soil respiration through accelerated tree growth in a mesocosm experiment. *Pedobiologia* **2017**, *65*, 24–28. [CrossRef]

15. Chen, J.; Luo, Y.; Li, J.; Zhou, X.; Cao, J.; Wang, R.W.; Wang, Y.; Shelton, S.; Jin, Z.; Walker, L.M.; et al. Costimulation of soil glycosidase activity and soil respiration by nitrogen addition. *Glob. Chang. Biol.* **2017**, *23*, 1328–1337. [CrossRef] [PubMed]

16. Chen, J.; Zhou, X.; Hruska, T.; Cao, J.; Zhang, B.; Liu, C.; Liu, M.; Shelton, S.; Guo, L.; Wei, Y.; et al. Asymmetric Diurnal and Monthly Responses of Ecosystem Carbon Fluxes to Experimental Warming. *CLEAN-Soil Air Water* **2017**, *45*, 1600557. [CrossRef]

17. Waldrop, M.P.; Zak, D.R.; Sinsabaugh, R.L. Microbial community response to nitrogen deposition in northern forest ecosystems. *Soil Biol. Biochem.* **2004**, *36*, 1443–1451. [CrossRef]

18. Shahzad, T.; Chenu, C.; Genet, P.; Barot, S.; Perveen, N.; Mougin, C.; Fontaine, S. Contribution of exudates, arbuscular mycorrhizal fungi and litter depositions to the rhizosphere priming effect induced by grassland species. *Soil Biol. Biochem.* **2015**, *80*, 146–155. [CrossRef]

19. Liang, C.; Schimel, J.P.; Jastrow, J.D. The importance of anabolism in microbial control over soil carbon storage. *Nat. Microbiol.* **2017**, *2*, 17105. [CrossRef] [PubMed]

20. Devaraju, N.; Bala, G.; Caldeira, K.; Nemani, R. A model based investigation of the relative importance of CO_2-fertilization, climate warming, nitrogen deposition and land use change on the global terrestrial carbon uptake in the historical period. *Clim. Dynam.* **2015**, *8*, 1–18. [CrossRef]

21. Zhou, L.; Zhou, X.; Zhang, B.; Lu, M.; Luo, Y.; Liu, L.; Li, B. Different responses of soil respiration and its components to nitrogen addition among biomes: A meta-analysis. *Glob. Chang. Biol.* **2014**, *20*, 2332–2343. [CrossRef] [PubMed]

22. Ren, C.; Wang, T.; Xu, Y.; Deng, J.; Zhao, F.; Yang, G.; Han, X.; Feng, Y.; Ren, G. Differential soil microbial community responses to the linkage of soil organic carbon fractions with respiration across land-use changes. *For. Ecol. Manag.* **2018**, *409*, 170–178. [CrossRef]

23. Raich, J.W.; Tufekciogul, A. Vegetation and soil respiration: Correlations and controls. *Biogeochemistry* **2000**, *48*, 71–90. [CrossRef]

24. Igor, N.; Aliaksra, R.; Alyona, S.; Evgeniy, L.; Vadim, Z.; Vladimir, P.; Rakovich, Y.P.; Donegan, J.F.; Rubin, A.B.; Govorov, A.O. Consequences of biodiversity loss for litter decomposition across biomes. *Nature* **2014**, *509*, 218–221.

25. Eisenhauer, N.; Bessler, H.; Engels, C.; Gleixner, G.; Habekost, M.; Milcu, A.; Partsch, S.; Sabais, A.C.; Scherber, C.; Steinbeiss, S. Plant diversity effects on soil microorganisms support the singular hypothesis. *Ecology* **2010**, *91*, 485–496. [CrossRef] [PubMed]

26. Aponte, C.; Marañón, T.; García, L.V. Microbial C, N and P in soils of Mediterranean oak forests: influence of season, canopy cover and soil depth. *Biogeochemistry* **2010**, *101*, 77–92. [CrossRef]

27. Heuck, C.; Weig, A.; Spohn, M. Soil microbial biomass C:N:P stoichiometry and microbial use of organic phosphorus. *Soil Biol. Biochem.* **2015**, *85*, 119–129. [CrossRef]

28. Cleveland, C.C.; Liptzin, D. C:N:P stoichiometry in soil: is there a "Redfield ratio" for the microbial biomass? *Biogeochemistry* **2007**, *85*, 235–252. [CrossRef]

29. Achbergerová, L.; Nahálka, J. Polyphosphate-an ancient energy source and active metabolic regulator. *Microb. Cell Fact.* **2011**, *10*, 63. [CrossRef] [PubMed]

30. Elser, J.J.; Sterner, R.W.; Gorokhova, E.; Fagan, W.F.; Markow, T.A.; Cotner, J.B.; Harrison, J.F.; Hobbie, S.E.; Odell, G.M.; Weider, L.W. Biological stoichiometry from genes to ecosystems. *Ecol. Lett.* **2010**, *3*, 540–550. [CrossRef]

31. Bai, Z.; Dent, D. Recent land degradation and improvement in China. *Ambio* **2009**, *38*, 150–156. [CrossRef] [PubMed]

32. Zhao, F.; Yang, G.; Han, X.; Feng, Y.; Ren, G. Stratification of carbon fractions and carbon management index in deep soil affected by the Grain-to-Green program in China. *PLoS ONE* **2014**, *9*, e99657. [CrossRef] [PubMed]

33. Zhang, C.; Xue, S.; Liu, G.B.; Song, Z.L. A comparison of soil qualities of different revegetation types in the Loess Plateau, China. *Plant Soil* **2011**, *347*, 163–178. [CrossRef]

34. Zhang, C.; Liu, G.; Xue, S.; Wang, G. Soil bacterial community dynamics reflect changes in plant community and soil properties during the secondary succession of abandoned farmland in the Loess Plateau. *Soil Biol. Biochem.* **2016**, *97*, 40–49. [CrossRef]

35. Tian, Q.; Taniguchi, T.; Shi, W.Y.; Li, G.; Yamanaka, N.; Du, S. Land-use types and soil chemical properties influence soil microbial communities in the semiarid Loess Plateau region in China. *Sci. Rep.* **2017**, *7*, 45289. [CrossRef] [PubMed]

36. Ren, C.; Sun, P.; Kang, D.; Zhao, F.; Feng, Y.; Ren, G.; Han, X.; Yang, G. Responsiveness of soil nitrogen fractions and bacterial communities to afforestation in the Loess Hilly Region (LHR) of China. *Sci. Rep.* **2016**, *6*, 28469. [CrossRef] [PubMed]

37. Du, S.; Wang, Y.L.; Kume, T.; Zhang, J.G.; Otsuki, K.; Yamanaka, N.; Liu, G.B. Sapflow characteristics and climatic responses in three forest species in the semiarid Loess Plateau region of China. *Agric. For. Meteorol.* **2011**, *151*, 1–10. [CrossRef]

38. Zhao, F.Z.; Zhang, L.; Ren, C.J.; Sun, J.; Han, X.H.; Yang, G.H.; Wang, J. Effect of microbial carbon, nitrogen, and phosphorus stoichiometry on soil carbon fractions under a black locust Forest within the central loess plateau of China. *Soil Sci. Soc. Am. J.* **2016**, *80*, 1520–1530. [CrossRef]

39. Ren, C.; Zhao, F.; Kang, D.; Yang, G.; Han, X.; Tong, X.; Feng, Y.; Ren, G. Linkages of C:N:P stoichiometry and bacterial community in soil following afforestation of former farmland. *For. Ecol. Manag.* **2016**, *376*, 59–66. [CrossRef]

40. Ren, C.; Kang, D.; Wu, J.P.; Zhao, F.; Yang, G.; Han, X.; Feng, Y.; Ren, G. Temporal variation in soil enzyme activities after afforestation in the Loess Plateau, China. *Geoderma* **2016**, *282*, 103–111. [CrossRef]

41. Zhao, F.; Kang, D.; Han, X.; Yang, G.; Yang, G.; Feng, Y.; Ren, G. Soil stoichiometry and carbon storage in long-term afforestation soil affected by understory vegetation diversity. *Ecol. Eng.* **2015**, *74*, 415–422. [CrossRef]

42. Hooper, D.U.; Chapin, F.S.; Ewel, J.J.; Hector, A.; Inchausti, P.; Lavorel, S.; Lawton, J.H.; Lodge, D.M.; Loreau, M.; Naeem, S. Effects of biodiversity on ecosystem functioning: A consensus of current knowledge. *Ecol. Monogr.* **2005**, *75*, 3–35. [CrossRef]

43. Davidson, E.A.; Janssens, I.A. Temperature sensitivity of soil carbon decomposition and feedbacks to climate change. *Nature* **2006**, *440*, 165–173. [CrossRef] [PubMed]

44. Chen, J.; Luo, Y.; Xia, J.; Shi, Z.; Jiang, L.; Niu, S.; Zhou, X.; Cao, J. Differential responses of ecosystem respiration components to experimental warming in a meadow grassland on the Tibetan Plateau. *Agric. For. Meteorol.* **2016**, *220*, 21–29. [CrossRef]

45. Breshears, D.D.; Rich, P.M.; Barnes, F.J.; Campbell, K. Overstory-imposed heterogeneity in solar radiation and soil moisture in a semiarid woodland. *Ecol. Appl.* **1997**, *7*, 1201–1215. [CrossRef]

46. Sinsabaugh, R.L.; Hill, B.H.; Follstad Shah, J.J. Ecoenzymatic stoichiometry of microbial organic nutrient acquisition in soil and sediment. *Nature* **2009**, *462*, 795–798. [CrossRef] [PubMed]

47. Allison, S.D.; Wallenstein, M.D.; Bradford, M.A. Soil-carbon response to warming dependent on microbial physiology. *Nat. Geosci.* **2010**, *3*, 336–340. [CrossRef]

48. Ali, R.S.; Ingwersen, J.; Demyan, M.S.; Funkuin, Y.N.; Wizemann, H.D.; Kandeler, E.; Poll, C. Modelling in situ activities of enzymes as a tool to explain seasonal variation of soil respiration from agro-ecosystems. *Soil. Biol. Biochem.* **2015**, *81*, 291–303. [CrossRef]

49. Sistla, S.A.; Schimel, J.P. Seasonal patterns of microbial extracellular enzyme activities in an arctic tundra soil: Identifying direct and indirect effects of long-term summer warming. *Soil Biol. Biochem.* **2013**, *66*, 119–129. [CrossRef]

50. Stone, M.M.; Deforest, J.L.; Plante, A.F. Changes in extracellular enzyme activity and microbial community structure with soil depth at the Luquillo Critical Zone Observatory. *Soil Biol. Biochem.* **2014**, *75*, 237–247. [CrossRef]

51. Spehn, E.M.; Joshi, J.; Schmid, B.; Diemer, M.; Korner, C. Above-ground resource use increases with plant species richness in experimental grassland ecosystems. *Funct. Ecol.* **2000**, *14*, 326–337.

52. Song, B.; Niu, S.; Zhang, Z.; Yang, H.; Li, L.; Wan, S. Light and heavy fractions of soil organic matter in response to climate warming and increased precipitation in a temperate steppe. *PLoS ONE* **2012**, *7*, e33217. [CrossRef] [PubMed]

53. Tucker, C.L.; Bell, J.; Pendall, E.; Ogle, K. Does declining carbon-use efficiency explain thermal acclimation of soil respiration with warming? *Glob. Chang. Biol.* **2013**, *19*, 252–263. [CrossRef] [PubMed]

54. Lloret, F.; Mattana, S.; Curiel Yuste, J. Climate-induced die-off affects plant-soil-microbe ecological relationship and functioning. *FEMS Microbiol. Ecol.* **2015**, *91*, 1–12. [CrossRef] [PubMed]

55. Spohn, M.; Chodak, M. Microbial respiration per unit biomass increases with carbon-to-nutrient ratios in forest soils. *Soil Biol. Biochem.* **2015**, *81*, 128–133. [CrossRef]

56. Schimel, J. Microbial ecology: Linking omics to biogeochemistry. *Nat. Microbiol.* **2016**, *1*, 15028. [CrossRef] [PubMed]

57. Shoemaker, W.R.; Locey, K.J.; Lennon, J.T. A macroecological theory of microbial biodiversity. *Nat. Ecol. Evol.* **2017**, *1*, 0107. [CrossRef] [PubMed]

58. Zhou, H.; Zhang, D.; Wang, P.; Liu, X.; Cheng, K.; Li, L.; Zheng, J.; Zhang, X.; Zheng, J.; Crowley, D.; et al. Changes in microbial biomass and the metabolic quotient with biochar addition to agricultural soils: A Meta-analysis. *Agric. Ecosyst. Environ.* **2017**, *239*, 80–89. [CrossRef]

59. Zhou, Z.; Wang, C.; Jiang, L.; Luo, Y. Trends in soil microbial communities during secondary succession. *Soil Biol. Biochem.* **2017**, *115*, 92–99. [CrossRef]

 *forests*

Article

Modeling Hydrological Appraisal of Potential Land Cover Change and Vegetation Dynamics under Environmental Changes in a Forest Basin

Jie Wang [1,*], Shaowei Ning [2] and Timur Khujanazarov [3]

[1] College of Hydrometeorology, Nanjing University of Information Science & Technology, Nanjing 210044, China
[2] School of Civil Engineering, Hefei University of Technology, Hefei 230009, China; ning@hfut.edu.cn
[3] Water Resources Research Center, Disaster Prevention Research Institute, Kyoto University, Uji, Gokasho 611-0011, Japan; khu.temur@gmail.com
* Correspondence: wj@nuist.edu.cn or wangjie0775@163.com; Tel.: +86-25-5869-5622

Received: 9 May 2018; Accepted: 24 July 2018; Published: 26 July 2018

Abstract: An integrated multi-model approach to predict future land cover in the Da River Basin in Vietnam was developed to analyze future impacts of land cover change on streamflow and sediment load. The framework applied a land cover change model and an ecological model to forecast future land cover and leaf area index (LAI) based on the historical land cover change, and these data were then used in a calibrated distributed hydrological model and a new sediment rating curve model to assess hydrological changes and sediment load in the river basin. Results showed that deforestation would likely continue, and that forest area would decrease by up to 21.3% by 2050, while croplands and shrublands would replace forests and increase by over 11.7% and 10%, respectively. Streamflow and sediment load would generally increase due to deforestation in the Da River Basin in the 2050s, in both the wet and dry seasons, but especially in the wet season. In this case, the predicted annual sediment load was expected to increase by about 9.7% at the Lai Chau station. As deforestation increased, sediment load and reservoir siltation could likely shorten the lifespan of the recently constructed Son La Reservoir. The applied integrated modeling approach provides a comprehensive evaluation of land/forest cover change effects on the river discharge and sediment load, which is essential in understanding human impacts on the river environment and in designing watershed management policies.

Keywords: forest cover; ecosystem model; BTOPMC model; LAI; streamflow; sediment load

---

1. Introduction

Land cover change is a global phenomenon and has dramatically altered river basin environments in the past decades [1–6]. Although land cover pattern changes certainly provide many socioeconomic benefits, they are also a major driving force in the degradation of natural environments, local river inundation [7], and serious soil degradation [6,8,9]. Land cover change is directly linked to the watershed hydrological cycle, altering the balance between rainfall and evapotranspiration [2,3,5,10], and additionally affecting flow velocity [6], whether in the form of streams or overland flows, due to changes in slope or gradient [11]. Land cover change also modifies the overland flow resistance and soil erodibility, and consequently impacts the sediment load in river basins.

Many studies have been carried out to evaluate impacts of the land cover change or vegetation dynamics on streamflow [1–3,10] and sediment load [6,9,12,13] at different spatial and temporal scales. Generally, researchers have agreed that land cover change or vegetation dynamics can alter streamflow or sediment load [1–7,9–12]. Many studies also agreed on the vegetation effect on streamflow, i.e., the streamflow should increase along with deforestation and vegetation degradation [3], or decrease following afforestation and vegetation restoration [2]. However, a few case studies have shown that forest/vegetation could have some positive effects on river discharge [14,15]. Although it is generally agreed that land cover change has negative effects on river streams, its role and importance in altering streamflow is still argued about, especially for large basins [2,5,14,16].

Deforestation or vegetation degradation is considered to be a main risk factor for riverbanks and surface soil erosion, as it results in the increase of sediment transportation in surface runoff and streamflow velocity [8,9]. For example, Schmidt et al., using an isotope tracer technique, concluded that deforestation caused by agricultural expansion significantly increased the sediment yield of six rivers in western China [6]. Moreover, the response of sediment load to land cover change appears to be more complex than that of streamflow response [17–19]. Although streamflow and sediment load are strongly correlated in response to land cover change [20,21], the level or severity of the land cover change impacts on both of these parameters are still arguable [2,5,15,16,22]. Land cover change has a big impact on the river basin's hydrological cycle, including many environmental issues, and thus it should be assessed as one unit for a better understanding of the overall impact to the catchment.

The Red River in the northwest of Vietnam is classified as one of the top ten rivers in terms of sedimentation load, and it is a well-known major agricultural area in Vietnam. Rapid land development in the upstream area and dam constrictions have changed the hydrological cycle, sediment load, and biodiversity in the area [23]. Land cover change was recognized as the major factor to alter streamflow and sediment flow in the upstream area of the Red River compared to climate change [21]. Satellite observation and model simulations suggested that the decrease in historical vegetation cover raised sediment load by 13.7% in the Red River Basin [24]. In addition, land cover change has caused significant variations in the peak river flow and velocity, increasing flood risk in the downstream area [23]. On the other hand, the sediment load has decreased, starting from the downstream area of the HoaBinh Dam since the construction of the dam, indicating that HoaBinh Reservoir has trapped half of the annual sediment [23,25]. Such a sediment load reduction and reservoir siltation issues have caused a re-evaluation of the HoaBinh Dam life span and its flooding prevention functions [26]. This example shows how sediment load evaluation in connection to land cover change is an important task for future dam construction design under rapidly changing conditions. Most studies have been focused separately on historical changes in streamflow or sediment load analysis in the Red River Basin, and few studies have combined river discharge and sediment load responses to analyses of future land cover change. However, streamflow and sediment load responses to land cover change are highly correlated, and it is essential to evaluate both variables simultaneously in order to provide useful information that will optimize the use of water resources and reservoir operation in the Red River Basin.

Defining future land cover change plays an important role in the impact assessment and it is crucial for future analyses of streamflow and sediment load. Currently, there are many studies that address this issue, and land use/cover change simulations based on the historical land change analysis are widely applied and well described. Several land use/cover change models have been developed and applied in various parts of the world, including CLUE-S [27], Dinamica EGO [28], GEOMOD [29], and the Land Change Model (LCM) [30]. Among these, the LCM was selected; this model could predict land cover more accurately in several sub-tropic forest-dominated areas, such as Bolivia [30], Mexico [31], and others. Although many studies have combined such land use/cover change models with hydrological models to predict future streamflow [1,32] or sediment load [13], only a few cases were carried out to predict future streamflow and sediment load simultaneously [33]. It is also important to point out that vegetation characteristics would change dynamically along with land cover conversion [5,34], and it is crucial not to overlook vegetation dynamics in the hydrological simulation. Hydrological response to land cover change or dynamic vegetation has therefore received increasing attention from both field observations and model simulations in the catchment area [1–3,17–19]. However, many studies have been focused on the individual effect of land cover change [2,13,17] or vegetation dynamics [12,24] on the hydrological cycle. Hence, both land cover change and dynamic vegetation impacts were considered in this research.

The main goal of this study was to evaluate the streamflow and the sediment load response to the changes in future land cover and vegetation dynamics in the Da River Basin, one of the largest sub-basins of the Red River Basin. We developed an integrated framework that included a land change model and an ecological model, coupled with a calibrated distributed hydrological model and a new sediment rating curve model. The system was applied in order to quantitatively assess future land cover change impacts in the Da River Basin.

2. Study Area and Data Description

2.1. Study Area Description

The Da River (also known as the Black River) originates in the southeast of China and is one of the largest tributaries of the Red River. Both the Da River and the Red River are trans-boundary rivers shared between Vietnam and China (Figure 1). The catchment area of the Da River is approximately 55,000 km^2. The upstream landscape is characterized with narrow and steep slopes with high erosion rates [23]. Climate is dominated by tropical monsoons, with an annual average precipitation of about 1320 mm, falling in two seasons, the rainy season (May to October), with 85% of total precipitation, and the cool dry season, with 15% of total precipitation [23,25]. The annual mean runoff from 1988 to 2004 was about 1168 m^3/s at the Lai Chau station, and there was a total annual sediment load of about 40.1 × 10^6 ton.

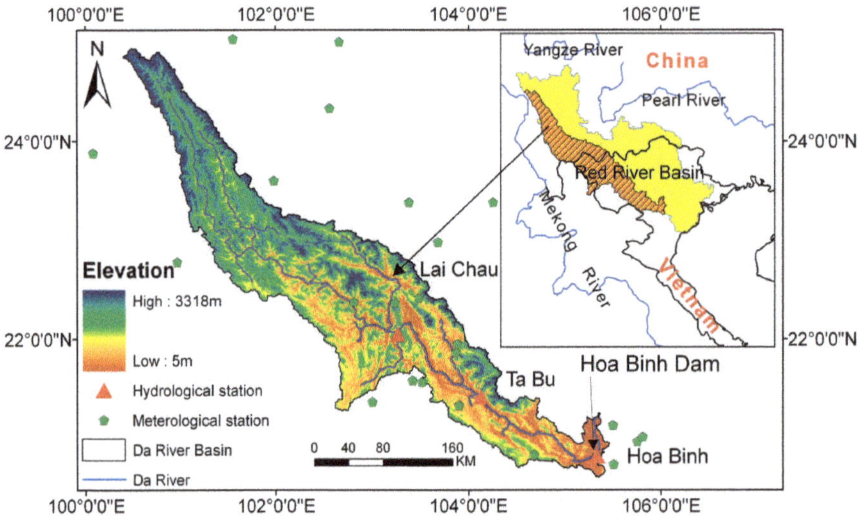

Figure 1. Location of the Da River Basin (DRB) and the hydro-meteorological stations.

Sediment load of the river is important, as the Vietnamese government has decided to build a cascade of five dams and hydropower facilities in the Da River Basin (DRB) (Figure 1) for hydroelectric power, flood prevention, and agricultural irrigation. Up to now, two dams, the Son La and the HoaBinh, were constructed in the DRB. The Son La reservoir in the DRB, with an effective storage of 16.2 km^3 [23], has just been completed in October 2012, and it has become the largest dam in Vietnam (Figure 1). Construction of the HoaBinh dam (V = 9.5 km^3), located in the downstream area of the Son La dam, was completed in 1993 [23].

Deforestation has been a growing concern in the region, and for the upstream area of the DRB in particular. The region was originally dominated by forest: 70% of the entire DRB was covered by evergreen broadleaf forests, and the remaining area mainly comprised croplands and shrublands. The forest cover of the Chinese portion of the river basin has declined by half from 1950 to 1990 [35], the most severe decline being in 1993 due to unbridled logging and agricultural expansion activities [36]. The past forest cover decline in Vietnam has been even more rapid over the same time period, especially in the mountainous regions of the Sino-Vietnamese border, where forest cover had been reduced by more than 70% of the previous forest area [37]. Since 1995, several programs for forest rehabilitation have been established and the total forest area of Vietnam has been continuously increased; however, the Da River Basin is still limited in terms of forest plantations due to poor accessibility [38]. Compared with the original natural forests, young man-made forests have lower canopy density and shallower rooting depth, and they cannot play an equal role in soil conservation. Over the last 500 years, deforestation have raised the soil erosion rate by 15-fold, resulting in increased sediment load in the DRB [23].

2.2. Dataset Description

The hydrological and meteorological datasets for 1991–2000 were provided by the Vietnam Academy of Science and Technology and the China Meteorological Data Sharing Service Center. These datasets were used as inputs for hydrological modeling and sediment concentration calculations. Daily streamflow data collected at the Lai Chau (LC) and the Ta Bu (TB) stations in the DRB (Figure 1) were used in calibrating and validating the distributed hydrologic model. Monthly suspended sediment concentration (SSC) data were also available at LC station. Available meteorological stations were well distributed in the study area (Figure 1), providing daily precipitation, wind speed, relative humidity, and hours of sunshine, as well as maximum, minimum, and mean air temperatures of the DRB. To run the ecological model, observation data of precipitation, mean air temperature, relative humidity, and vapor pressure were interpolated to gridded datasets using the ordinary Kriging interpolation method [39,40]. Particularly, the altitude effect on air temperature (a lapse rate of 0.65 °C per 100 m) was considered in the process of the air temperature interpolation [5].

Geographical information (e.g., topography, land use, soil type, vegetation, population density) was used as inputs to the distributed hydrological model, the land change model, and the ecological model. Digital elevation data (GTOPO30) with a 30-arc-second spatial resolution was obtained from the U.S. Geological Survey (USGS) [41]. Soil type and soil properties were extracted from the FAO global soil map [42] with a spatial resolution of 5-arc minutes. Land cover data (from 2001 and 2011) were obtained from Moderate Resolution Imaging Spectroradiometer (MODIS) product [43]. Land cover has been regrouped into eight categories: water body, urban, bare land, forest, cropland, grassland, wetland, and shrublands. The global gridded population density in 2000, the global roads dataset, and the global Human Footprint (HF), provided by the data center in NASA's Earth Observing System Data and Information System, the digital river network from the Vietnam Academy of Science and Technology, and basin rainfall trends as calculated from TRMM satellite data (TRMM_3A12), were also employed in this study (Table 1).

In addition, a 31-year long global dataset of vegetation leaf area index (LAI3g) from 1981 to 2012, derived from the third generation GIMMS NDVI3g dataset, was introduced to reflect the current LAI (vegetation cover), with a 1/12 degree resolution and 15-day composites [44].

Table 1. Dataset summary for different models in the study.

Models	Spatial Dataset	Point Dataset
Hydrological model BTOPMC	Elevation, land cover, soil map, LAI	Rainfall, streamflow meteorological data (e.g., precipitation, maximum/minimum air temperature, vapor pressure)
New sediment rating curve NSRC	LAI	Streamflow, SSC

Table 1. Cont.

Models	Spatial Dataset	Point Dataset
Land change model LCM	Elevation, land cover, soil map, population density, road and river networks, slope, human footprint, rainfall trends	
Ecological model Biome-BGC	Elevation, land cover, soil map	precipitation, mean air temperature, mean air temperature, vapor pressure

BTOPMC: Block wise use of TOPMODEL with Muskingum–Cunge routing model; LAI: Leaf Area Index; SSC: Suspended Sediment Concentration; Biome-BGC: Biome-BioGeochemical Cycle.

3. Methodology

Four models were used to evaluate change in land cover and its impact on streamflow and sediment load in the DRB: the LCM, Biome-BioGeochemical Cycle (Biome-BGC) Model, Block wise use of TOPMODEL with Muskingum–Cunge routing model (BTOPMC) and New Sediment Rating Curve (NSRC). The models were coupled in a 'one-way' manner framework shown on Figure 2 and datasets used in these models were summarized in Table 1. In the initial step, the LCM was used to forecast scenarios of future potential land cover change (LCC); simulated land cover types from the LCM were used as inputs into the Biome-BGC to estimate future LAI. The scenarios of future potential land use types and LAI were then used in the validated BTOPMC for future streamflow analysis. Then, scenarios of future LAI and streamflow were used in the NSRC for future sediment load evaluation. A feedback loop was not included, as spatial changes in land cover and temporal dynamic changes in the LAI were reflected in the basin scale streamflow and sediment load simulations.

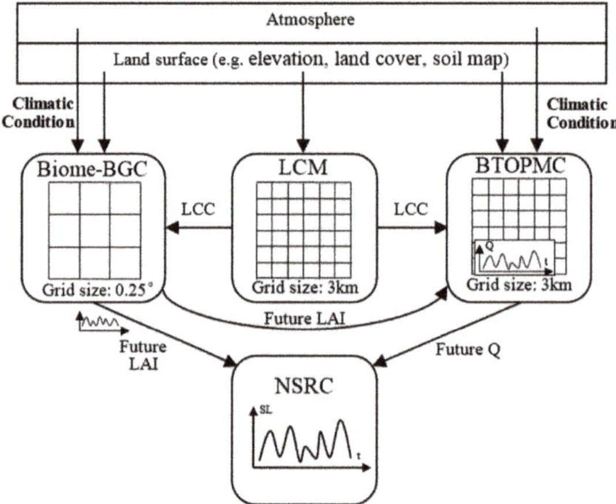

Figure 2. Model integration and data flow within models. Biome-BGC: Biome-BioGeochemical Cycle; LCM: Land Change Model; BTOPMC: Block wise use of TOPMODEL with Muskingum–Cunge routing model; NSRC: New Sediment Rating Curve.

3.1. Land Cover Change Model

The Land Change Model, developed at Clark University [45], is a widely-used effective tool to simulate and forecast land cover change [46]. It allows for the determination of the transition potential of land use from one category to a different one, taking into account static or dynamic variables that may explain the change. For predicting the land cover change, protected areas (e.g., natural parks) that can limit the changes are also included. Through a three-stage process, the LCM can produce potential scenarios of future land cover changes based on an assessment of two time periods in the past [47].

Key features of the LCM used in this study were: (1) Land change analysis—to assess the net change and differences between the separated land classes; (2) Land transition potential modeling—to identify driving factors of land cover transition possibilities between different land cover types. In this study, the driving factors (e.g., distance to river network, distance to road, distance to urban area, distance to protected area, annual rainfall trend, basin slopes, human footprint, and population density) were selected in order to generate potential transition maps; and (3) Land change prediction—to calculate transition potential by a multi-layer perception neural network (MLP) [45]. The LCM model output was validated by quantitative comparison of the forecasted land cover map of 2011, obtained through the LCM with the actual map from MODIS. The strength of the model was evaluated by the kappa index (k) and mean error between simulated and observed land cover maps.

The kappa index can be used as a measure of agreement for land cover map comparisons, or to determine whether the values contained in an error matrix represent a result that is significantly better than random [48]. A kappa index ranging from 1 (accurate) to 0 (inaccurate) is computed as:

$$k = \frac{N \sum_{i=1}^{r} x_{ii} - \sum_{i=1}^{r} (x_{i+} \times x_{+i})}{N^2 - \sum_{i=1}^{r} (x_{i+} \times x_{+i})} \tag{1}$$

where N is the total number of pixels in the matrix, r is the number of rows in the matrix, x_{ii} is the number in row i and column i, x_{+i} is the total for row i, and x_{i+} is the total for column i [49]. According to the suggestion from Monserud and Leemans [49], kappa index values that are greater than 0.85 indicate that the agreement between two maps is excellent.

In this study, actual land cover layers in 2001 and 2008 were used as inputs in assessing land cover change in the DRB. Land cover change was first analyzed, the transition potential of eight land classes (water body, urban area, bare land, forest, cropland, grassland, wetland, and shrublands) were then determined, and a future land cover likelihood map was simulated by the MLP neural network.

3.2. Ecological Model (Biome-BGC)

The Biome-BGC is a biogeochemical point simulation model developed by the University of Montana to estimate the storage and fluxes of carbon, nitrogen, and water within terrestrial ecosystems [49]. The Biome-BGC has been employed in calculating the future LAI combined with the LCM.

The Biome-BGC requires daily climatic data, soil types, vegetation types, site conditions, and parameters describing the eco-physiological characteristics of the vegetation. To achieve equilibrium conditions when the initial soil and plant compartment pools actually match the mass balance equations, Biome-BGC needs "spin-up" simulations. Among the numerous flux and state variables calculated by this model, the LAI was used in this study [5,21]. Ichii et al. [50] applied this model to simulate the carbon fluxes and gross primary productivity in Amazonian, African, and Asian terrestrial ecosystems. In order to obtain LAI values for all grids, we applied the point-based Biome-BGC model to each grid cell, with a spatial resolution of 0.25 degrees [21]. The model has been well validated in simulating potential LAI in the DRB [21].

3.3. Hydrological Model (BTOPMC)

The BTOPMC is a grid-based blockwise distributed model [51,52]. The model extends the TOPMODEL concept [53] by adopting the Muskingum-Cunge method for flow routing components on a grid basis, with sub-catchments serving as blocks [51]. This concept helps to address TOPMODEL's limitation in flow timing and heterogeneity for modeling large river basins in warm humid regions [52,54]. For each grid, four vertical zones are considered: the vegetation zone, root zone, unsaturated zone and the saturated zone [51]. The model has been validated in several river basins with various scale resolutions using geographical data, remote sensing data, and global datasets on ungauged catchments, showing good performance [21,51,52,54]. For a more detailed description of the BTOPMC model and its underlying conceptualizations and parameters, the readers are referred to [51].

In this paper, the BTOPMC was applied to predict future streamflow under future land cover change and LAI change scenarios. The maximum water capacity of the root zone and LAI related to land cover changes were key parameters for evaluating land cover change effects on the streamflow. Following recommendations [55], three statistics were applied to assess the performance of the hydrological model:

(i) Nash-Sutcliffe efficiency (NSE): NSE ranges between $-\infty$ and 1.0 (1 inclusive), with NSE = 1 being the optimal value. Values between 0.0 and 1.0 are generally viewed as being at a satisfactory performance level:

$$\text{NSE} = 1 - \frac{\sum_{i=1}^{n} \left(\text{Qobs}_i - \text{Qsim}_i\right)^2}{\sum_{i=1}^{n} \left(\text{Qobs}_i - \overline{\text{Qobs}}\right)^2} \tag{2}$$

(ii) The ratio of the root mean squared error to the observations standard deviation (RSR): RSR varies from the optimal value of 0, which indicates perfect performance in the simulation, to a large positive value. A lower RSR represents better the model simulation performance:

$$\text{RSR} = \frac{\text{RMSE}}{\text{STDEV}_{\text{obs}}} = \frac{\sqrt{\sum_{i=1}^{n} \left(\text{Qobs}_i - \text{Qsim}_i\right)^2}}{\sqrt{\sum_{i=1}^{n} \left(\text{Qobs}_i - \overline{\text{Qobs}}\right)^2}} \tag{3}$$

(iii) Percent bias (PBIAS): The PBIAS value should be close to zero. Positive values indicate the model contains underestimation bias and vice versa:

$$\text{PBIAS} = \frac{\left|\overline{\text{Qsim}} - \overline{\text{Qobs}}\right|}{\overline{\text{Qobs}}} \times 100\% \tag{4}$$

where Qsim is the simulated value, Qobs is the observed value, $\overline{\text{Qsim}}$ is the average simulated value, and $\overline{\text{Qobs}}$ is the average observed value.

An RSR less than 0.7 and a NSE greater than 0.5 suggest that there is good model performance for both streamflow and sediment load calculations [55]. However, values of PBIAS for streamflow and sediment load vary significantly. Moriasi [55] defined the PBIAS of less than 25% as being of satisfactory indication for the streamflow simulation, and PBIAS results of less than 55% were considered to be acceptable for the sediment load.

3.4. New Sediment Rating Curve (NSRC)

The calculation of sediment load (SL) requires both streamflow and sediment concentration data in river basins. Sediment concentration data are rare since data collection requires manual individual sampling taken at fixed temporal intervals. This type of data is still absent at most hydrological stations, especially in developing countries. Instead, physically-based models or sediment rating curves have been used to estimate the SSC. Physically-based models that are used to simulate the SSC tend to suffer from problems associated with the difficulty of a huge dataset and the identifiability of parameter values. Conversely, traditional sediment rating curves [56] generally represent a simple power functional relationship relating the SSC to streamflow; unfortunately they do not consider the temporal dynamic changes in vegetation cover. Vegetation cover, as discussed above, should have an important effect on soil erosion and sediment transport capacity by slowing the flow through friction losses [57]. Hence low intensity vegetation cover conditions should provide greater sediment flux for the same streamflow. The NSRC with vegetation dynamics information such as normalized difference vegetation index (NDVI) or LAI, could simulate SL well in Southeast Asia [23,58]. The SL was calculated by:

$$SL = Q \times SSC \tag{5}$$

$$SSC = a(1 - M_{NDVI/LAI}{}^c)Q^b \tag{6}$$

$$M_{NDVI} = (NDVI - NDVI_{min})/(NDVI_{max} - NDVI_{min}) \tag{7}$$

$$MLAI = (LAI - LAI_{min})/(LAI_{max} - LAI_{min}) \tag{8}$$

in which a, b, and c are model parameters for a particular stream, Q (m^3/s) is streamflow, SSC (g/m^3) is suspended sediment concentration, $M_{NDVI/LAI}$ is standardized NDVI or LAI, and $NDVI_{min/max}$ and $LAI_{min/max}$ are the minimum and maximum NDVI or LAI values respectively.

To estimate sediment load from the streamflow and land cover change, the NSRC was developed based on the time series of M_{LAI} and the streamflows from 1991 to 2000. To provide a comprehensive assessment of this sediment model performance and to indicate the accuracy of the calculated curve, the same statistics and evaluation rules as in the BTOPMC model were used. Based on the well-fitted NRSC, the impact of future land cover change on sediment load was evaluated.

4. Results

4.1. Future Land Cover Change Prediction

In order to predict future land cover, historical remote sensing maps of 2001 and 2008 for land cover change were investigated. Figure 3a presented gains and losses of the land cover between 2001 and 2008, showing that croplands, forest, shrublands, and grasslands had changed more than others. The net land cover change (Figure 3b) indicated that gains of croplands (+4.3%) mainly transferred from forest (−2.9%) and grassland (−1.1%) losses, other types of land cover showed less noticeable changes.

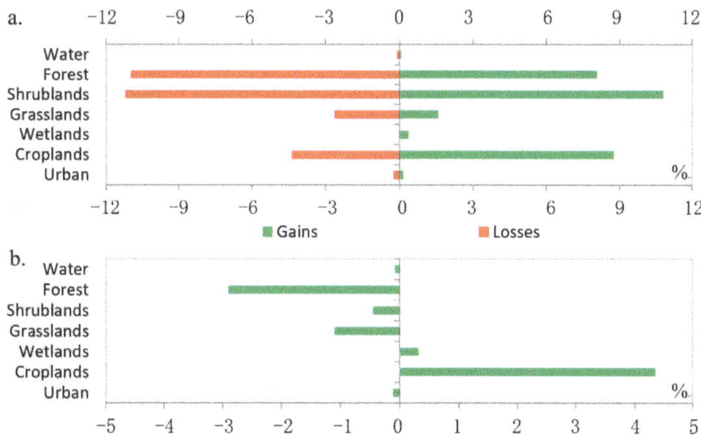

Figure 3. (a) Gains and losses in land cover between 2001 and 2008; (b) Net land cover change between 2001 and 2008.

Results presented in Figure 4a showed consistency in spatial simulations and greater similarity in the water, forest, croplands, and urban areas; though areas covered by shrublands had been overestimated (14.8%) to a certain extent over the area of the grassland (−15%) (Table 2). Comparison of the observed and simulated value showed an acceptable level of likelihood and a very high kappa index of 0.94, which guaranteed good adjustment of the simulated map to the reference map. All these facts assured that the LCM may provide highly agreeable results for creating the future map of land cover in the DRB.

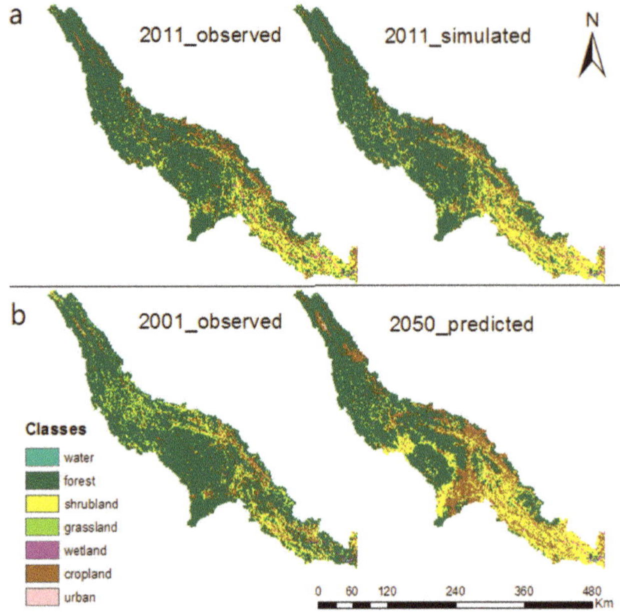

Figure 4. (**a**) Observed (left) and simulated (right) land cover maps of 2011; (**b**) Land cover maps of 2001(basin line) and 2050 (predicted).

Table 2. Statistical parameters between the reference map and the simulated map of 2011.

Land Cover	Reference 2011 (pix)	Simulated 2011 (pix)	Mean Error	Percent (%)	Kappa Index
Water	869	842	−27	−3.1	
Forest	184,701	177,657	−7044	−3.8	
Shrublands	60,433	69,411	8978	14.8	
Grasslands	10,743	9126	−1617	−15.0	0.94
Wetlands	825	896	71	8.6	
Croplands	58,871	58,403	−468	−0.8	
Urban	1618	1725	107	6.6	

Finally, the land cover map for 2050 was generated by the LCM (Figure 4b). Land cover changes from 2001 to 2050 were found mainly in the downstream areas, with an obvious decrease in forest and an increase in the shrublands and cropland (Table 3). Results showed an increase in croplands to over 11.7% of the total area, as well as an increase in shrublands to almost 10%, replacing forest areas, which decreased to about 21% of the total area.

Table 3. Potential land cover change simulated by LCM in the Da River Basin (area percent: %).

Land Cover	Baseline 2001 (%)	Predicted 2050 (%)	Changes (%)
Water	0.06	0.05	−0.01
Forest	70.91	49.57	−21.34
Shrublands	15.5	25.47	9.97
Grasslands	0.72	0.16	−0.56
Wetlands	0.19	0.34	0.15
Croplands	12.57	24.27	11.7
Urban	0.06	0.14	0.08

4.2. Future LAI Predictions

The Biome-BGC ecological model was driven by the current climate data from 1991 to 2000, and future land cover change scenarios, to calculate the future leaf area index for the decade of 2046 to 2055. The model extended the future land cover change from point-in-time to the temporal dynamic scale, as a result, the predicted future LAI stood for future vegetation cover without impacts of future climate change. Deforestation was the main factor in decrease in LAI. Annual maximum LAI decrease was estimated for the most grids in the DRB, especially in the middle and lower reaches with the highest percentage change of −43% (Figure 5). The dominant land cover change was a transfer from forest cover to croplands or shrublands, which would cause lower vegetation productivity. Seasonal vegetation cover (LAI) changes in this case would decrease by around 30% on average during the wet season (Figure 5), while the most drastic change, with a decrease of up to more than 60%, would happen during dry season as it was the best season for felling operations.

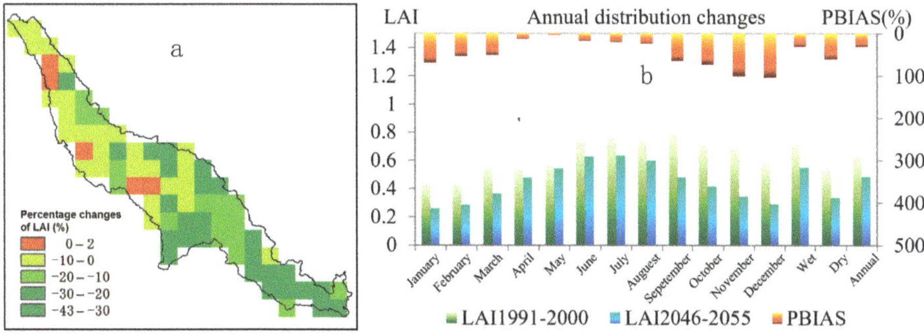

Figure 5. Spatial distribution of percentage changes in LAI (**a**) and annual distribution changes (**b**) in LAI.

4.3. Hydrological Model and NSRC Model Simulation

In this study, the daily streamflow data for 1991–1995 were used for model calibration, and the data for 1996–2000 were used for validation of the BTOPMC hydrological model. The statistics for the evaluation of the BTOPMC model gave consistent results and good accuracy according to the established criteria (Table 4). Direct comparison of the simulated and observed daily streamflow in the baseline period showed a reasonable match within the established criteria at the Lai Chau and Ta Bu stations (Figure 6).

The NSRC model was also calculated for the period from 1991 to 2000, as by the following equation:

$$SSC = 0.41(1 - M_{LAI}{}^{6.5})Q^{1.05} \tag{9}$$

Our results showed a well-noted correlation between the simulated and observed monthly SSCs (Figure 6). For the low and medium values, the model showed very good results; the peak values for year of 1996 and 1999 were overestimated, but overall performance was considered to be good (Figure 6 and Table 4). In addition, the same three statistic criteria used to evaluate the new sediment rating curve were in good agreement with the established validation technique. The high NSE (0.85), low RSR, and PBIAS (Table 4) suggested that the NSRC could evaluate the SSC accurately at the Lai Chau station, and it could be used to evaluate future land cover change effects on the sediment load.

Table 4. Evaluation of BTOPMC and NSRC model simulations during the baseline period for the catchments controlled by the Lai Chau and Ta Bu stations in the DRB.

| | Streamflow | | | | SSC | |
| | Lai Chau | | Ta Bu | | Lai Chau | |
	Calibration	Validation	Calibration	Validation	Calibration	Validation
NSE	0.75	0.70	0.70	0.65	0.85	0.82
RSR	0.48	0.57	0.51	0.65	0.33	0.45
PBIAS (%)	6.2	7.2	7.0	7.8	0.53	1.22

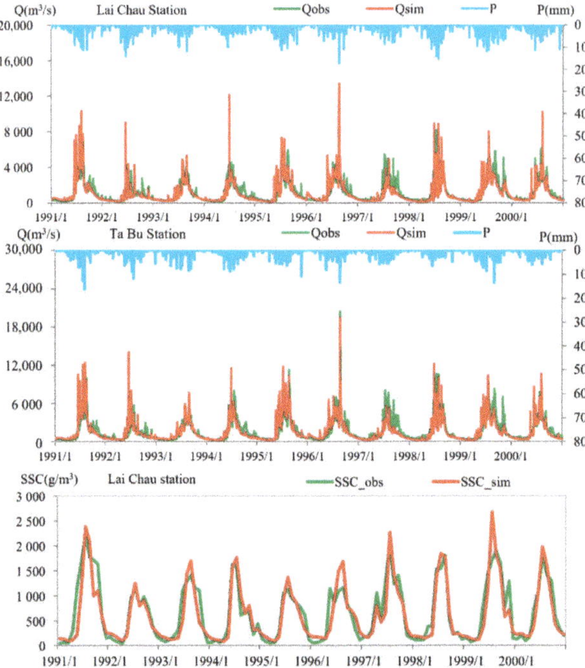

Figure 6. Comparison of observed and simulated daily streamflow (the Lai Chau and Ta Bu first and second figures, respectively) and monthly SSCs (bottom figure) in the DRB (calibration period: 1991–1995, validation period: 1996–2000).

4.4. Future Land Cover Change Impacts on Streamflow and Sediment Load

Streamflow under potential land cover change increased in both the wet season and the dry season with different magnitudes. Annual streamflow in the Lai Chau catchment and the Ta Bu catchment increased by 5.8% and 6.6% respectively. The future monthly streamflow changes (Figure 7a) varied from 4.0% to 10.3%, the highest being in May (wet season). Spatial distribution of the average annual runoff depth change at the sub-basin scale was also analyzed to explore the impacts of spatial variations due to land cover change. As shown in Figure 7b, the annual runoff depth was expected to increase by 5–150 mm/year due to the different percentages of the changed land covers. Most of the sub-basins in the middle and the southeast parts of the watershed showed higher increases in runoff depth, caused by strong land cover changes from forest to croplands or shrublands (Figures 4b and 5a). Moreover, land cover change impacts on individual water balance components, as given in Table 5, indicated an increase in surface runoff and total runoff, but a decrease in actual evapotranspiration and ground water in both upstream catchments (LC) and downstream catchments (TB) in the future. A reduction in the forest cover could decline the infiltration rate and lower the evapotranspiration rate, and so the potential forest cover changes mentioned above were considered as the main reason for water balance changes. Among all four water balance components, the surface runoff showed a maximum increase of more than 10%, which was the main contribution to total runoff growth. Compared with year of 2001, the future evapotranspiration in 2050 also exhibited a distinct decrease, ranging from −5.0% to −8.0% over different seasons and catchments. The projected evapotranspiration in the wet season decreased more than in the dry season in the LC and TB catchments. Spatially, all four water balance components showed higher changes in the downstream area, which agreed with the serious potential for future deforestation in this area.

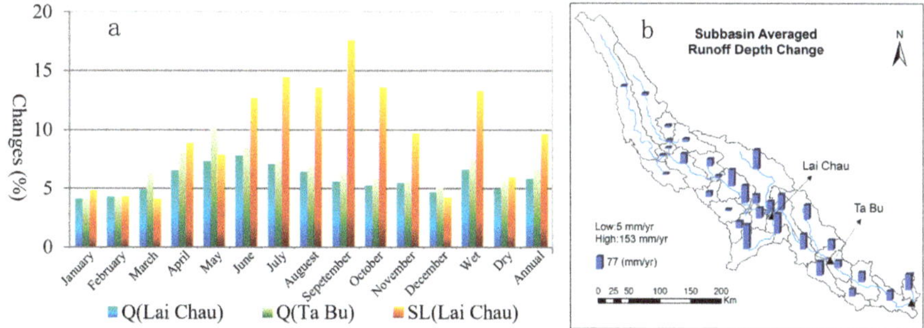

Figure 7. (**a**) Annual distribution changes in streamflow and sediment load under land cover change scenario at the Lai Chau and Ta Bu (**b**) Spatial distribution of averaged annual runoff depth change.

Table 5. Water balance changes due to land cover change between year of 2001 and 2050: Precipitation (P, mm), actual evapotranspiration (AET, mm), surface runoff (SR, mm), ground water (GW, mm), total runoff (TR, mm).

Catchment	Time Scale	P	AET	SR	GW	TR
LC	wet season	2523.1	630.1→586.0 (−7.0%)	1343.3→1534.1 (14.2%)	548.7→521.2 (−5.0%)	1892.0→2011.2 (6.3%)
	dry season	540.2	135.1→128.3 (−5.0%)	241.8→273.5 (13.1%)	161.2→156.4 (−2.9%)	403.0→423.2 (5.0%)
	annual	1608.5	401.1→377.1 (−6.0%)	832.1→940.3 (13.0%)	373.9→358.9 (−4.0%)	1206.0→1275.9 (5.8%)
TB	wet season	3202.3	796.9→733.1 (−8.0%)	1706.1→1989.3 (16.6%)	696.9→648.1 (−7.0%)	2403.0→2573.6 (7.1%)
	dry season	760.6	188.2→176.9 (−6.0%)	342.0→394.0 (15.2%)	228.0→216.6 (−5.0%)	570.0→600.2 (5.3%)
	annual	1989.1	495.3→460.7 (−7.0%)	1030.2→1192.9 (15.8%)	462.8→435.1 (−6.0%)	1493.0→1593.0 (6.7%)

Sediment load would increase significantly in the wet season and only slightly in the dry season, by 13.3% and 6.0%, respectively. The total annual sediment load rate was expected to increase to 9.7% at the Lai Chau station (Figure 7a). The future sediment load change varied from 4.0% to 14.5% over different months, with very high percentage changes from June to October. High growth rates of the sediment load compared with streamflow changes clearly indicated that the sediment load was more easily altered by the means of the land cover pattern changes. Overall, both streamflow and sediment load showed a tendency to increase as an outcome of the land cover change impacts in future.

4.5. Discussion

4.5.1. Combination of Future Land Cover and LAI Changes

In this study, the effects of future land cover changes and corresponding LAI changes were investigated. A future land cover change scenario for 2050 was predicted by the LCM, based on the historical trend of land cover changes from 2001 to 2008. Main variables [1,30,31,59] for land cover predictions were included in the model, such as basin slopes, distance to the river network, distance to urban areas, distance to protected area, human footprint, population density, rainfall trends, etc. The model was able to produce convincing maps of future land cover in the river basin as supported by established criteria (Table 2).

Although a series of driving factors reflecting current human activities were used to predict potential land cover change, future changes in government policy [33] (construction of new dam, afforestation efforts, and others) were hard to predict. It is also impossible to consider future climate change impacts and responses to these impacts on the policy level. These limitations are quite important and should have major impacts on the river stream operation, but they are more uncertain than factors such as relatively moderate population growth, urbanization, and population density, which were also used as input data for the study. There were some areas on the simulated map that were overestimated or underestimated; these errors were hard to neglect. For example, areas covered by shrublands were overestimated (14.8%) to a certain extent, and grasslands were underestimated (−15%) (Table 2); predictions over a large basin with coarse resolution can produce such outcomes due to the low resolution, the lack of government policy, and other factors. The most important factor that was also captured well in the scenario was the change from natural forest cover to human-induced forest in the DRB. This impact was observed as deforestation near the urban areas. In this study, the trend in the reduction of forest cover over the basin was predicted well and it was supported by comparative analysis. Deforestation occurred easily in areas with high population densities and a strong human footprint [13]. In our prediction, analysis of the satellite maps showed quite clearly that a great portion of forest cover was converted to shrublands and croplands in the downstream area of the DRB as a response to its rapid development and population growth, while almost no changes occurred in the middle-west portion of the basin, due to the presence of national nature reserves. Simulated maps and real data from the MODIS satellite in 2011 matched quite well and were accepted by established criteria. Results showed that the forest area in 2050 would decrease by 21% compared to the forest cover of 2001 (Figure 3); meanwhile, shrublands and cropland areas were expected to increase by 10% and 12% respectively. To improve the evaluation of the future land cover change impacts, historical land cover change maps and LAI changes were crucial [13,32,33].

Land cover changes produce further changes in vegetation coverage [21], and many reports have pointed that evaluation of the land cover changes on the hydrology of the basin should also include the impact of the vegetation cover dynamics [5,21,34]. In this study, LAI was considered to be a suitable parameter to reflect vegetation cover dynamics [5,21]. The results of the ecological model have been validated previously and they showed a good match with satellite observed data [21]. The outcome had also pointed towards deforestation as a main cause of the LAI decreasing trends. Such correlations were observed in the downstream area of the DRB, which was severely affected by deforestation, where the LAI values also declined significantly. Areas with unchanged land cover have shown positive vegetation dynamics trends. For example, the land cover in the western area of the DRB remained mostly intact, and LAI values showed an increased trend in response to the increasing rainfall that promoted vegetation growth [34]. The accuracy of the land cover map and LAI predictions was important for further evaluating the impacts of land cover changes on streamflow and sediment load [1,13,33].

4.5.2. Impacts of Forest Cover Changes on Streamflow

Impacts of deforestation on streamflow were analyzed by the BTOPMC hydrological model simulation. Our results have shown that both land cover and LAI have contributed to changes in streamflow.

There have been many arguments over the impacts of forest cover on streamflow in the large basins/regions. Though most researchers have agreed that forest degradation increases the streamflow or water yield [5,60–62], some have argued that the same forest cover changes had weak effects on streamflow in large basins due to their high water retention capacity [16]. Liu et al. [5] concluded that afforestation could reduce streamflow, while deforestation could increase streamflow at both grid and river basin scales in China. Moreover, Wang et al. [15] on the other hand indicated that forest cover had positive impacts on runoff generation in northeast China. Our results agreed with the mainstream view [60–62]; deforestation (21.3%) was the main reason for an increase in annual streamflow (6.7%) and surface runoff (15.8%), along with the decrease in ET (7%) and ground water (6%) through the whole basin. The consistency of spatial variation among runoff depth (Figure 7b), land cover changes (Figure 4b), and LAI changes (Figure 5) supported the same opinion as the majority of other studies [5,61].

LAI values mainly affect the ET [2,5,51,52] and the canopy interception [51,52]; the maximum storage capacity of the root zone (Srmax) is also important as it reflects the available water in the root zone, which is used for evapotranspiration and infiltration to the ground water [2,48,63]. In our hydrological simulation, deforestation decreased LAI values, and this would consequentially produce less ET [5] and canopy interception [51] on different scales. Meanwhile, Srmax was reduced from 0.05 m (forest) to 0.3 m (shrublands) or 0.2 m (croplands) respectively, which would further decrease ET and groundwater values. All of these behaviors were caused by deforestation, affected the water balance of the river basin, and increased streamflow in the river basin.

Additionally, many researchers [7,64] have revealed that land cover changes have significant influences on flood discharge. An increase in the streamflow due to forest degradation would be highly likely to increase flood risk in the Red River [64]; however, a shortage of hourly flood event data limited detailed analysis into the flood risk in this study.

4.5.3. Impacts of Increased Sediment Load on Reservoir Lifetime

Deforestation is known for increasing soil erosion and suspended sediment transportation in rivers of southeast Asian countries [21,23,24,65]. A decrease in LAI and an increase in streamflow due to deforestation would further increase sediment loads [21,25]. This study indicated that the predicted annual sediment load was expected to increase by 9.7% by the 2050s at the Lai Chau station of the DRB. It is also well known that large reservoirs trap suspended sediment from upstream areas, and so increasing sediment load would aggravate reservoir siltation and shorten the life span of the reservoir. The HoaBinh Reservoir has decreased its total suspended load to almost 70% [23]. Wang et al. [21] concluded that land cover change was the main factor for the sedimentation load in the reservoir inflow. This effect has caused a reevaluation of the HoaBinh Dam's operation, reducing its lifetime from more than 100 years to about 50 years [26]. Many observers have feared that the newly constructed the Son La Dam could suffer the same situation.

The Son La Reservoir's sediment trapping efficiency (TE), used to estimate the siltation of the dam, can be evaluated by the hydraulic retention time (HRT) [66,67]. The HRT allows the calculation of the average length period of sediment compounds remaining in the reservoir; it is calculated as a volume of the effective storage divided by the average streamflow into the reservoir. The TE is the percentage of the sediment compounds retained by reservoir, calculated as follows [66]:

$$\mathrm{TE} = 1 - \frac{0.05}{\sqrt{\mathrm{HRT}}} \tag{10}$$

where HRT is water residence time (year) and TE is the trapping efficiency (%). The Son La Reservoir effective storage is equal to 16.2 km^3; the calculated HRT is averaged to about 103 days, that is, 28.3% of a year. Thus, the annual average TE of the Son La Reservoir was estimated to be equal to 90.6%. SSC at the Lai Chau station was used to calculate sediment load into the Son La Reservoir, as SSC was not available at the Son La Dam. The mean annual sediment settlement of the Son La Dam was evaluated to be equal to 17 Mt/year without land cover changes, and 19 Mt/year with land cover changes. The increase in sediment load by about 10% as a result of land cover change would reduce the useful lifetime of the Son La Reservoir ahead of the designed time. The subsequent vital problems of possible sedimentation in reservoirs are declines in dam lifetime, increases in flood risk, and reduction in hydropower generation.

4.5.4. Uncertainties Analysis

Impacts of land cover change and vegetation dynamics were considered at the same time. However, as only aggregated results were generated and there were no mechanism-based combinations, the prediction models were also full of uncertainties. Although a series of driving factors reflecting both human activities and climate were used to drive the land cover change model to predict potential changes, changes in government policy were hard to oversee and predict. Overall uncertainties may come from potential errors in input datasets and model parameters, and simplification of ecological or hydrological process [59,60]. In addition, the future increase in CO_2 concentration was not used to simulate future dynamics of LAI changes, and this should be addressed in the future [5]. All of the uncertainties above would bring prediction errors in land cover change and this may cause a deceptive evaluation of impacts. Therefore, in-depth surveys of government policy and physical-based ecohydrological processes should be investigated, in order to reduce uncertainties in future research.

5. Conclusions

In this study, modeling hydrological appraisal of future potential land cover change and vegetation dynamics under environmental changes was carried out. This study indicated that deforestation pressure in the study basin would continue to rise by 2050. The forest area was expected to decrease to about 22% of the total area by 2050, with a lower LAI during the dry season. The streamflow and sediment yield in the DRB would generally increase in the 2050s in both the wet season and dry season, due to deforestation. Streamflow changes in the wet season showed higher rates than that in the dry season, and this poses flood risks in the future. The increase of the sediment load to 2 million tons per year and the high sediment trapping efficiency of the Son La Reservoir would likely shorten the designed reservoir operation time. It is important to analyze the impacts of the land cover change on all characteristics of the river, including impacts on sediment load and streamflow change, for a better understanding of this phenomenon. This information could be helpful for policy makers and water resource managers with regards to dam operation and vegetation cover protection, in a rapidly changing environment.

Author Contributions: This manuscript was primarily designed and written by J.W.; S.N. contributed to its model simulation and calculation; T.K. contributed to the results analysis and polished the English writing of this draft.

Funding: This research was supported by the National Natural Science Foundation of China (Grant No. 41501029, No. 41671022, No. 51709071, No. 51709148), the Huaihe Meteorology Research Open Foundation (HRM201703), and the Startup Foundation for Introducing Talent of NUIST (Jie Wang).

Acknowledgments: The authors are also thankful to the Vietnam Academy of Science and Technology and the China Meteorological Data Sharing Service Center for providing the hydro-meteorological data. We would like to express our thanks to Hiroshi Ishidaira for his guidance in the writing process of this paper. We also express our thanks to the academic editor and the reviewers for their useful comments.

Conflicts of Interest: The authors declare no conflict of interest.

Abbreviations

The following abbreviations are used in this manuscript:

DRB   Da River Basin
LCM   Land Change Model
Biome-BGC   Biome-BioGeochemical Cycle Model
BTOPMC   Block wise use of TOPMODEL with Muskingum–Cunge routing model
NSRC   New Sediment Rating Curve
SSC   Suspended Sediment Concentration
SL   Sediment Load
LAI   Leaf Area Index
LC   Lai Chau
TB   Ta Bu

References

1. Gashaw, T.; Tulu, T.; Argaw, M.; Worqlul, A.W. Modeling the hydrological impacts of land use/land cover changes in the Andassa watershed, Blue Nile Basin, Ethiopia. *Sci. Total Environ.* **2018**, *619–620*, 1394–1408. [CrossRef] [PubMed]
2. Zhang, L.; Hickel, K.; Shao, Q.X. Predicting afforestation impacts on monthly streamflow using the DWBM model. *Ecohydrology* **2017**, *10*. [CrossRef]
3. Wiekenkamp, I.; Huisman, J.A.; Bogena, H.R.; Graf, A.; Lin, H.S.; Drüe, C.; Vereecken, H. Changes in measured spatiotemporal patterns of hydrological response after partial deforestation in a headwater catchment. *J. Hydrol.* **2016**, *542*, 648–661. [CrossRef]
4. Sun, S.L.; Chen, H.S.; Ju, W.M.; Hua, W.J.; Yu, M.; Yin, Y.X. On the attribution of changing hydrological cycle in Poyang Lake Basin, China. *J. Hydrol.* **2014**, *514*, 214–225. [CrossRef]
5. Liu, Y.B.; Xiao, J.F.; Ju, W.M.; Xu, K.; Zhou, Y.L.; Zhao, Y.T. Recent trends in vegetation greenness in China significantly altered annual evapotranspiration and water yield. *Environ. Res. Lett.* **2016**, *11*, 094010. [CrossRef]
6. Schmidt, A.H.; Gonzalez, V.S.; Bierman, P.R.; Thomas, B.N.; Dylan, H.R. Agricultural land use doubled sediment loads in western China's rivers. *Anthropocene* **2017**, *21*, 95–106. [CrossRef]
7. Ward, P.J.; Renssen, H.; Aerts, J.C.J.H.; Van Balen, R.T.; Vandenberghe, J. Strong increases in flood frequency and discharge of the River Meuse over the late Holocene: Impacts of long-term anthropogenic land use change and climate variability. *Hydrol. Earth Syst. Sci.* **2008**, *12*, 159–175. [CrossRef]
8. Doyle, M.; Harbor, J.; Rich, C.; Spacie, A. Examining the effects of urbanization on streams using indicators of geomorphic stability. *Phys. Geogr.* **2000**, *21*, 155–181.
9. Garcia-Ruiz, J.M.; Lana-Renault, N. Hydrological and erosive consequences of farmland abandonment in Europe, with special reference to the Mediterranean region—A review. *Agric. Ecosyst. Environ.* **2011**, *140*, 317–338. [CrossRef]
10. Costa, M.H.; Botta, A.; Cardille, J.A. Effects of large-scale changes in land cover on the discharge of the Tocantins Rivers, Southeastern Amazonia. *J. Hydrol.* **2003**, *283*, 206–217. [CrossRef]
11. Wardrop, D.H.; Brooks, R.P. The Occurrence and impact of sedimentation in central Pennsylvania wetlands. *Environ. Monit. Assess.* **1998**, *51*, 119–130. [CrossRef]
12. María, J.M.; Ramón, B.; Luis, J.; Raquel, P. Effect of vegetal cover on runoff and soil erosion under light intensity events. Rainfall simulation over USLE. *Sci. Total Environ.* **2007**, *378*, 161–165.
13. Leh, M.; Bajwa, S.; Chaubey, I. Impact of land use change on erosion risk: An integrated remote sensing, geographic information system and modeling methodology. *Land Degrad. Dev.* **2013**, *24*, 409–421. [CrossRef]
14. Zhou, G.X.; Wei, X.; Luo, Y.; Zhang, M.; Li, Y.; Qiao, Y.; Liu, H.G.; Wang, C.L. Forest recovery and river discharge at the regional scale of Guangdong province, China. *Water Resour. Res.* **2010**, *46*, w09503. [CrossRef]
15. Wang, S.; Fu, B.J.; He, C.S.; Sun, G.; Gao, G.Y. A comparative analysis of forest cover and catchment water yield relationships in northern China. *For. Ecol. Manag.* **2011**, *262*, 1189–1198. [CrossRef]
16. Zhou, G.; Wei, X.; Chen, X.; Zhou, P.; Liu, X.; Xiao, Y. Global pattern for the effect of climate and land cover on water yield. *Nat. Commun.* **2015**, *6*, 5918. [CrossRef] [PubMed]

17. Tang, L.H.; Yang, D.W.; Hu, H.P.; Gao, B. Detecting the effect of land-use change on streamflow, sediment and nutrient losses by distributed hydrological simulation. *J. Hydrol.* **2011**, *409*, 172–182. [CrossRef]

18. Dao, N.K.; Suetsugi, T. The responses of hydrological processes and sediment yield to land-use and climate change in the Be River Catchment, Vietnam. *Hydrol. Process.* **2014**, *28*, 640–652.

19. Zorzal-Almeida, S.; Salim, A.; Andrade, M.R.M.; Nascimento, M.N.; Bini, L.M.; Bicudo, D.C. Effects of land use and spatial processes in water and surface sediment of tropical reservoirs at local and regional scales. *Sci. Total Environ.* **2018**, *644*, 237–246. [CrossRef] [PubMed]

20. Gao, G.Y.; Fu, B.J.; Zhang, J.J.; Ma, Y.; Sivapalan, M. Multiscale temporal variability of flow-sediment relationships during the 1950s–2014 in the loess plateau, china. *J. Hydrol.* **2018**, *563*, 609–619. [CrossRef]

21. Wang, J.; Ishidaira, H.; Ning, S.; Khujanazarov, T.; Yin, G.; Guo, L. Attribution Analyses of Impacts of Environmental Changes on Streamflow and Sediment Load in a Mountainous Basin, Vietnam. *Forests* **2016**, *7*, 30. [CrossRef]

22. Kellner, E.; Hubbart, J.A. Improving understanding of mixed-land-use watershed suspended sediment regimes: Mechanistic progress through high-frequency sampling. *Sci. Total Environ.* **2017**, *598*, 228–238. [CrossRef] [PubMed]

23. Le, T.P.Q.; Garnier, J.; Gilles, B.; Sylvain, T.; Van, M.C. The changing flow regime and sediment load of the Red River, VietNam. *J. Hydrol.* **2007**, *334*, 199–214. [CrossRef]

24. Wang, J.; Ishidaira, H. Effects of human-induced vegetation cover change on sediment flow using satellite observations and terrestrial ecosystem model. *J. Jpn. Soc. Civ. Eng. Ser. B1* **2013**, *69*, I_205–I_210. [CrossRef]

25. Dang, T.H.; Coynel, A.; Orange, D.; Blanc, G.; Etcheber, H.; Le, L.A. Long-term monitoring (1960–2008) of the river-sediment transport in the Red River Watershed (Vietnam): Temporal variability and dam-reservoir impact. *Sci. Total Environ.* **2010**, *408*, 4654–4664. [CrossRef] [PubMed]

26. Imhof, A. Vietnam dam to cause hardship for ethnic minorities. *World Rivers Rev.* **1998**, *13*, 3–4.

27. Verburg, P.H.; Soepboer, W.; Veldkamp, A.; Limpiada, R.; Espaldon, V.; Mastura, S.S. Modeling the spatial dynamics of regional land use: The CLUE-S model. *Environ. Manag.* **2002**, *30*, 391–405. [CrossRef] [PubMed]

28. Soares, B.S.; Cerqueira, G.C.; Pennachin, C.L. DINAMICA—A stochastic cellular automata model designed to simulate the landscape dynamics in an Amazonian colonization frontier. *Ecol. Model.* **2002**, *154*, 217–235. [CrossRef]

29. Pontius, R.G.; Cornell, J.D.; Hall, C.A.S. Modeling the spatial pattern of land-use change with GEOMOD2: Application and validation for Costa Rica. *Agric. Ecosyst. Environ.* **2001**, *85*, 191–203. [CrossRef]

30. Kim, O.S. An assessment of deforestation models for reducing emissions from deforestation and forest degradation (REDD). *Trans. GIS* **2010**, *14*, 631–654. [CrossRef]

31. Azucena, P.V.; Mas, J.F.; Arika, L.Z. Comparing two approaches to land use/cover change modeling and their implications for the assessment of biodiversity loss in a deciduous tropical forest. *Environ. Model. Softw.* **2011**, *29*, 11–23.

32. Lin, Y.P.; Hong, N.M.; Wu, P.J.; Lin, C.J. Modeling and assessing land-use and hydrological processes to future land-use and climate change scenarios in watershed land-use planning. *Environ. Geol.* **2007**, *53*, 623–634. [CrossRef]

33. Romano, G.; Abdelwahab, O.M.M.; Gentile, F. Modeling land use changes and their impact on sediment load in a Mediterranean watershed. *Catena* **2018**, *163*, 342–353. [CrossRef]

34. Xiao, J.F.; Zhou, Y.; Zhang, L. Contributions of natural and human factors to increases in vegetation productivity in China. *Ecosphere* **2016**, *6*, 1–20. [CrossRef]

35. United Nations Environment Programme (UNEP). *China Conservation Strategy*; UNEP & China Environmental Science Press: Beijing, China, 1990.

36. Ye, C.Q.; Gan, S.; Wang, W.L.; Deng, Q.Y.; Chen, W.H.; Li, Y.G. Analysis on the runoff distribution and the variability in the downstream of Honghe River. *Resour. Environ. Yangtze Basin* **2008**, *17*, 886–891.

37. World Bank. *Vietnam Water Resources Sector Review: Selected Working Papers of the World Bank, ADB, FAO/UNP and NGO Water Resources Sectoral Group*; World Bank Group: Hanoi, Vietnam, 1996; p. 340.

38. Forest Science Institute of Vietnam (FSIV). *Vietnam Forestry Outlook Study, Asia-Pacific Forestry Sector Outlook Study II*; Working Paper Series; Forest Science Institute of Vietnam: Hanoi, Vietnam, 2009.

39. Li, X.; Cheng, G.; Lu, L. Spatial analysis of air temperature in the Qinghai-Tibet Plateau. *Arct. Antarct. Alp. Res.* **2005**, *37*, 246–252. [CrossRef]

40. Mahdian, M.H.; Bandarabady, S.R.; Sokouti, R.; Banis, Y.N. Appraisal of the geostatistical methods to estimate monthly and annual temperature. *J. Appl. Sci.* **2009**, *9*, 128–134. [CrossRef]

41. Available online: https://lta.cr.usgs.gov/GTOPO30 (accessed on 16 May 2017).

42. Available online: http://www.fao.org/geonetwork/srv/en/metadata.show?id=14116 (accessed on 15 December 2017).

43. Available online: https://lpdaac.usgs.gov/products/modis_products_table/mcd12q1 (accessed on 10 June 2017).

44. Zhu, Z.C.; Bi, J.; Pan, Y.Z.; Ganguly, S.; Anav, A.; Xu, L.; Samanta, A.; Piao, S.L.; Nemani, R.R.; Myneni, R.B. Global Data Sets of Vegetation Leaf Area Index (LAI) 3g and Fraction of Photosynthetically Active Radiation (FPAR) 3g Derived from Global Inventory Modeling and Mapping Studies (GIMMS) Normalized Difference Vegetation Index (NDVI3g) for the Period 1981 to 2011. *Remote Sens.* **2013**, *5*, 927–948.

45. Clark Labs. *The Land Change Modeler for Ecological Sustainability*; IDRISI Focus Paper; Clark University: Worcester, MA, USA, 2009.

46. Oñate-Valdivieso, F.; Sendra, J.B. Application of GIS and remote sensing techniques in generation of land use scenarios for hydrological modeling. *J. Hydrol.* **2010**, *395*, 256–263. [CrossRef]

47. Eastman, J.R. *IDRISI Andes Tutorial*; Clark-Labs, Clark University: Worcester, MA, USA, 2006.

48. Monserud, R.A.; Leemans, R. Comparing global vegetation maps with the Kappa statistic. *Ecol. Model.* **1992**, *62*, 275–293. [CrossRef]

49. Running, S.W.; Gower, S.T. FOREST-BGC, A general model of forest ecosystem processes for regional applications. *Tree Physiol.* **1991**, *9*, 147–160. [CrossRef] [PubMed]

50. Ichii, K.; Hashimoto, H.; Nemani, R.; White, M. Modeling the interannual variability and trends in gross and net primary productivity of tropical forests from 1982 to 1999. *Glob. Planet. Chang.* **2005**, *48*, 274–286. [CrossRef]

51. Takeuchi, K.; Ao, T.Q.; Ishidaira, H. Introduction of blockwise use of TOPMODEL and Muskingum–Cunge method for the hydro-environmental simulation of a large ungauged catchment. *Hydrol. Sci. J.* **1999**, *44*, 633–646. [CrossRef]

52. Ishidaira, H.; Takeuchi, K.; Ao, T. Hydrological simulation of large river basins in Southeast Asia. In Proceedings of the Fresh Perspectives on Hydrology and Water Resources in Southeast Asia and the Pacific, Christchurch, New Zealand, 21–24 November 2000; pp. 53–54.

53. Beven, K. Linking parameters across scales-sub grid parameterizations and scale-dependent hydrological models. *Hydrol. Process.* **1995**, *9*, 507–525. [CrossRef]

54. Shrestha, S.; Bastola, S.; Babel, M.S.; Dulal, K.N.; Magome, J.; Hapuarachchi, H.A.P.; Kazama, F.; Ishidaira, H.; Takeuchi, K. The assessment of spatial and temporal transferability of a physically based distributed hydrological model parameters in different physiographic regions of Nepal. *J. Hydrol.* **2007**, *347*, 153–172. [CrossRef]

55. Moriasi, D.N.; Arnold, J.G.; Van Liew, M.W.; Bingner, R.L.; Harmel, R.D.; Veith, T.L. Model evaluation guidelines for systematic quantification of accuracy in watershed simulations. *Trans. ASABE* **2007**, *50*, 885–900. [CrossRef]

56. Asselman, N.E.M. Fitting and interpretation of sediment rating curves. *J. Hydrol.* **2000**, *234*, 228–248. [CrossRef]

57. Howe, A.; Rodriguez, J.; MacFarlane, G. Vegetation Sediment Flow Interactions in Estuarine Wetlands. In Proceedings of the International Congress on Modelling and Simulation (MODSIM05), Melbourne, Australia, 12–15 December 2005; pp. 332–338.

58. Wang, J.; Ishidaira, H.; Sun, W.C.; Ning, S.W. Development and Interpretation of New Sediment Rating Curve Considering the Effect of Vegetation Cover for Asian Basins. *Sci. World J.* **2013**, *2013*, 154375. [CrossRef] [PubMed]

59. Aguejdad, R.; Houet, T.; Hubert-Moy, L. Spatial Validation of Land Use Change Models Using Multiple Assessment Techniques: A Case Study of Transition Potential Models. *Environ. Model. Assess.* **2017**, *22*, 591–606. [CrossRef]

60. Panday, P.K.; Coe, M.T.; Macedo, M.N.; Lefebvre, P.; Castanho, A.D.D.A. Deforestation offsets water balance changes due to climate variability in the Xingu river in Eastern Amazonia. *J. Hydrol.* **2015**, *523*, 822–829. [CrossRef]

61. Alvarenga, L.A.; Mello, C.R.D.; Colombo, A.; Cuartas, L.A.; Bowling, L.C. Assessment of land cover change on the hydrology of a brazilian headwater watershed using the distributed hydrology-soil-vegetation model. *Catena* **2016**, *143*, 7–17. [CrossRef]

62. Coe, M.T.; Costa, M.H.; Soares-Filho, B.S. The influence of historical and potential future deforestation on the stream flow of the Amazon River—Land surface processes and atmospheric feedbacks. *J. Hydrol.* **2009**, *369*, 165–174. [CrossRef]

63. Zhao, J.; Xu, Z.X.; Singh, V.P. Estimation of root zone storage capacity at the catchment scale using improved Mass Curve Technique. *J. Hydrol.* **2016**, *540*, 959–972. [CrossRef]

64. Sanyal, J.; Densmore, A.L.; Carbonneau, P. Analysing the effect of land-use/cover changes at sub-catchment levels on downstream flood peaks: A semi-distributed modelling approach with sparse data. *Catena* **2014**, *118*, 28–40. [CrossRef]

65. Dudgeon, D.; Choowaew, S.; Ho, S.C. River conservation in South-East Asia. In *Global Perspectives on River Conservation: Science, Policy and Practice*; Boon, P.J., Davies, B.R., Petts, G.E., Eds.; John Wiley & Sons Ltd.: Chichester, UK, 2000; pp. 279–308.

66. Maneux, E.; Probst, J.L.; Veyssy, E.; Etcheber, H. Assessment of dam trapping efficiency from water residence time: Application to fluvial sediment transport in the Adour, Dordogne, and Garonne River basins (France). *Water Resour. Res.* **2001**, *37*, 801–811. [CrossRef]

67. Vörösmarty, C.J.; Meybeck, M.; Fekete, B.; Sharma, K.; Green, P.; Syvitski, J.P.M. Anthropogenic sediment retention: Major global impact from registered river impoundments. *Glob. Planet. Chang.* **2003**, *39*, 169–190. [CrossRef]

Article

Spatio-Temporal Dynamic Architecture of Living Brush Mattress: Root System and Soil Shear Strength in Riverbanks

Dong Zhang [1], Jinhua Cheng [1,*], Ying Liu [2], Hongjiang Zhang [1], Lan Ma [1], Xuemei Mei [1] and Yihui Sun [1]

[1] School of Soil and Water Conservation, Beijing Forestry University, Beijing 100083, China; allianz_d@bjfu.edu.cn (D.Z.); zhanghj@bjfu.edu.cn (H.Z.); mlpcz@bjfu.edu.cn (L.M.); mxm0115@bjfu.edu.cn (X.M.); yihui_sun@bjfu.edu.cn (Y.S.)

[2] Hubei Key Laboratory of Ecological Restoration of Rivers and Lakes and Algae Use, Hubei University of Technology, Wuhan 430068, China; liuy@hbut.edu.cn

* Correspondence: jinhua_cheng@bjfu.edu.cn; Tel.: +86-10-6233-6612

Received: 15 June 2018; Accepted: 8 August 2018; Published: 13 August 2018

Abstract: As a basal measure of soil bioengineering, the living brush mattress has been widely applied in riparian ecological protection forest construction. The living brush mattress shows favorable protective effects on riverbanks. However, there are few reports on the root structure and the soil strengthening benefit of the living brush mattress. The present work reports a series of experiments on root morphology and soil shear strength enhancement at the temporal and spatial scales. The object of the study is 24 living brush mattress trees constructed with *Salix alba* L. 'Tristis' (LBS hereafter). Traditional root morphology and mechanical measurement methods were used to collect the parameters. The results showed that the root systems of LBS had the characteristics of symmetry and upslope growth. The roots were mainly distributed in a cylindrical region of the soil (radius × thickness: 0.4 m × 0.5 m) and their biomass increased with different growth rates for the periods from 1 to 5 and from 5 to 7 years. Both age and slope position were factors that influence root growth. The root diameter falls within 0–5 mm, has a significant effect on the soil shear strength and provides a conical-shape potentiation zone to ensure the efficient protection of a riverbank. The results of this study demonstrate that LBS is an efficient and feasible engineering measure in the field of riverbank protection.

Keywords: living brush mattress; root distribution; shear strength; spatio-temporal scales; soil bioengineering; riverbank

---

1. Introduction

Riverbank ecosystem functions play important roles in human society and economic development [1]. The stability of a riverbank is related not only to property losses on immediately adjoining lands and in adjacent areas but also to the downstream deposition phenomenon [2,3]. Although the concept of ecological riparian shelterbelts has been proposed in China [4], the traditional hardening protection methods, such as dry-stone masonry, concrete, and precast block revetments, are still applied in riverbank reinforcement projects due to a lack of theoretical research. In July 2016, a heavy rainfall event occurred in Beijing, the maximum peak flow of which was approximately 7.26 times that of the daily average water discharge in the flood season, and the event destroyed the bare bank and many riverbank hardening projects. According to the investigation of the soil bioengineering measure damage rate in the demonstration area after the flood, 100 m of a high-density planting living brush mattress that was installed along a straight riverbank was approximately 75% well preserved (Figure 1).

Figure 1. The protection measures damaged by the flood. (**a**) Destroyed retaining wall; (**b**) collapsed riverbank; (**c**) the living brush mattress measure before the flood; (**d**) the living brush mattress measure during the flood.

The living brush mattress is widely used as an asexual reproductive technique in soil bioengineering projects because of its characteristics of simple construction, low cost and quick effect [5–7]. In the past decade, soil bioengineering researchers have explored the contributions of certain measures to the stability of slopes and riverbanks. Bischetti et al. [8] calculated the safety factor of slopes reinforced by brush layering, which is a common soil bioengineering technique, based on the principle of limit equilibrium, and demonstrated that the safety factor will increase over time as the root system grows. Dhital and Tang [9] compared the control efficiency of wire net check dams and soil bioengineering vegetative check dams on riverbank erosion in three growth cycles. The results indicated that the effect of vegetative check dams on bank slope stability was more obvious. Fernandes and Guiomar [10] used a SLIP4EX model and a normal Coulomb model to evaluate the slope protection efficacies of different types of soil bioengineering interventions (including certain complex structures constructed using wooden materials in combination with plants) and suggested that although a wooden structure will decay with time, the plants in the structure can ensure the safety of the slope. These studies indicate that soil bioengineering techniques that employ living plant materials for civil engineering structures [11] depend on the plants to stabilize riverbanks [12].

Plants stabilize slopes through mechanical mechanisms between their root systems and the soil [13]. The root system morphology affects the mechanical properties of root-soil composites [14,15]. It is important to describe the root distribution, especially on the time scales used to evaluate the effects of root systems on soil enhancement [16,17]. Many studies have been performed on the morphological characteristics of plant roots in gentle and steep slopes and have suggested that site conditions (e.g., the slope, soil composition and soil moisture content) significantly affect root development and that tree species can also cause different root distributions [18,19]. In addition, McIvor et al. [19]

found that the root distribution has downslope growth characteristics, meaning that the roots in the downslope direction are closer to the surface than are the roots that follow an upslope path. In the studies of Di et al. and Nicoll et al., an asymmetric growth characteristic on both sides of the slope of the root distribution was found [20,21]. Furthermore, the root biomass is increased with the growth of seeding but showing different increments in the same time interval; that is, the growth rate of root biomass is not uniform [22,23]. However, in the juvenile stages of soil bioengineering plants, the proportion of adventitious roots is prominent [24], and there is still an absence of knowledge on whether a living brush mattress planted in riparian zones has these features.

The contribution of plants to soil safety lies in the complex mechanical interactions at the root-soil interface, mainly due to the resistance to soil shear deformation through root strength [25–30]. The soil shear strength of root-soil composites, which can directly reflect the effect of the root system on the soil [14,31], is influenced by many factors, such as the root strength, root branches, root hairs and secretions that are difficult to reconstruct in laboratory experiments [32]. Furthermore, in assessment models of root-to-soil reinforcement effects, such as the simple force equilibrium theory [33,34], the fiber bundle model [35] and the root bundle model [36], the shear strength of the rooted soil has been used as a factor to evaluate slope stabilization. Therefore, clearly understanding the soil shear strength distribution with roots is the most effective way to assess the safety of ecological engineering measures.

In recent years, soil bioengineering techniques have been increasingly applied in China [5,37]. Liu [38] and Qian [39] analyzed the suitability of eight dominant plant species in the riparian zone in Beijing and identified that living brush mattress constructed with *Salix alba* L. 'Tristis' (hereafter referred to as LBS) has a better rooting ability, survival rate, and tolerance to waterlogging than other plants.

The key to predicting the outcomes of soil bioengineering interventions is to quantify the evolution of the structures [10]. LBS has been applied with favorable results in a riparian ecological restoration project in Beijing during the flood event. However, little information exists regarding LBS root development and its shear strength distribution characteristics in different zones of the bank and in its growth stages, which are important for characterizing its bank stabilization ability. Therefore, this study aimed to (1) explore the root distribution characteristics of LBS in different time intervals; and (2) identify the variation in the soil shear strength enhanced by the LBS root system at varying depths, lateral distances from the stem, and development times. Furthermore, (3) the relationships between the root morphology (the number of root and root cross-sectional area) and shear strength were considered to estimate the soil-reinforcing effects of LBS.

2. Materials and Methods

2.1. Construction Methods for Living Brush Mattress and Salix alba L. 'Tristis' (LBS)

The live brush mattress comprises living shoots of plants (approximately 20 mm in diameter) that have been spread in horizontal rows on the surface of the bank such that the thick end is soaked in water and the soil is covered above the shoots. The interspaces between each living shoot are as small and uniform as possible to ensure greater bank protection benefits [40]. After construction, a number of newborn plants will grow on the living shoots. Our research is aimed at these individual newborn plants.

The structural material of LBS is *Salix alba* L. 'Tristis' living shoots with a diameter of 20 mm. The interspace between adjacent shoots is approximately 0.5 m. A schematic of the LBS structure is shown in Figure 2. The LBS was constructed before the end of the plant dormancy period in 2009, 2010, 2011, 2012, 2015 and 2016, with planting areas of approximately 300 m^2, 300 m^2, 180 m^2, 180 m^2, 300 m^2 and 300 m^2, respectively. In the first year after construction, a large number of herbs were found growing in the interspaces between the living shoots, and weeding and watering were undertaken twice a month for the first three months. In the following year after construction, few herbs were found, and only a weeding operation was undertaken in May. After five years of growth, few plants were growing in the interspaces, and the LBS had colonized.

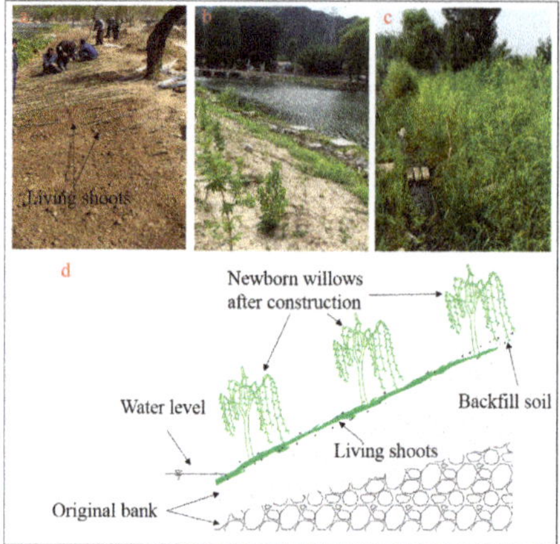

Figure 2. Construction effect diagram of a living brush mattress. (**a**) Under construction; (**b**) one month after completion; (**c**) six months after completion; (**d**) side view of the living brush mattress.

2.2. Site Details

The experiment was conducted at a bank along the Liuli River in the Huairou District, Beijing, China (alt.: 243 m, lat.: 40°39′ N, long.: 116°40′ E) (Figure 3). Following artificial transformation in accordance with the original riverbank terrain, the soil distribution 0–0.4 m below the surface at the experiment site is sandy loam (measured by the hydrometer method) with small gravel, and below 0.4 m, loamy sand (measured by the hydrometer method) with gravel is found. After land preparation, the riverbank slope was transformed to be consistently 15°.

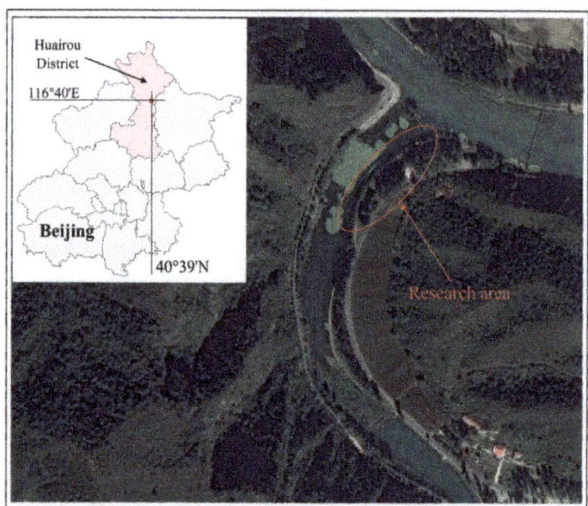

Figure 3. The location of the research area.

2.3. Tree Selection and Extraction

Twenty-four LBS trees, eight each of 1-year-old, 5-year-old and 7-year-old trees, were chosen for the experiment. Five trees were used for the morphological study, and three trees were used for shear strength tests of each age. Before sample collection, the basic condition of the LBS in the plot was investigated. The results are shown in Table 1. According to the arithmetic mean of the basis diameter of the survey, a well-grown single tree that had no neighbors [20] within a radius of nearly 1 m was regarded as the standard wood.

Table 1. Basic conditions of various growth years.

Age	1			5			7		
Plot Area (m²)	**3 m × 2 m**			**3 m × 4 m**			**3 m × 4 m**		
Plot Number	1	2	3	1	2	3	1	2	3
Plot Slope	15°	15°	15°	15°	15°	15°	15°	15°	15°
Number of Trees	8	8	8	13	9	11	12	7	8
Mean Basal Diameter (mm)	10.7	10.2	10.6	75.4	82.7	86.8	115.6	130.4	113.2

The root system exposure method, which requires a heavy workload but can provide more comprehensive root information, was used in this study. Excavation was conducted from 15 August to 17 September 2016, and from 3 August to 25 August 2017. Before excavation, each tree was cut horizontally at the point immediately above where the top-most root emerged from the stem, and the aspect was marked using a permanent pen. A trench with the stem at the center was dug by hand, trowel and water rinsing until all roots were exposed. After excavation, efforts were made to ensure that the root system remained unchanged, and the sample was transported to the laboratory for digitizing. Inevitably, roots were broken during excavation, and these broken roots were collected, marked, and reconstructed in the laboratory.

2.4. Root System Parameter Measurements

Four test zones were established for each sample, divided into the upslope-upstream face (hereafter referred to as U-U), the downslope-upstream face (hereafter referred to as D-U), the upslope-downstream face (hereafter referred to as U-D) and the downslope-downstream face (hereafter referred to as D-D). In each test zone, the space was divided into many test units composed of various special geometric shapes. In the vertical direction, the thickness of the geometric shape was 0.1 m, and in the horizontal direction, demi-semi-circles or annuluses were divided with individual radii of 0.1 m (Figure 4).

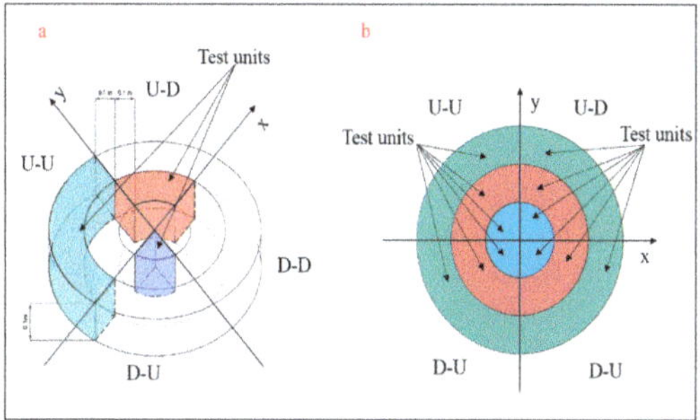

Figure 4. Schematic diagram of the test units. (**a**) 3D schematic of the test units; (**b**) top view of the test units. The intersection of the X-axis and the Y-axis is the center of the stem. This figure shows only the shape of the test unit. The morphological measurement range of the root includes the whole root system, and the test depth for the soil shear strength is 0.6 m below the surface.

The maximum vertical depth (MVD) from the surface, the maximum lateral distance (MLD), and the diameter of every root in each test unit from the stem were measured. The root biomass values were recorded after oven-drying at 70 °C for 48 h [41]. The number of roots and the root cross-sectional area were recorded during shear strength tests, and the content was converted into unit volume.

2.5. Shear Strength Determination

The shear strength was measured using a field inspection vane tester (Figure 5), which was used to measure the in situ undrained shear strength of the soil [13]. The lower and upper parts of the instrument are connected by a threaded joint. The scale-ring is also supplied with threads and follows the upper part of the instruments by means of two lugs. The 0-point is indicated by a line on the upper part. When the handle is turned, the spring deforms, and a mutual angular displacement formed between the upper and lower parts of the instrument, with the scale-ring following the upper part of the instrument. When failure is obtained, the scale-ring will remain in its position due to the friction in the threads. The size of the displacement depends on the torque that is necessary to turn the vane. The product of the scale reading and the vane coefficient is the measured value of the soil shear strength. The measuring range of the instrument is from 0 to 260 kPa when three differently sized vanes (16 × 32, 20 × 40 and 25.4 × 50.8 mm) are used, and the corresponding vane coefficients for each size are 2, 1 and 0.5, respectively. The measurement accuracy is within 10% of the reading.

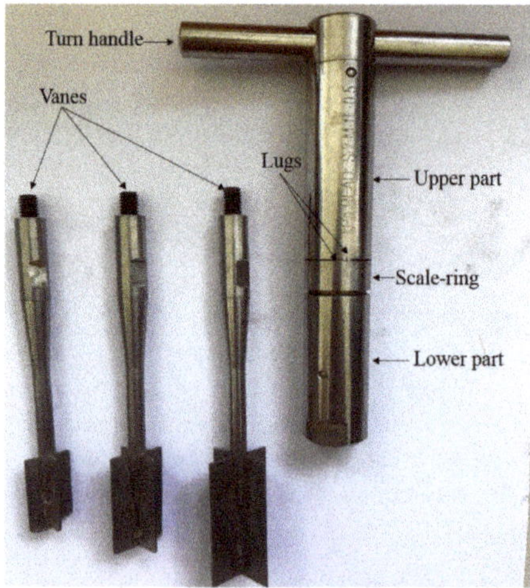

Figure 5. Field inspection vane tester device used for the shear strength tests.

Profile quantities parallel to the bank slope were excavated, and the shear strength was recorded for each unit (similar to Section 2.4). Certain outliers (e.g., samples with no obvious signs of root rupture or samples for which boulders were encountered during the test) were eliminated. Although it cannot be denied that the soil shear strength test method used in this study has some shortcomings for soils permeated by large roots, the sample plot was on the riverbank, and the soil contained large amounts of gravel of an uneven distribution, which can greatly affect the measurements obtained with other methods such as the shear box test [18,42]. We attempted to increase the number of tests in each unit as much as possible to compensate for the defects caused by the methods used in this study.

During the test, at greater distances from the stem, no roots were found in a number of test units. The shear strength of these units was considered to be the shear strength of the pure soil. For each root-soil composite unit, the shear strength enhancement value was determined by a subtraction operation of the shear strength between this unit and the pure soil.

The fact that coarse roots tend to act as separate anchors rather than as a component reinforcing soil strength has been confirmed in past studies [43]. Compared with fine roots, coarse roots make no significant contribution to improving the shear strength of root-soil composites [44]. In this study, the relationships between the number and cross-sectional area of fine roots and the soil shear strength were revealed. In addition to hair roots, the diameter of the fine root was defined as less than 5 mm [45].

2.6. Statistical Analysis

The data were analyzed using the statistics package SPSS 18.0 (SPSS Inc., Chicago, IL, USA). Multiple comparisons of the tree height, stem diameter, root biomass, maximum lateral root extent, and maximum root depth were performed using Tukey's honestly significant difference test for the three tree ages. Analysis of variance (ANOVA) was performed to test the effect of age on root biomass in different slope positions (upslope and downslope) and a paired *t*-test was performed to test the effect of slope position on root biomass for each age. Regression analyses were performed to investigate the relationships between the root biomass and depth, between the root biomass and the distance from the stem, between the shear strength and the number of roots, and between the shear strength and cross-sectional area.

3. Results

3.1. LBS Dimensions

Table 2 shows selected basic parameters of the excavated samples. The basal diameters of all trees were close to the average of the plot.

Table 2. General measurements of the excavated *Salix alba* L. 'Tristis' (LBS).

Age (Year)	N	H (m)	D (mm)	RB (g)	MLD (m)	MVD (m)
1	5	2.01 ± 0.35 [a]	9.16 ± 1.36 [a]	24.5 ± 7.02 [a]	0.45 ± 0.06 [a]	0.20 ± 0.03 [a]
5	5	9.79 ± 1.00 [b]	79.92 ± 11.46 [b]	507.28 ± 49.00 [b]	1.00 ± 0.20 [b]	1.46 ± 0.11 [b]
7	5	14.11 ± 1.22 [c]	122.08 ± 11.66 [c]	1006.25 ± 157.85 [c]	1.27 ± 0.11 [c]	1.73 ± 0.07 [c]

Note: Data are represented as means and standard deviations (mean ± SD). N = sample size; H = tree height; D = stem diameter at ground level; RB = root biomass; MLD = maximum lateral root extent; MVD = maximum root depth. Different letters in the same column indicate a significant difference at $\alpha = 0.05$.

After 5 and 7 years of growth, the root biomass was 20 and 41 times the biomass of the 1-year-old LBS, respectively. As observed during the extraction process, the LBS had a large number of adventitious roots, and the main root was not formed in the early stage of planting. After 5 and 7 years of growth, the main root was formed and extended to above 1.46 m and 1.73 m, respectively. In terms of the basic parameters over time, both the tree height and the basal diameter increased steadily. The MLD development was relatively fast in the first 5 years and then became slower in the next two years, in contrast with the development of the MVD.

The approximate relationships between the tree height (H) and the MLD and MVD are presented in Table 3. Note that with the exception of the MLD, that was measured for the 1-year-old LBS, both the MVD and MLD are approximately 10% of the tree height.

Table 3. Relationships between the tree height (H) and the maximum lateral distance (MLD) and maximum vertical depth (MVD) of the root system.

Age (Year)	MLD/H	MVD/H
1	0.22	0.1
5	0.1	0.15
7	0.09	0.12

3.2. Spatio-Temporal Root Biomass Distribution with the Depth below the Ground Surface

The spatio-temporal root biomass distribution is shown in Figure 6. The root biomass was less on the upslope face than on the downslope face, and the biomass was larger on the downstream face than on the upstream face, except in the 1-year-old LBS. As observed during the excavation, the roots of the LBS developed on both sides of the living shoot buried in the soil. Therefore, the root growth symmetry of the LBS was determined from the biomass on both sides of the living shoot (namely the upstream and the downstream faces of the slope, respectively). When the colors of the color blocks in the figure (representing the root biomass content of each test unit) at the same slope position are compared, similar colors are observed on the upstream and downstream faces for the 1-year-old LBS, indicating that the root system showed obvious symmetry in these two zones. After 5 years of growth, this symmetry was preserved only in the soil above 0.4 m in the upslope area. After the 7th year, this symmetry was no longer detectable.

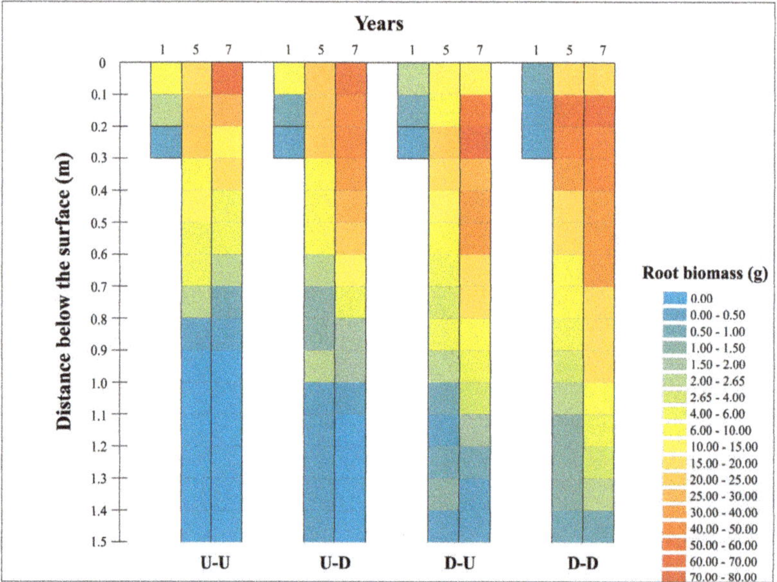

Figure 6. Spatio-temporal distribution of root biomass with depth in every unit. U-U, D-U, U-D and D-D represent the upslope-upstream face, the downslope-upstream face, the upslope-downstream face and the downslope-downstream face of the sample, respectively. Each color block represents the range of root biomass in each test unit. The value of each color block is the arithmetic mean corresponds of the 5 morphological research objects at each age.

In general, roots located in the upslope zone were concentrated primarily in the 0–0.4 m soil layer or in the 0.1–0.7 m downslope layer. In the 0–0.1 m soil layer, the root biomass was much larger in the upslope zone than in the downslope zone; however, at an increased depth, the result was the opposite. The root biomass distributed in the downslope zone was deeper than that in the upslope zone, except for the 1-year-old LBS. For example, for the 5-year-old LBS, color blocks representing less than 10 g of biomass are found below 0.5 m in the U-U zone and 0.3 m in the U-D zone, while for the downslope zone, such color blocks are found below 0.5 m in the D-U zone and 0.6 m in the D-D zone. The LBS root system exhibited characteristics upslope growth, meaning that the root system in the upslope zone grew close to the surface, whereas the growth behavior in the downslope zone was the opposite.

The root biomass distributions with depth in the upslope and downslope zones were analyzed (Figure 7). The root biomasses of the 5- and 7-year-old LBS, with approximately 71% and 63% of the total residing in the depth range of 0–0.4 m, first increased with depth and then decreased while exhibiting larger fluctuations. Therefore, this soil layer could be identified as the active layer of root growth. When the depth reached below 0.4 m, the root biomass, accounting for approximately 29% and 37% of the total, showed a steady decrease. Thus, this area could be defined as the stable layer of root growth [17]. Since the root system of the 1-year-old LBS was concentrated in the 0–0.3 m soil layer, the stable layer and the active layer could not be separately identified. The change in root biomass with depth was described by regression analysis. The optimal curve fit was identified by evaluating linear, power, logarithmic, exponential and polynomial equations in terms of their R^2 values.

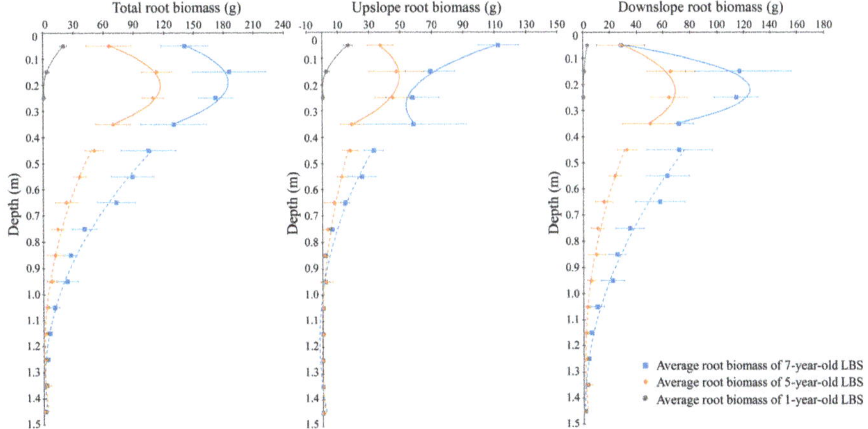

Figure 7. Average root biomass distribution with soil depth for 1, 5 and 7-year-old LBS. The fit curves of root biomass with depth by regression analysis are also shown. The solid lines represent the fitted curves of the active layers. The dotted lines represent the fitted curves of the stable layers.

Based on the fitting results, the relationships are best represented by the 2nd-order polynomial equation (Table 4). From these equations, it is possible to estimate the root biomass at any depth under similar site conditions.

Table 4. Regression equations and R^2 values of the root material by biomass with depth in the axial direction.

		1-Year-Old	5-Year-Old	7-Year-Old
Active Layer	Total	$B = 727.9d^2 - 311.9d + 34.03$ $R^2 = 0.9148$	$B = -2193.2d^2 + 885.03d + 26.90$ $R^2 = 0.6731$	$B = -2174.9d^2 + 823.17d + 106.31$ $R^2 = 0.4272$
	Upslope	$B = 623.2d^2 - 265.68d + 28.68$ $R^2 = 0.921$	$B = -902.55d^2 + 304.34d + 23.41$ $R^2 = 0.5164$	$B = 1091.8d^2 - 610.72d + 139.09$ $R^2 = 0.5717$
	Downslope	$B = 104.7d^2 - 46.22d + 5.35$ $R^2 = 0.711$	$B = -1290.6d^2 + 580.69d + 3.50$ $R^2 = 0.4486$	$B = -3266.7d^2 + 1433.9d - 32.78$ $R^2 = 0.7217$
Stable Layer	Total		$B = 77.72d^2 - 190.25d + 116.66$ $R^2 = 0.8665$	$B = 140.34d^2 - 371.09d + 246.37$ $R^2 = 0.8891$
	Upslope		$B = 32.01d^2 - 75.92d + 44.20$ $R^2 = 0.7618$	$B = 64.561d^2 - 152.06d + 87.29$ $R^2 = 0.8966$
	Downslope		$B = 47.71d^2 - 114.33d + 72.47$ $R^2 = 0.8418$	$B = 75.78d^2 - 219.03d + 159.08$ $R^2 = 0.8368$

Note: B = root biomass (g); d = distance below the surface (m).

The growth rates of the total root biomass from 1 to 5 years and from 5 to 7 years in the vertical direction are shown in Table 5. The growth rates in the first four years were significantly smaller than those from 5 to 7 years in the soil layers 0–1 m below the surface. In the 1–1.3 m layers, the growth rates from 5 to 7 years were still greater than those from 1 to 5 years, but the growth trend was gradual. When the soil depth reached 1.3 m or more, the growth rates from 5 to 7 years were no longer greater than those from 1 to 5 years.

Table 5. The growth rate of total root biomass in each soil layer in the vertical direction.

Vertical Depth (m)	Growth Rate	
	1—5 Years	5—7 Years
0.0–0.1	11.16	37.79
0.1–0.2	27.44	36.42
0.2–0.3	27.05	31.38
0.3–0.4	17.19	30.38
0.4–0.5	12.50	27.23
0.5–0.6	8.92	26.31
0.6–0.7	5.64	24.75
0.7–0.8	3.40	13.48
0.8–0.9	2.77	7.86
0.9–1.0	1.98	7.74
1.0–1.1	0.80	3.74
1.1–1.2	0.57	1.80
1.2–1.3	0.56	0.85
1.3–1.4	0.71	−0.08
1.4–1.5	0.22	0.13

The paired *t*-test was performed to test the effect of slope position (upslope and downslope) on root biomass for each age and the ANOVA was performed to test the effect of age on root biomass in different slope positions (Table 6). Considering the effect of slope position on the root biomass, in the 1st year, the slope position influenced the root biomass at the soil depths of 0–0.2 m, and in the 5th year, impacted the root biomass at the depths of 0.4–0.6 m, 0.7–0.9 m, 1.0–1.1 m and 1.2–1.3 m. In the 7th year, the slope position affected the root biomass throughout the whole system except at the depths of 0.1–0.2 m and 0.3–0.4 m. Considering the effect of age on the root biomass, in the upslope area, the age influenced the root biomass at the soil depths of 0–0.8 m, and in the downslope area, it impacted the root biomass at the depths of 0–1.3 m.

Table 6. Results of paired *t*-test and analysis of variance (ANOVA) for evaluating the effects of slope position and age on the root biomass across different ranges of soil depth.

Soil Depth (m)	*p*-Value				
	Paired *t*-Test			ANOVA	
	Slope Position			Age	
	1-Year-Old	5-Year-Old	7-Year-Old	Upslope	Downslope
0.0–0.1	**<0.01**	0.308	**<0.01**	**<0.01**	**0.030**
0.1–0.2	**0.038**	0.271	0.090	**<0.01**	**<0.01**
0.2–0.3	0.398	0.130	**0.012**	**<0.01**	**<0.01**
0.3–0.4	-	0.062	0.477	**<0.01**	**<0.01**
0.4–0.5	-	**0.019**	**0.018**	**<0.01**	**<0.01**
0.5–0.6	-	**<0.01**	**<0.01**	**<0.01**	**<0.01**
0.6–0.7	-	0.060	**<0.01**	**<0.01**	**<0.01**
0.7–0.8	-	**0.045**	**<0.01**	**<0.01**	**<0.01**
0.8–0.9	-	**0.020**	**<0.01**	0.061	**<0.01**
0.9–1.0	-	0.227	**<0.01**	0.428	**<0.01**
1.0–1.1	-	**0.037**	**0.011**	0.471	**<0.01**
1.1–1.2	-	0.122	**<0.01**	0.397	**<0.01**
1.2–1.3	-	**0.035**	**<0.01**	0.397	**<0.01**
1.3–1.4	-	0.160	**0.021**	0.397	0.142
1.4–1.5	-	0.174	**0.012**	0.397	0.078

Note: The boldface values indicate significant differences ($p < 0.05$). The symbol "-" indicates that there was no root in this soil layer.

3.3. Spatio-Temporal Root Biomass Distribution with Lateral Distance from the Tree Stem

The root biomass distributions with lateral distance from the tree stem were analyzed (Figure 8). The root biomasses of the 5- and 7-year-old LBS, with approximately 93% and 96% of the total residing in the distance range of 0–0.6 m, first increased with distance and then decreased while exhibiting larger fluctuations. Therefore, this soil layer could be identified as the active layer of root growth. When the distance reached 0.6 m, the root biomass, accounting for approximately 7% and 4% of the total, showed a steady decrease. Thus, this area could be defined as a stable layer of root growth. For the 1-year-old LBS, the roots were distributed entirely within 0–0.6 m of the lateral distance. The root biomass in the downslope was slightly larger than that in the upslope area, and approximately 80% of the root materials were concentrated in the 0–0.4 m lateral distance. Contrary to expectations [17], the root biomass variation with the horizontal distance in certain growth periods did not show a decreasing trend at the beginning but decreased after the first increase in the 0.1–0.2 m soil layer. The variation of root biomass and the horizontal distance from the tree stem were fitted using regression analysis (Table 7). The results showed that the root biomass of 1-year-old LBS was best fitted by the 2rd-order polynomial equation, and that of the 5- and 7-year-old LBS by the 3rd-order polynomial equation.

Similar to the presentation of the results in Section 3.2, the growth rates of the total root biomass from 1 to 5 years and from 5 to 7 years in the horizontal direction are shown in Table 8.

Table 7. Regression equations and R^2 values for the variation of the root biomass with the distance from the stem in the horizontal direction.

		1-Year-Old	5-Year-Old	7-Year-Old
Active layer	Total	$B = -5.2l^2 - 14.70l + 9.11$ $R^2 = 0.8109$	$B = 2909.4l^3 - 2713.2l^2 + 443.9l + 116.4$ $R^2 = 0.8426$	$B = 4651.7l^3 - 4162l^2 + 573.2l + 232.5$ $R^2 = 0.895$
	Upslope	$B = -15.80l^2 - 3.94l + 6.44$ $R^2 = 0.7758$	$B = 368.7l^3 - 173.1l^2 - 147.8l + 76.4$ $R^2 = 0.8612$	$B = 2666.1l^3 - 2587.3l^2 + 535.9l + 64.2$ $R^2 = 0.8406$
	Downslope	$B = -35.7l^2 + 3.72l + 1.89$ $R^2 = 0.3868$	$B = 2540.6l^3 - 2540.1l^2 + 591.7l + 40$ $R^2 = 0.6702$	$B = 1985.6l^3 - 1574.7l^2 + 37.2l + 168.3$ $R^2 = 0.8358$
Stable layer	Total		$B = -38.6l^3 + 147.2l^2 - 187.1l + 79.3$ $R^2 = 0.8225$	$B = -218.8l^3 + 783.6l^2 - 930.7l + 368$ $R^2 = 0.6232$
	Upslope		$B = 3.9l^3 - 7.93l^2 + 0.4l + 4.3$ $R^2 = 0.2126$	$B = -48.1l^3 + 186.3l^2 - 240.7l + 103.9$ $R^2 = 0.5382$
	Downslope		$B = -42.5l^3 + 155.2l^2 - 187.5l + 75.1$ $R^2 = 0.5966$	$B = -170.7l^3 + 597.3l^2 - 689.9l + 264.2$ $R^2 = 0.6001$

Note: B = root biomass (g); l = lateral distance from the tree stem (m).

Table 8. The growth rate of the total root biomass in each soil layer in the horizontal direction.

Distance from the Stem (m)	Growth Rate	
	1–5 Years	5–7 Years
0.0–0.1	30.26	60.19
0.1–0.2	33.67	52.53
0.2–0.3	22.32	43.65
0.3–0.4	14.33	30.71
0.4–0.5	9.83	17.81
0.5–0.6	5.26	20.05
0.6–0.7	2.30	13.36
0.7–0.8	1.48	4.43
0.8–0.9	0.58	3.11
0.9–1.0	0.36	2.22
1.0–1.1	0.21	0.69
1.1–1.2	0.03	0.56
1.2–1.3	0.01	0.37
1.3–1.4	0.01	0.21

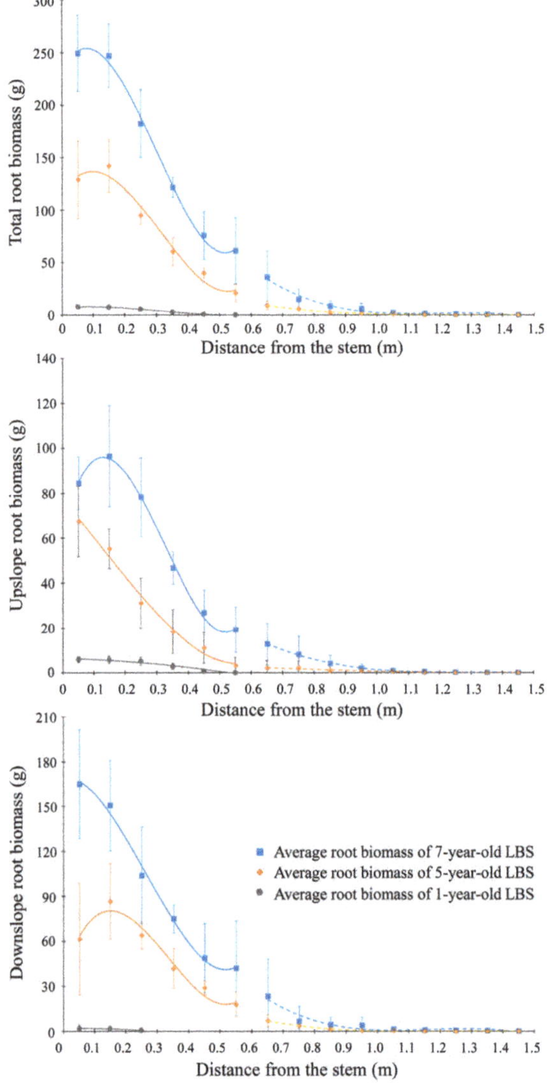

Figure 8. Average root biomass distribution with distance from the stem for 1-, 5- and 7-year-old LBS. The fit curves of root biomass with depth by regression analysis are also shown. The solid lines represent the fitted curves of the active layers. The dotted lines represent the fitted curves of the stable layers.

The growth rates of the root biomass in the first four years were smaller than that in the 5–7 years in the 0–0.7 m soil layer from the stem. In the 1–1.4 m layer, the growth rates over 5 to 7 years were still greater than that over 1 to 5 years, but the growth trend was gradual.

The paired *t*-test was performed to test the effect of slope positions (upslope and downslope) on root biomass for each age and the ANOVA was performed to test the effects of age on root biomass in different slope position (Table 9). Considering the effect of slope position on the root biomass, in the

1st year, the slope position influenced the root biomass at the lateral distances of 0–0.4 m, and in the 5th year, it impacted the root biomass at the lateral distances of 0.2–0.5 m. In the 7th year, the slope position influenced the root biomass at the lateral distances of 0–0.4 m. Considering the effect of age on the root biomass, in the upslope area, the age influenced the root biomass at the lateral distances of 0–0.7 m and 0.8–0.9 m, and in the downslope area, it impacted the root biomass at the lateral distances of 0–0.9 m.

Table 9. Results of paired *t*-test and ANOVA for evaluating the effects of slope position and age on the root biomass across different distances from the stem.

Distance from the Stem (m)	Paired *t*-Test			ANOVA	
	Slope Position			Age	
	1-Year-Old	5-Year-Old	7-Year-Old	Upslope	Downslope
0.0–0.1	<0.01	0.376	0.012	<0.01	<0.01
0.1–0.2	<0.01	0.132	<0.01	<0.01	<0.01
0.2–0.3	<0.01	0.018	0.043	<0.01	<0.01
0.3–0.4	0.010	0.016	0.038	<0.01	<0.01
0.4–0.5	0.052	0.017	0.092	<0.01	<0.01
0.5–0.6	0.374	0.065	0.051	<0.01	<0.01
0.6–0.7	-	0.219	0.054	<0.01	<0.01
0.7–0.8	-	0.584	0.671	0.062	<0.01
0.8–0.9	-	0.755	0.857	0.049	<0.01
0.9–1.0	-	0.693	0.331	0.108	0.070
1.0–1.1	-	0.898	0.689	0.231	0.230
1.1–1.2	-	0.374	0.565	0.292	0.111
1.2–1.3	-	0.374	0.404	0.175	0.117
1.3–1.4	-	0.374	0.293	0.619	0.257
1.4–1.5	-	-	-	-	-

Note: The boldface values indicate significant differences ($p < 0.05$). The symbol "-" indicates that there was no root in this soil layer.

3.4. Spatio-Temporal Distribution of the Soil Shear Strength with LBS Roots

Figure 9 describes the shear strength distribution of the soil with LBS roots. Because the water level exceeded the riverbank height, it was impossible to determine the shear strength of certain deeper soil layers; thus, testing was conducted only up to 0.6 m below the surface. An unambiguous stratification of shear strength caused by the unique distribution characteristics of the riparian soil structure occurred 0.4 m underground. The maximum shear strength of the root-soil composite was approximately 159.61 kPa in the upper layer (0–0.4 m below the surface) and 80.94 kPa in the lower layer (0.4–0.6 m below the surface) (Table 10). In the upper and lower layers, the shear strength of pure soil was 67.61 kPa and 36.78 kPa, respectively (Table 10). The reinforcement effect of the LBS roots on the soil was mainly concentrated in the horizontal layer 0.6 m from the stem. When the distance was more than 0.6 m, the strengthening effect was not easily detectable. In the upslope zone, the distribution range of the shear strength in the horizontal direction markedly decreased with an increase in soil depth. By contrast, in the downslope zone, the distribution range of the shear strength in the horizontal direction initially increased and then markedly decreased with increasing soil depth. In general, the range of the shear strength increased with the tree age. For a given unit, however, the shear strength did not necessarily increase with age; for example, in the U-U region, the horizontal 0.1–0.2 m layer, and the vertical 0.1–0.2 m layer, the soil shear strength of the 7-year-old LBS was less than that of the 5-year-old LBS. The shear strength of the 1-year-old LBS in the 0–0.1 m soil layer varied between 20 kPa and 31 kPa, which was obviously less than the shear strength of the pure soil. In order to demonstrate that the use of LBS has an advantage over other plant species in improving soil shear strength, the average soil shear strength enhancement values of all root-soil composite units, 24.05 kPa and 11.4 kPa in the upper and lower layers, respectively, were calculated (Table 10).

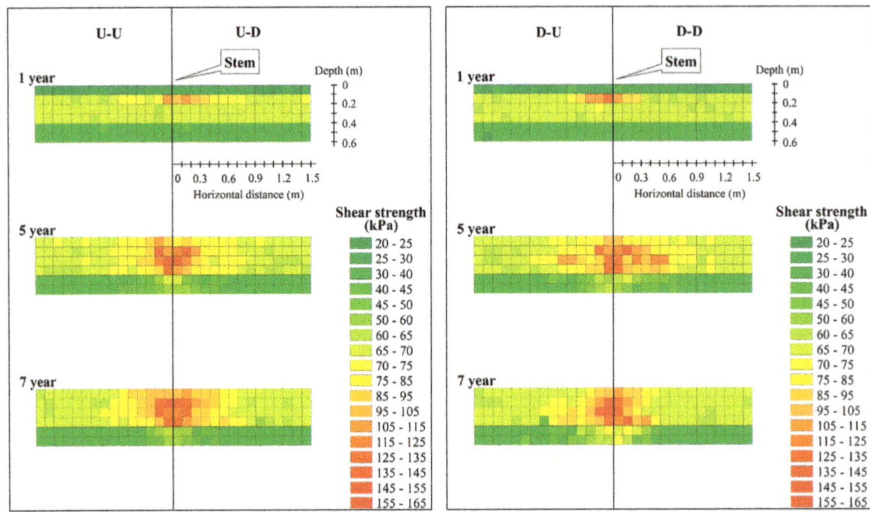

Figure 9. Spatio-temporal distribution of the soil shear strength. U-U, D-U, U-D and D-D represent the upslope-upstream face, the downslope-upstream face, the upslope-downstream face and the downslope-downstream face of the riverbank, respectively.

Table 10. The shear strengths of the pure soil and the root-soil composite.

Depth (m)	SSP (kPa)	MSSC (kPa)	ESS (kPa)
0–0.4	67.61 ± 4.41	159.61	24.05 ± 22.27
0.4–0.6	36.78 ± 2.33	80.94	11.74 ± 10.53

Note: SSP and ESS data are represented as means and standard deviations (mean ± SD). SSP = shear strength of the pure soil; MSSC = maximum shear strength of the root-soil composite; ESS = average soil shear strength enhancement values of all root-soil composite units.

The regression relationships between the shear strength enhancement and the root materials for the 5- and 7-year-old LBS are shown in Figure 10. The results show that both the number of roots and the root cross-sectional area per unit volume were highly significantly correlated with the enhancement values of the soil shear strength ($p < 0.01$).

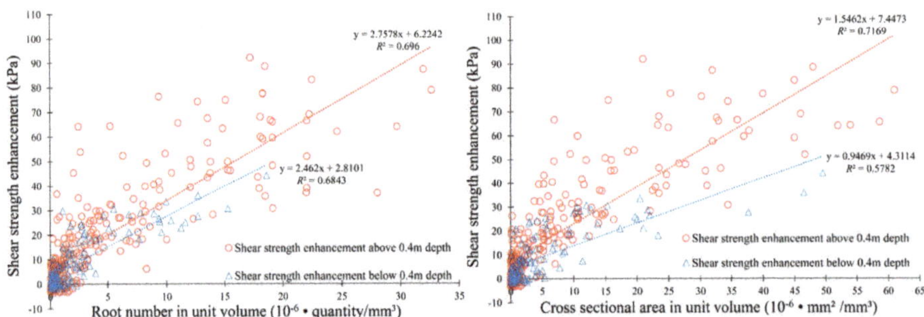

Figure 10. Best-fit curves of the shear strength enhancement with the number of roots and the root cross-sectional area per unit volume.

4. Discussion

4.1. Spatio-Temporal Dynamic Distribution Characteristics of the LBS Root System

Among all tree ages, the MVD and MLD corresponded to a relatively small proportion of the tree height. The relationships were 0.1 H < MVD < 0.15 H and 0.09 H < MLD < 0.22 H. These ratios are obviously less than the approximate relationships of H < Dr < 2 H reported by Greenway [46], Wu [47], and Docker and Hubble [17], where Dr = the root diameter = 2MLD. These findings show that the extents of the LBS root system in the vertical and lateral slope directions were small relative to the shoot growth aboveground. Furthermore, the MLDs of the 5- and 7-year-old LBS were 1 m and 1.27 m, respectively, corresponding to only 1/8 and 1/10 of the MLD of 5- and 7-year-old Veronese poplars, respectively [19]. Therefore, the LBS had a small root system in general. Although the lateral root extension of the LBS was small, the living shoots that were used to construct the LBS limited its growth position, and the interspaces between the living shoots were also small to ensure that the LBS would form a root grid in the interspace of shoots and soil. According to the spatio-temporal dynamic distribution of the root biomass, the roots were distributed primarily in a cylindrical soil pattern with an approximate radius of 0.4 m and a thickness of 0.5 m. This characteristic arose because the root system was mainly distributed in the shallow soil layer, as reported by Mclvor et al. [19]. For the 1-year-old LBS, 95% of the total root biomass was distributed in the first 0.1 m of depth; thus, this shallow root biomass was obviously larger than the root biomass in the 0.1–0.3 m layer underground. This result indicates that in the early planting stage, the roots of the LBS were concentrated in the 0–0.1 m layer and developed primarily in the horizontal direction. Thus, the ratio between the MLD and the tree height for the 1-year-old trees is greater than that for the LBS at other ages, as shown in Table 3. The total root biomasses at the depth of 0.1–0.2 m for the 5- and 7-year-old LBS and at 0–0.1 m for the 1-year-old LBS were the largest of any soil layer (Figure 7). These findings indicate that the roots of the LBS were able to develop even in the thinner soil layer on the lower part of the slope. For original bank slope structures such as a rocky bank slope, root development would be difficult. However, an artificial soil covering could create a growing environment for the roots of LBS.

Symmetry of the 1-year-old LBS root system was found between the upstream and downstream faces. This finding is consistent with the symmetries of other plant species' root systems during the seedling period [48,49]. However, with increasing time, the root system symmetry was no longer evident. Root growth reacts to mechanical impedances and soil nutrients [50]. During extraction, gravels were found in the deep soil, and the random distribution of gravel and soil nutrients was probably the cause of the disappearance of symmetry. An upslope growth characteristic of LBS was also found in this study; the roots in the upslope zone were closer to the surface than the roots in the downslope zone, in contrast to the growth trend reported by Mclvor et al. [19]. In fact, however, the amount of root biomass in the downslope zone was obviously higher than that in the upslope zone; thus, it should be noted that the upslope growth explains only the direction of root growth. Our results also exhibited that, except at depths below 1.3 m, the growth rate in the last two years was greater than that in the first four years. From this discrepancy in root biomass growth rate, the root development of LBS shows a non-uniform trend in time scale over the first 7 years.

Both slope position and age had an effect on the LBS root biomass (Tables 6 and 9). In the vertical direction, the effect of the slope position on the root biomass at different ages were different, and with aging, the influence range gradually increased. In the 1st year, the effect of slope position on root biomass was significant at depths of 0–0.2 m; in the 5th year, the effect of slope position on root biomass was mainly concentrated in deeper soil layers; in the 7th year, slope position almost influenced the root biomass in all soil layers. In addition, the influence range of age on the root biomass in the downslope area was deeper than the upslope area. This difference may be due to the upslope growth characteristic of the root of LBS. In the horizontal direction, the influence of the slope position on the root biomass at different ages was mainly concentrated in the range of 0–0.5 m. This concentration may be due to the fact that the LBS roots were mainly distributed in this soil layer at these three ages. Notably, for 5-year-old LBS, no effect of slope position on root biomass was observed in the 0–0.1 m and 0.1–0.2 m soil layers. This may indicate that there were no remarkable differences in the increments of root biomass in upslope and downslope areas in these two soil layers from 1 to 5 years. On the other hand, the effects of age on the root biomass in the upslope and downslope areas were found in the 0–0.9 m soil layers. This result may indicate that although the root biomass increases with age, in the horizontal direction, the LBS roots frequently grew only in the soil layers at 0–0.9 m.

4.2. Soil Shear Strength Enhancement by the LBS Root System

From the distribution of the soil shear strength (Figure 9), it was found that the shear strength enhancement was distributed in the soil with a conical shape, and with an increase in the lateral distance, the enhancement became no longer obvious; this result is consistent with the conclusion offered by Fan and Lai [14] and indicates that plant roots provide a cone-shaped soil shear strength enhancement zone in the soil below that is centered on the stem [17,44,51]. In the first five years, the plant age has an obvious influence on the range of the root-induced enhancement of the soil shear strength. With aging, the soil shear strength gradually expands in both the horizontal and vertical directions. However, the effect of age on the soil shear strength expansion was not obvious between the 5th and 7th years. Furthermore, in some soil layers closer to the stem, the shear strength did not increase but rather decreased. This result may demonstrate that the contribution of plants to riverbank stability cannot be determined solely from the growth period of the plants but mainly from the root distribution, which has also been demonstrated by previous studies [17]. In the 1-year growth stage after construction, the backfill was loose, and a large number of adventitious roots were growing in the first 0.1 m depth layer. Most of the roots were pulled out of the soil rather than shearing during the test. This effect caused the soil shear strength provided by the 1-year-old LBS root system in the 0–0.1 m depth layer to be less than that of other units.

The significant linear relationship between the shear strength enhancement by roots in the diameter class from 0 to 5 mm and the number of roots or the root cross-sectional area was similar to that reported in earlier studies [43,52]. These results demonstrate that fine roots mainly increase soil shear strength, while thicker roots anchor the soil in the vertical direction [21,44]. The number of roots in the 0–5 mm diameter class decreased with vegetation development in some soil layers close to the stem. This effect may explain why the shear strength did not increase but decreased.

The maximum enhancement values in the upper and lower layers were approximately 1.3 and 1.2 times with respect to the pure soil (calculated from Table 10). This finding indicates that the LBS roots had an obvious strengthening effect on the soil. The average soil shear strength enhancement values of LBS (Table 10) were 24.05 kPa and 11.4 kPa in the upper and lower soil layers, respectively. Compared with the living brush mattress constructed using *Salix babylonica* L. [5], the enhancement value of LBS is larger. Liu et al. used the pull-out test to compare the capability of *Populus canadensis* Moench, *Amorpha fruticosa* L., and LBS, which are dominant plant species suitable for soil bioengineering technology, to reinforce soil. The capabilities of *Populus canadensis* Moench and LBS were similar [37]. However, the vertical distribution of the *Populus canadensis* Moench root system [39] was smaller than that of LBS. This caused enhanced range of soil shear strength for *Populus canadensis* Moench that was less than that of LBS. Although the capability of individual *Amorpha fruticosa* L. was the largest [37], the survival rate of planting is much lower than that of LBS using soil bioengineering technology [38]. Therefore, it is not guaranteed that *Amorpha fruticosa* L. can form a dense population on the slope to improve the stability of the riverbank. Moreover, Guo [53] measured the tensile strength of the LBS roots using tensile tests. Compared with *Picea abies* (L.) H. Karst. and *Ostrya carpinifolia* Scop. [54], the roots of LBS had greater tensile strength. Based on the Wu's model theory [29,47], this finding indirectly indicated that the contribution of LBS roots to enhance soil shear strength is greater than the contributions of *Picea abies* (L.) H. Karst. and *Ostrya carpinifolia* Scop. Based on these results, we speculate that LBS has a higher efficiency for soil protection, which further confirms the advantages of LBS for riverbank restoration engineering.

In this study, we investigated the root distribution characteristics of LBS and its bank protection capacity. However, considering the results may be affected by site-specific factors, such as river morphology, river flow, riverbank structure, etc., further investigations should be conducted to test the generalization of this measure. Despite this limitation, our results suggest that LBS plays an important role in riverbank protection and represents an efficient and easily applicable engineering measure for riverbank protection.

5. Conclusions

In this study, we revealed the root distribution characteristics and bank protection capacity of LBS by investigating the root biomass and soil shear strength. Our results show that the roots of LBS exhibited characteristics of symmetry and upslope growth and were mainly distributed in a cylindrical region of the soil. Moreover, the LBS provided a conically shaped shear strength enhancement zone in the soil by means of the root system. The amount of fine roots (the number and cross sectional areas here) per unit volume significantly affected the soil shear strength enhancement value. On the basis of these results, LBS protected the riverbank via improving the soil shear strength by fine roots. Compared with other plant species, LBS showed advantages in improving the soil shear strength. Although this is a case study, our results indicate that LBS is an efficient and feasible engineering measure in the field of riverbank protection.

Author Contributions: D.Z., Y.L., J.C., L.M. and H.Z. conceived and designed the experiments; D.Z., X.M. and Y.S. performed the experiments; D.Z. analyzed the data; D.Z. wrote the paper.

Funding: This research was funded by [the Beijing Municipal Science and Technology Commission] grant number [Z151100001115001] and [the Chinese Natural Science Foundation Program] grant number [41501297].

Acknowledgments: The authors would like to express their appreciation to the students who contributed their time, labor and effort to the data collection effort. Thanks are given particularly to Peiyan Chen, Mengjun Xue, Congyu Xiao, Chunyang Chen, Lu Yuan and Yi Cui. The assistance provided by Zhen Liu while analyzing the data is also appreciated.

References

1. Cheng, X.; Chen, L.; Sun, R.; Kong, P. Land use changes and socio-economic development strongly deteriorate river ecosystem health in one of the largest basins in China. *Sci. Total Environ.* **2018**, *616–617*, 376–385. [CrossRef] [PubMed]
2. Samadi, A.; Amiri-Tokaldany, E.; Darby, S.E. Identifying the effects of parameter uncertainty on the reliability of riverbank stability modelling. *Geomorphology* **2009**, *106*, 219–230. [CrossRef]
3. Darby, S.E.; Thorne, C.R. Prediction of tension crack location and riverbank erosion hazards along destabilized channels. *Earth Surf. Process. Landf.* **1994**, *19*, 233–245. [CrossRef]
4. Yao, S.; Yue, H.; Li, L. Analysis on Current Situation and Development Trend of Ecological Revetment Works in Middle and Lower Reaches of Yangtze River. *Procedia Eng.* **2012**, *28*, 307–313. [CrossRef]
5. Li, X.; Zhang, L.; Zhang, Z. Soil bioengineering and the ecological restoration of riverbanks at the airport Town, Shanghai, China. *Ecol. Eng.* **2006**, *26*, 304–314. [CrossRef]
6. Li, M.H.; Eddleman, K.E. Biotechnical engineering as an alternative to traditional engineering methods: A biotechnical streambank stabilization design approach. *Landsc. Urban Plan.* **2002**, *60*, 225–242. [CrossRef]
7. Wang, P.; Chen, G.Q. Contaminant transport in wetland flows with bulk degradation and bed absorption. *J. Hydrol.* **2017**, *552*, 674–683. [CrossRef]
8. Bischetti, G.B.; Chiaradia, E.A.; D'Agostino, V.; Simonato, T. Quantifying the effect of brush layering on slope stability. *Ecol. Eng.* **2010**, *36*, 258–264. [CrossRef]
9. Dhital, Y.P.; Tang, Q. Soil bioengineering application for flood hazard minimization in the foothills of siwaliks, Nepal. *Ecol. Eng.* **2015**, *74*, 458–462. [CrossRef]
10. Fernandes, J.P.; Guiomar, N. Simulating the stabilization effect of soil bioengineering interventions in Mediterranean environments using limit equilibrium stability models and combinations of plant species. *Ecol. Eng.* **2016**, *88*, 122–142. [CrossRef]

11. Rauch, H.P.; Sutili, F.; Hörbinger, S. Installation of a riparian forest by means of soil bio engineering techniques—Monitoring results from a river restoration work in southern Brazil. *Open J. For.* **2014**, *4*, 161–169. [CrossRef]

12. Capilleri, P.P.; Motta, E.; Raciti, E. Experimental study on native plant root tensile strength for slope stabilization. *Procedia Eng.* **2016**, *158*, 116–121. [CrossRef]

13. Li, Y.; Wang, Y.; Ma, C.; Zhang, H.; Wang, Y.; Song, S.; Zhu, J. Influence of the spatial layout of plant roots on slope stability. *Ecol. Eng.* **2016**, *91*, 477–486. [CrossRef]

14. Fan, C.C.; Lai, Y.F. Influence of the spatial layout of vegetation on the stability of slopes. *Plant Soil* **2014**, *377*, 83–95. [CrossRef]

15. Stokes, A.; Atger, C.; Bengough, A.G.; Fourcaud, T.; Sidle, R.C. Desirable plant root traits for protecting natural and engineered slopes against landslides. *Plant Soil* **2009**, *324*, 1–30. [CrossRef]

16. Giadrossich, F.; Cohen, D.; Schwarz, M.; Seddaiu, G.; Contran, N.; Lubino, M.; Valdes-Rodriguez, O.A.; Niedda, M. Modeling bio-engineering traits of *Jatropha curcas* L. *Ecol. Eng.* **2016**, *89*, 40–48. [CrossRef]

17. Docker, B.B.; Hubble, T.C.T. Modelling the distribution of enhanced soil shear strength beneath riparian trees of south-eastern Australia. *Ecol. Eng.* **2009**, *35*, 921–934. [CrossRef]

18. Fan, C.C.; Tsai, M.H. Spatial distribution of plant root forces in root-permeated soils subject to shear. *Soil Tillage Res.* **2016**, *156*, 1–15. [CrossRef]

19. McIvor, I.R.; Douglas, G.B.; Hurst, S.E.; Hussain, Z.; Foote, A.G. Structural root growth of young Veronese poplars on erodible slopes in the southern north island, New Zealand. *Agrofor. Syst.* **2008**, *72*, 75–86. [CrossRef]

20. Di, I.A.; Lasserre, B.; Scippa, G.S.; Chiatante, D. Root system architecture of Quercus pubescens trees growing on different sloping conditions. *Ann. Bot.* **2005**, *95*, 351–361. [CrossRef]

21. Nicoll, B.C.; Berthier, S.; Achim, A.; Gouskou, K.; Danjon, F.; Beek, L.P.H.V. The architecture of Picea sitchensis, structural root systems on horizontal and sloping terrain. *Trees* **2006**, *20*, 701–712. [CrossRef]

22. Yang, X.; Blagodatsky, S.; Liu, F.; Beckschäfer, P.; Xu, J.; Cadisch, G. Rubber tree allometry, biomass partitioning and carbon stocks in mountainous landscapes of sub-tropical China. *For. Ecol. Manag.* **2017**, *404*, 84–99. [CrossRef]

23. Luo, T.; Luo, J.; Pan, Y. Leaf traits and associated ecosystem characteristics across subtropical and timberline forests in the Gongga Mountains, Eastern Tibetan Plateau. *Oecologia* **2005**, *142*, 261–273. [CrossRef] [PubMed]

24. Czernin, A.; Phillips, C. Below-ground morphology of Cordyline australis (New Zealand cabbage tree) and its suitability for river bank stabilization. *N. Z. J. Bot.* **2005**, *43*, 851–864. [CrossRef]

25. Stokes, A.; Douglas, G.B.; Fourcaud, T.; Giadrossich, F.; Gillies, C.; Hubble, T.; Walker, L.R. Ecological mitigation of hillslope instability: Ten key issues facing researchers and practitioners. *Plant Soil* **2014**, *377*, 1–23. [CrossRef]

26. Schwarz, M.; Preti, F.; Giadrossich, F.; Lehmann, P.; Or, D. Quantifying the role of vegetation in slope stability: A casestudy in Tuscany (Italy). *Ecol. Eng.* **2010**, *36*, 285–291. [CrossRef]

27. Schwarz, M.; Lehmann, P.; Or, D. Quantifying lateral root reinforcement in steep slopes-From a bundle of roots to tree stand. *Earth Surf. Process. Landf.* **2010**, *35*, 354–367. [CrossRef]

28. Cohen, D.; Schwarz, M.; Or, D. An analytical fiber bundle model for pullout mechanics of root bundles. *J. Geophys. Res. Earth Surf.* **2011**, *116*, F03010. [CrossRef]

29. Wu, T.H.; Iii, M.K.; Swanston, D.N. Strength of tree roots and landslides on Prince of Wales Island, Alaska. *Can. Geotech. J.* **1979**, *16*, 19–33. [CrossRef]

30. Wang, P.; Chen, G.Q. Solute dispersion in open channel flow with bed absorption. *J. Hydrol.* **2016**, *543*, 208–217. [CrossRef]

31. Zhang, C.B.; Chen, L.H.; Liu, Y.P.; Ji, X.D.; Liu, X.P. Triaxial compression test of soil-root composites to evaluate influence of roots on soil shear strength. *Ecol. Eng.* **2010**, *36*, 19–26. [CrossRef]

32. Giadrossich, F.; Schwarz, M.; Cohen, D.; Preti, F.; Or, D. Mechanical interactions between neighbouring roots during pullout tests. *Plant Soil* **2013**, *367*, 391–406. [CrossRef]

33. Wu, T.H. *Investigation of landslides on Prince of Wales Island, Alska*; Geotechnical Engineering Report 5; Ohio State University: Columbus, OH, USA, 1976.

34. Waldron, L.J. The shear resistance of root-permeated homogeneous and stratified soil. *Soil Sci. Soc. Am. J.* **1977**, *41*, 843–849. [CrossRef]

35. Pollen, N.; Simon, A. Estimating the mechanical effects of riparian vegetation on stream bank stability using a fiber bundle model. *Water Resour. Res.* **2005**, *41*, 226–244. [CrossRef]

36. Schwarz, M.; Cohen, D.; Or, D. Pullout tests of root analogs and natural root bundles in soil: Experiments and modeling. *J. Geophys. Res. Earth Surf.* **2011**, *116*, F02007(1-14). [CrossRef]

37. Liu, Y.; Rauch, H.P.; Zhang, J.; Yang, X.; Gao, J.R. Development and soil reinforcement characteristics of five native species planted as cuttings in local area of Beijing. *Ecol. Eng.* **2014**, *71*, 190–196. [CrossRef]

38. Liu, Y. Effects of Soil bioengineering Techniques Applied in Riverbank Ecological Restoration. Ph.D. Thesis, Beijing Forestry University, Beijing, China, 2011.

39. Qian, B.T. Research on Plant Materials Selection and Construction Methods of Soil Bio-Engineering. Master's Thesis, Beijing Forestry University, Beijing, China, 2013.

40. Studer, R.; Zeh, H.; Givoanni, D.C. *SOIL BIOENGINEERING—Construction Type Manual*, 5th ed.; vdf Hochschulverlag AG der ETH Zürich: Zürich, Switzerland, 2014; pp. 218–250, ISBN 978-3-7281-3642-8.

41. Böhm, W. *Methods of Studying Root Systems*, 12th ed.; Springer: Berlin/Heidelberg, Germany, 1979; pp. 125–138, ISBN 978-3-642-67284-2.

42. Fan, C.C.; Chen, Y.W. The effect of root architecture on the shearing resistance of root-permeated soils. *Ecol. Eng.* **2010**, *36*, 813–826. [CrossRef]

43. Ziemer, R.R. Roots and the stability of forested slopes. In *Erosion and Sediment Transport in Pacific Rim Steeplands*, 3rd ed.; Davies, T.R.H., Pearce, A.J., Eds.; International Association of Hydrological Sciences: Christchurch, New Zealand, 1981; Publication 132, pp. 343–361.

44. Abernethy, B.; Rutherfurd, I.D. The distribution and strength of riparian tree roots in relation to riverbank reinforcement. *Hydrol. Process.* **2001**, *15*, 63–79. [CrossRef]

45. Kajimoto, T.; Matsuura, Y.; Osawa, A.; Abaimov, A.P.; Zyryanova, O.A.; Isaev, A.P.; Yefremov, D.P.; Mori, S.; Koike, T. Size-mass allometry and biomass allocation of two larch species growing on the continuous permafrost region in Siberia. *For. Ecol. Manag.* **2006**, *222*, 314–325. [CrossRef]

46. Greenway, D.R. Vegetation and slope stability. In *Slope Stability Geotechnical Engineering and Geomorphology*, 3rd ed.; Anderson, M.G., Richards, K.S., Eds.; John Wiley & Sons: Chichester, UK, 1987; pp. 187–230.

47. Wu, T.H. Slope stabilization. In *Slope Stabilization and Erosion Control: A Bioengineering Approach*, 7th ed.; Morgan, R.P.C., Rickson, R.J., Eds.; E & FN Spon: London, UK, 1995; pp. 221–248, ISBN 0-419-15630-5.

48. Reubens, B.; Achten, W.M.J.; Maes, W.H.; Danjon, F.; Aerts, R.; Poesen, J.; Muys, B. More than biofuel? *Jatropha curcas* root system symmetry and potential for soil erosion control. *J. Arid Environ.* **2011**, *75*, 201–205. [CrossRef]

49. Amichev, B.Y.; Bailey, B.E.; Rees, K.C.J.V. White spruce (*Picea glauca*) structural root system development and symmetry influenced by disc trenching site preparation. *For. Ecol. Manag.* **2014**, *326*, 1–8. [CrossRef]

50. Pasquale, N.; Perona, P.; Francis, R.; Burlando, P. Effects of streamflow variability on the vertical root density distribution of willow cutting experiments. *Ecol. Eng.* **2012**, *40*, 167–172. [CrossRef]

51. Shields, F.D.; Gray, D.H. Effects of woody vegetation on sandy levee integrity. *J. Am. Water Resour. Assoc.* **1992**, *28*, 917–931. [CrossRef]

52. Waldron, L.J.; Dakessian, S. Effect of grass, legume, and tree roots on soil shearing resistance. *Soil Sci. Soc. Am. J.* **1982**, *46*, 894–899. [CrossRef]

53. Guo, K.L. *Salix* × *aureo-pendula* Root System Distribution and Tensile Mechanical Properties in Soil Bioengineering Revetment. Master's Thesis, Beijing Forestry University, Beijing, China, 2016.

54. Vergani, C.; Chiaradia, E.A.; Bischetti, G.B. Variability in the tensile resistance of roots in alpine forest tree species. *Ecol. Eng.* **2012**, *46*, 43–56. [CrossRef]

Article

Quantitative Assessment of Surface Runoff and Base Flow Response to Multiple Factors in Pengchongjian Small Watershed

Lei Ouyang [1,2], Shiyu Liu [1,2,*], Jingping Ye [1,2], Zheng Liu [1,2], Fei Sheng [1,2], Rong Wang [1] and Zhihong Lu [1]

[1]   School of Land Resources and Environment, Jiangxi Agricultural University, Nanchang 330045, China; leioy1992@gmail.com (L.O.); jingpye@gmail.com (J.Y.); lzheng048@gmail.com (Z.L.); sf961001@gmail.com (F.S.); wrong7411@163.com (R.W.); luzhihong1@163.com (Z.L.)

[2]   Key Laboratory of Poyang Lake Watershed Agricultural Resources and Ecology of Jiangxi Province, Nanchang 330045, China

*   Correspondence: sliuy2@jxau.edu.cn; Tel.: +86-158-7061-1238

Received: 22 July 2018; Accepted: 6 September 2018; Published: 10 September 2018

Abstract: Quantifying the impacts of multiple factors on surface runoff and base flow is essential for understanding the mechanism of hydrological response and local water resources management as well as preventing floods and droughts. Despite previous studies on quantitative impacts of multiple factors on runoff, there is still a need for assessment of the influence of these factors on both surface runoff and base flow in different temporal scales at the watershed level. The main objective of this paper was to quantify the influence of precipitation variation, evapotranspiration (ET) and vegetation restoration on surface runoff and base flow using empirical statistics and slope change ratio of cumulative quantities (SCRCQ) methods in Pengchongjian small watershed ($116°25'48''$–$116°27'7''$ E, $29°31'44''$–$29°32'56''$ N, 2.9 km^2), China. The results indicated that, the contribution rates of precipitation variation, ET and vegetation restoration to surface runoff were 42.1%, 28.5%, 29.4% in spring; 45.0%, 37.1%, 17.9% in summer; 30.1%, 29.4%, 40.5% in autumn; 16.7%, 35.1%, 48.2% in winter; and 35.0%, 38.7%, 26.3% in annual scale, respectively. For base flow they were 33.1%, 41.9%, 25.0% in spring; 39.3%, 51.9%, 8.8% in summer; 40.2%, 38.2%, 21.6% in autumn; 24.3%, 39.4%, 36.3% in winter; and 24.4%, 47.9%, 27.7% in annual scale, respectively. Overall, climatic factors, including precipitation and ET change, affect surface runoff generation the most, while ET affects the dynamic change of annual base flowthe most. This study highlights the importance of optimizing forest management to protect the water resource.

Keywords: precipitation variation; vegetation restoration; evapotranspiration (ET); surface runoff; base flow; empirical statistics; slope change ratio of cumulative quantities (SCRCQ)

---

1. Introduction

At present, runoff is generally divided into two parts: surface runoff and base flow [1–3]. During a rainfall process in a basin, the flow duration curve formed at the outlet of a watershed consists of surface runoff and base flow (Figure 1) [4]. There are many quantitative studies on the effects of precipitation variation and vegetation restoration on surface runoff [5–11]. Distinct hydrological effects based on different time scales were reported [12]. For instance, Chaplot investigated water and soil resources response to changes in precipitation and air temperature in Iowa (USA), pointing to a considerable effect of precipitation changes with a 170% runoff increase as precipitation increases by 40% [11], while Zhang and Nearing predicted that decreasing precipitation by 7–14% would result in greater runoff (67–82%) [13]. The above two results showed the opposite effects on runoff caused by precipitation, which is mainly due to different topography, forest change, soil condition, temperature as well as other factors like NO_3–N loads [9]. Zhang conducted research over Poyang Lake basin and found that contributions of climate change and human activities to streamflow changes were 73.2% and 26.8%, respectively. However, human-induced and climate-induced influences on streamflow were different in different river basins [14]. Similarly, Gu demonstrated that human activities accounted for a low portion (5.5%) in the runoff inputs in Poyang Lake [15]. Interestingly, however, the decrease of streamflow in the Fuhe River (sub-catchment of Poyang Lake) in 2000s was mainly affected by human activities (156.9%), rather than climate change [6]. More recently, Rogger et al. reviewed research gaps in the field of land-use change impacts on floods at the catchment scale [16]. They proposed that land-use change impacts on floods are poorly understood at the catchment scale at present, then suggested possible ways forward for addressing these gaps. Despite numerous works, most research focused on calculating surface runoff in annual scale [8,17,18], but little attention was paid to the seasonal scale. However, a study on the effects of precipitation and evapotranspiration (ET) variation, vegetation restoration on the seasonal surface runoff is of great significance for rationally allocating water resources yearly, especially for watersheds with seasonal droughts. Thus, it is still necessary to further study the influences of precipitation and ET variation, vegetation restoration on seasonal surface runoff.

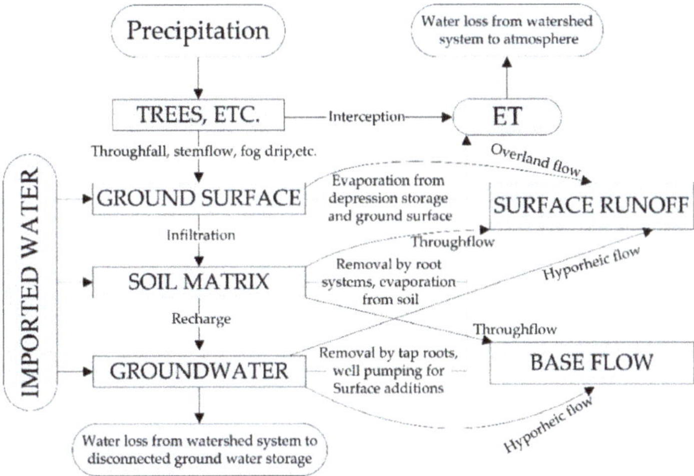

Figure 1. Process of water inputs, storage, and losses in watershed.

Base flow is a relatively stable part of streamflow. Base flow studies are not only helpful for reasonable regulation of water quantity and high-efficiency utilization of water resources, but also beneficial for hydrological analysis, evaluation and the management of water resources [19,20]. At present, studies on base flow mainly focus on regression models, improvement of base flow separation methods and influencing factors. The studies of base flow in China began in the 1960s, and mainly dealt with qualitative analysis of base flow evolution principles and its driving factors, base flow separation and estimation. Nonetheless, quantitative studies on the contribution rates of different factors to base flow are rarely reported.

Consequently, with the support of the National Natural Science Foundation of China, taking Pengchongjian small watershed with rich precipitation as a study area, we quantitatively assessed the impacts of multiple factors on both surface runoff and base flow (Figure 2). This study could help to understand how precipitation variation, ET and vegetation restoration affect runoff, and provide scientific foundations for the sustainable utilization of water resources and the evaluation of hydrological benefits for natural forest protection and afforestation projects.

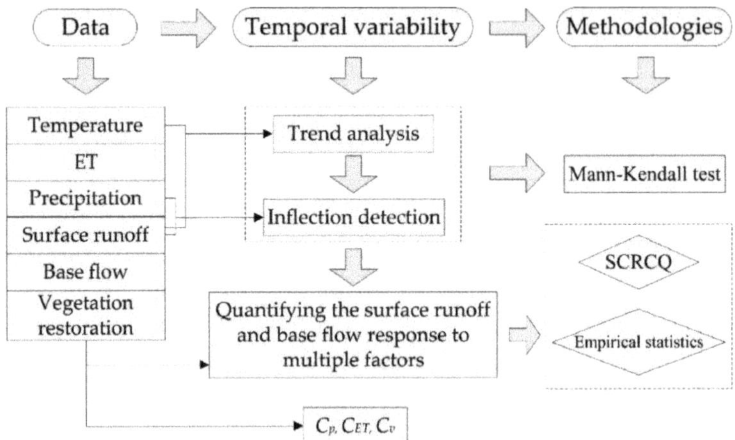

Figure 2. Graphical structure of this paper.

2. Study Area and Methods

2.1. Study Area

Pengchongjian small watershed is located between 116°25′48″–116°27′7″ E and 29°31′44″–29°32′56″ N in Duchang County, Jiujiang City, Jiangxi Province, covering a catchment area of 2.90 km² (Figure 3). It is high in the north-west and low in the south-east, whose altitude ranges from 80 to 560 meters, belonging to subtropical humid monsoon climate zone with average annual precipitation of 1560 mm. Major exposure strata here are epimetamorphic rock, granite and limestone. Moreover, the watershed is rich in vegetation types, with stand types of 70% Cunninghamia lanceolata (Lamb.) Hook., 20% *Quercus acutissima* Carruth. and 2% *Phyllostachys heterocycla* (Carr.) Mitford cv. Pubescens. The watershed is completely closed as there are no inhabitants, water conservancy facilities or soil and water conservation projects. Yet the China fir forest suffered from deforestation in early 1980s, after which the secondary forest has been recovering, which provides an ideal study area for this study. According to the investigating data of forest resources by Wushan forest farm in Duchang County, forest coverage increased from 80% in 1985 to 98% in 2016, and forest stocking volume rose from 1.2×10^4 m³ in 1985 to 2.5×10^4 m³ in 2016. In 1981, a hydrological station was built up at the outlet of the watershed by Jiangxi Province Bureau, and the precipitation and surface runoff have been continuously monitored till now.

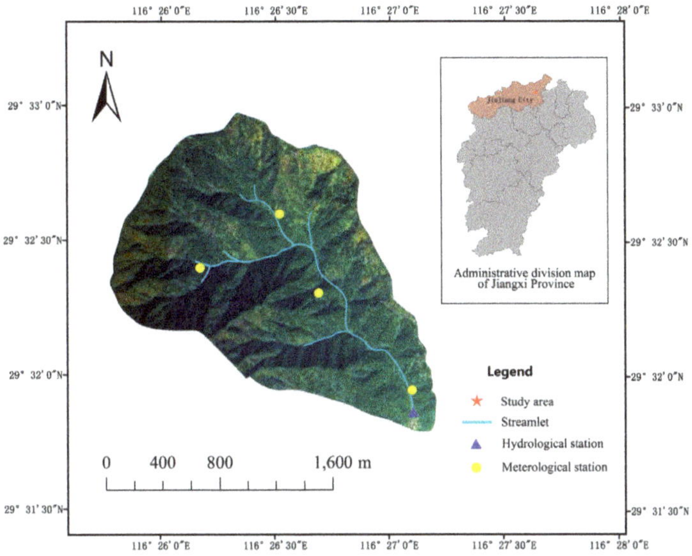

Figure 3. Location map and observation sites in Pengchongjian small watershed.

2.2. Methods

2.2.1. Data

Data used in this paper include daily precipitation, surface runoff and daily mean temperature at Pengchongjian hydrological station from 1983 to 2014, which was supplied by Jiangxi Province Hydrology Bureau. Base flow were separated from daily surface runoff datasets using the Kalinlin improving method [21]. Seasonal and annual precipitation, surface runoff and base flow during these 30 years were then calculated based on daily statistical data for subsequent trend analysis. Meanwhile, based on the water balance equation in a totally closed small watershed, mean ET is basically equal to mean precipitation minus mean surface runoff.

2.2.2. Trend Analysis

The Mann–Kendall test (MK) is a suitable method to determine the variation trend of hydrological time series [22,23], which has been widely applied for trend and inflection analysis of precipitation and runoff [24,25]. Thus, in this paper, we found a consistent inflection point of the year 2003 both for precipitation and surface runoff. The research period was the divided into baseline period (1983–2003) and changing period (2004–2014), which lays a solid foundation for separating the impacts of multiple factors on surface runoff and base flow in different time scales. The calculation procedures of an MK trend test are given as follows.

For a time series $\{r_1, r_2, \ldots, r_{n-1}, r_n\}$, the statistic index UF_m is calculated as,

$$UF_m = \frac{S_m - E(S_m)}{\sqrt{Var(S_m)}} \ (m = 1, 2, \ldots n) \tag{1}$$

where S_m is deemed as cumulative number of $r_i > r_j$ ($1 \leq i \leq j$), $E(S_m)$ is averaged S_m, and $Var(S_m)$ represents variance of S_m. The change of UF_m reflects the variation in hyro-climate variables. $UF_m > 0$ (<0) means the variables show an increasing (a decreasing) trend. If $|UF_m| > 1.64, 1.96, 2.58$, the change trend is significant at $p > 0.1, 0.05$ and 0.01, respectively. Contrarily, the value calculated with inverse series ($r_n, r_{n-1}, \ldots, r_2, r_1$) is termed as UB_m. If the exact intersection point of the two lines located within the critical limit line and the trend is significance, the point is deemed as the possible change point [22].

2.2.3. Separating the Effect of ET on Surface Runoff and Base Flow

The slope change ratio of cumulative quantities (SCRCQ) method was proposed to quantitatively assess the factors impacting the streamflow in the Huangfuchuan River Basin, a first-level tributary of the middle reaches of the Yellow River. It is believed that this method can be applied to the quantitative assessment of river runoff changes and its influencing factors in arid and semi-arid regions [26]. Subsequently, similar research on Songhua River basin [27], Miyun Reservoir watershed [28] and some areas of the southern humid areas such as the Yinjiang River watershed [29], Dongting Lake catchment [30], and Ning Jiang River basin [31] had been carried out in China. The cumulative quantities can eliminate the effects of inter-annual fluctuations to some extent, with a high correlation coefficient between year and cumulative quantities, which creates conditions for quantifying analysis [26]. Therefore, using the method of SCRCQ, we calculated the contribution rates of ET to surface runoff and base flow in seasonal and annual scale. Calculation equations are as follows:

$$C_{ET} = -100 \times \frac{(K_2 - K_1)/K_1}{(K_4 - K_3)/K_3} \tag{2}$$

where K (mm·year^{-1}) is the slope of the linear relationship between year and cumulative ET in baseline period (labeled by 1) and changing period (labeled by 2), cumulative surface runoff or base flow in baseline period (labeled by 3) and changing period (labeled by 4). C_{ET} is the relative contribution rate of ET to surface runoff or base flow. According to the water balance equation in a relatively closed watershed, the change in soil water storage is likely to be small, which is negligible in annual scale [26,28]. Thus, the contribution rates of precipitation variation and vegetation restoration to surface runoff and base flow is equal to $1 - C_{ET}$.

2.2.4. Excluding the Contribution Rates of Precipitation Variation and Vegetation Restoration

The formation of surface runoff is mainly affected by precipitation and underlying surface conditions. For the same small watershed with no inhabitants and water and soil conservancy projects, underlying surface change can be mainly regarded as vegetation restoration. In this paper, empirical statistics was used to quantitatively assess the two factors (precipitation variation and vegetation restoration) that have significant influence on surface runoff as well as base flow.

Firstly, regression analyses were conducted between observed precipitation, surface runoff and base flow in baseline period when vegetation was less changed, after which a linear Equation (1) could be built. Then, substituting observed mean precipitation in the changing period, when vegetation changes apparently in the equation, a value that represents surface runoff (base flow) produced in the condition of vegetation in baseline period can be obtained which is called simulated surface runoff (base flow). The value of simulated surface runoff (base flow) minus observed mean surface runoff (base flow) in the changing period represents surface runoff (base flow) variation caused by vegetation restoration. Similarly, the value of observed mean surface runoff (base flow) in the baseline period minus simulated surface runoff (base flow) represents surface runoff (base flow) variation caused by precipitation variation. In this way, the impacts of precipitation variation and vegetation restoration on surface runoff and base flow can be quantitatively distinguished. Calculation equations of the contribution rates of precipitation variation and vegetation restoration are listed as follows:

$$Q = aP + b \tag{3}$$

where Q is surface runoff (base flow) [mm], P is precipitation [mm], a and b are parameters.

$$\Delta Q_v = Q_{sim} - \overline{Q'_{obs}}, \ \Delta Q_P = \overline{Q_{obs}} - Q_{sim} \tag{4}$$

$$\Delta Q_{total} = \Delta Q_v + \Delta Q_P = \overline{Q_{obs}} - \overline{Q'_{obs}} \tag{5}$$

$$C_p = \frac{\Delta Q_P}{\Delta Q_{total}} \times (1 - C_{ET})\% \text{ and } C_v = \frac{\Delta Q_v}{\Delta Q_{total}} \times (1 - C_{ET})\% \tag{6}$$

where $\overline{Q_{obs}}$ and $\overline{Q'_{obs}}$ are observed mean surface runoff (base flow) (mm) in baseline period and changing period respectively; $Q_{sim} = a\overline{P'_{obs}} + b$, which means simulated surface runoff (base flow), where $\overline{P'_{obs}}$ is observed mean precipitation (mm) in the changing period; ΔQ_{total} is total change in observed mean surface runoff (base flow) comparing baseline period to changing period; ΔQ_v and ΔQ_P are surface runoff (base flow) changes caused by vegetation restoration and precipitation variation, respectively.

3. Results and Discussion

3.1. Seasonal Distribution of Precipitation, Evapotranspiration (ET), Surface Runoff and Base Flow

Based on daily precipitation, runoff data and base flow separation results of Pengchongjian hydrological station in 1983–2014, the monthly, seasonal (spring is March to May, summer is June to August, autumn is September to November, winter is December to the following February) and annual precipitation, ET, surface runoff and base flow were obtained. In this condition, it was possible to investigate the annual and inter-annual distribution of precipitation, ET and runoff, and the impacts of various factors on surface runoff and base flow in small watershed. Annual mean precipitation, ET, surface runoff and base flow from 1983 to 2014 were 1560.3, 811.9, 743.7 and 353.2 mm, respectively (Table 1). In spring and summer, both parameters except ET accounted for a large proportion in whole year, while a small ratio in autumn and winter. However, the seasonal distribution of ET was relatively even.

Table 1. Seasonal distribution of precipitation, ET, surface runoff and base flow in Pengchongjian small watershed from 1983 to 2014.

Parameters	Spring	Summer	Autumn	Winter	Year
Precipitation/mm	564.6	589.7	199.3	206.7	1560.3
Annual proportion/%	36.2%	37.8%	12.8%	13.2%	—
ET/mm	218.0	272.2	168.8	153.0	811.9
Annual proportion/%	26.9%	33.5%	20.8%	18.8%	—
Surface runoff/mm	354.8	300.9	34.3	53.7	743.7
Annual proportion/%	47.7%	40.5%	4.6%	7.2%	—
Base flow/mm	159.9	131.2	19.1	43.0	353.2
Annual proportion/%	45.3%	37.1%	5.4%	12.2%	—

3.2. Variations of Precipitation and Surface Runoff

According to the MK test, the annual variation trend of precipitation (P) in Pengchongjian small watershed during 1983–2014 was as follows: (i) precipitation generally showed a fluctuated downward trend (Figure 4a). For example, it declined significantly in 1983–1986, and reached the bottom in 1986 ($p > 0.05$), while in 1987–2003, it showed a fluctuated rise, reaching the peak in 1998. However, from 2004 to 2009, precipitation began to decline, at the beginning of 2010 a peak (2010) and a valley (2011) appeared, then gradually it increased. (ii) According to the intersection point of UF and UB curves, the t test method was applied. It was also determined that the inflection point of the small watershed was 2003 (Figure 4b).

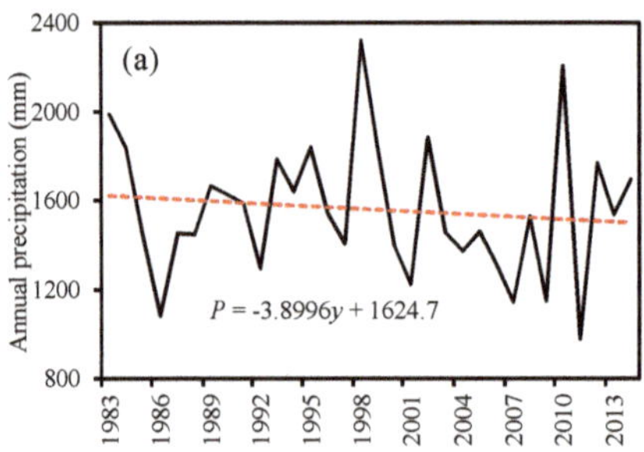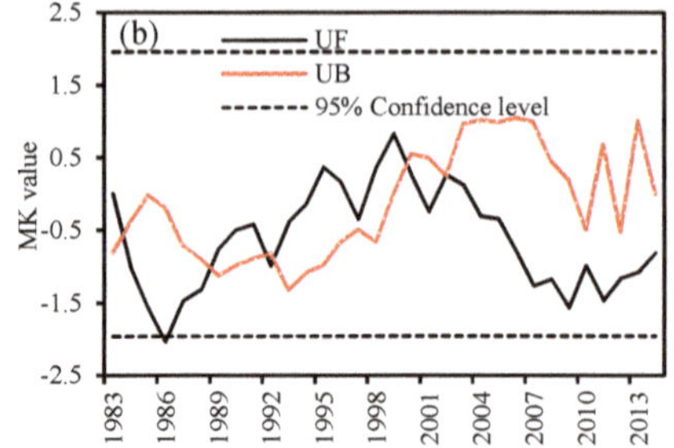

Figure 4. Inter-annual variation of observed precipitation (**a**) and its Mann–Kendall (MK) test results (**b**) in Pengchongjian small watershed. in 1983–2014.

According to the MK test, annual variation trend of surface runoff (SR) in 1983–2014 was as follows: (i) surface runoff showed a fluctuated downward trend (Figure 5a). It decreased significantly in 1983–1986 ($p > 0.05$), then increased gradually in 1987–2003 and reached the maximum value in 1999; since 2004, surface runoff was relatively small except for 2010. (ii) UF and UB curves have multiple intersections (Figure 5b). In order to remove the invalid mutation point, the t test method was used. Furthermore, surface runoff and precipitation shared the same inflection point of year 2003. In other words, surface runoff change was directly related to precipitation.

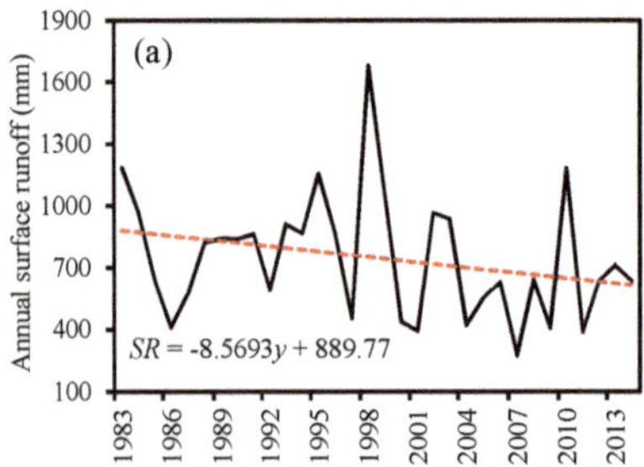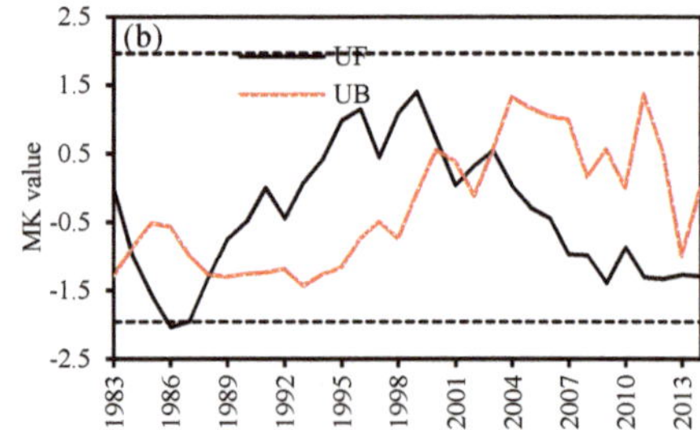

Figure 5. Inter-annual variation of observed surface runoff (**a**) and its MK results (**b**) in Pengchongjian small watershed in 1983–2014.

On the basis of the trend and inflection point analysis of precipitation and surface runoff, the research periods were divided into 1983–2003 and 2004–2014. The average annual precipitation and surface runoff in 1983–2003 were 1608.2 and 831.9 mm, while in 2004–2014, they were 1468.9 and 588.9 mm. Compared with the baseline period, precipitation and surface runoff in the changing period decreased by 8.7% and 29.2%, respectively, with mean annual reduction of 12.7 and 22.1 mm. The decreasing proportion of surface runoff was larger than that of precipitation. That was to say, besides precipitation, ET and vegetation factors also influenced surface runoff. How to quantitatively distinguish the contribution rates of various factors on surface runoff in a small watershed is a scientific problem that needs to be further investigated.

3.3. Contribution Rates of ET to Surface Runoff and Base Flow

3.3.1. Variation in Temperature and ET

Annual mean temperature from 1983–2014 overall presented an upward trend, with a rate of 0.0349 °C·year^{-1} (Figure 6a). It reached its highest level of 17.9 °C in year 2013, while in year 1984, its historical low level was 15.9 °C. Compared to the baseline period (1983–2003), annual mean temperature in the changing period (2004–2014) increased by 1 °C. The annual variation of ET is shown in Figure 6b. The fluctuation since the early 21st century was more remarkable than that in the 1980s, and showed an upward trend not as remarkable as that as temperature. However, the rising temperature caused the ET to go up, which would probably affect the process of the water cycle, directly leading to a decrease in surface runoff as well as base flow. Thus, we firstly calculated the contribution rate of ET to surface runoff and base flow decrease by SCRCQ.

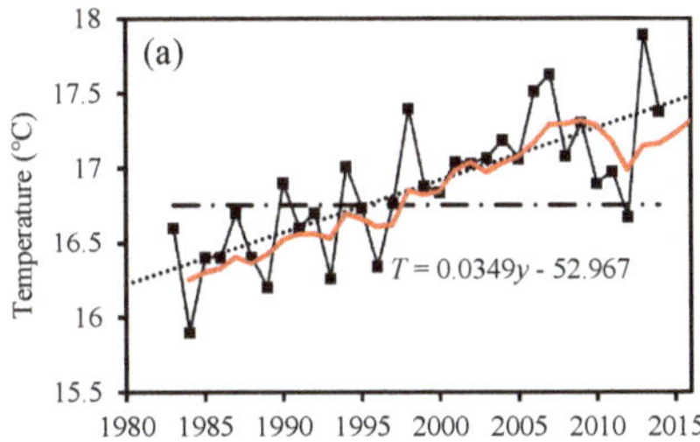

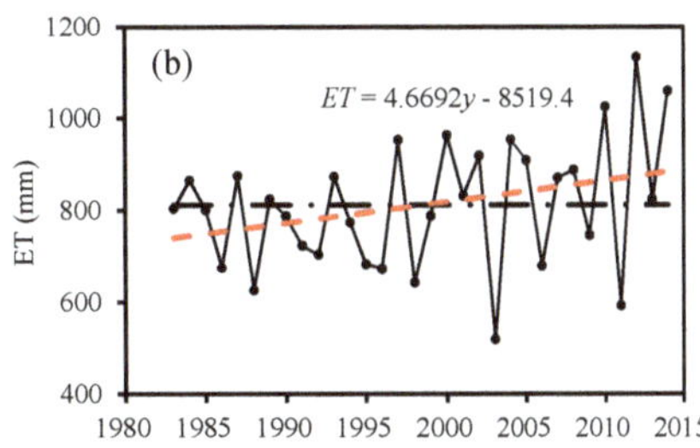

Figure 6. Variation trend of annual mean temperature (**a**) and ET (**b**) from 1983 to 2014.

3.3.2. Impacts of ET on Surface Runoff and Base Flow

The linear relationships between year and cumulative ET, surface runoff and base flow are shown in Figures 7 and 8. The calculation results of the contribution rates of ET to surface runoff and base flow are shown in Table 2. Annual average ET during the baseline period and changing period were 773.4 and 860.9 mm, while annual mean surface runoff were 831.9 and 588.9 mm, and average base flow were 384.3 and 293.7 mm, respectively. Compared with the baseline period, ET, surface runoff and base flow in changing period increased by 87.5, 243 and 90.6 mm, with corresponding change rates of 11.3%, −29.2% and −23.6%, respectively. This somewhat reflected the fact that surface runoff and base flow declined synchronously with the increase of ET. In annual scale, the contribution rates of ET to the decrease of surface runoff and base flow were 38.7% and 47.9%, which was basically consistent with the conclusion drawn in the Yinjiang basin [29], while quite distinct from that (less than 10%) in the Yellow River basin [32]. This fully illustrated that under the subtropical humid monsoon climate, ET could affect surface runoff and base flow to a greater extent, similar to the conclusion drawn by Chaplot [11]. On a seasonal scale, the contribution rates of ET to the decrease of surface runoff was between 28%~40%, while that to base flow reduction was between 38%~52%. We could deem that the sensitivity of response of base flow to ET is higher than that of surface runoff, as depicted by Lin [33]. Moreover, the contribution rate of ET to base flow in summer was obviously higher than that in other seasons, which may be due to more precipitation, higher temperature and higher ET in summer caused by forest restoration. On the other hand, the variation of average ET, surface runoff and base flow in winter were both 2~3 times that of other seasons compared to the baseline period (Table 2). Meanwhile, the contribution rates of ET to surface runoff and base flow in winter were quite close to that in other seasons, indicating that vegetation accounted for a larger part in water yield when encountering the dry season with less precipitation.

In short, ET impact on surface runoff and base flow could not be negligible whether in annual scale or seasonal scale, especially in winter. The response of dry season runoff to forest change is mainly determined by soil conditions, tree species and vegetation regeneration after disturbances, as well as topography [34], or rather, water consumption (ET) and subsequent changes in soil infiltration and water storage capacity.

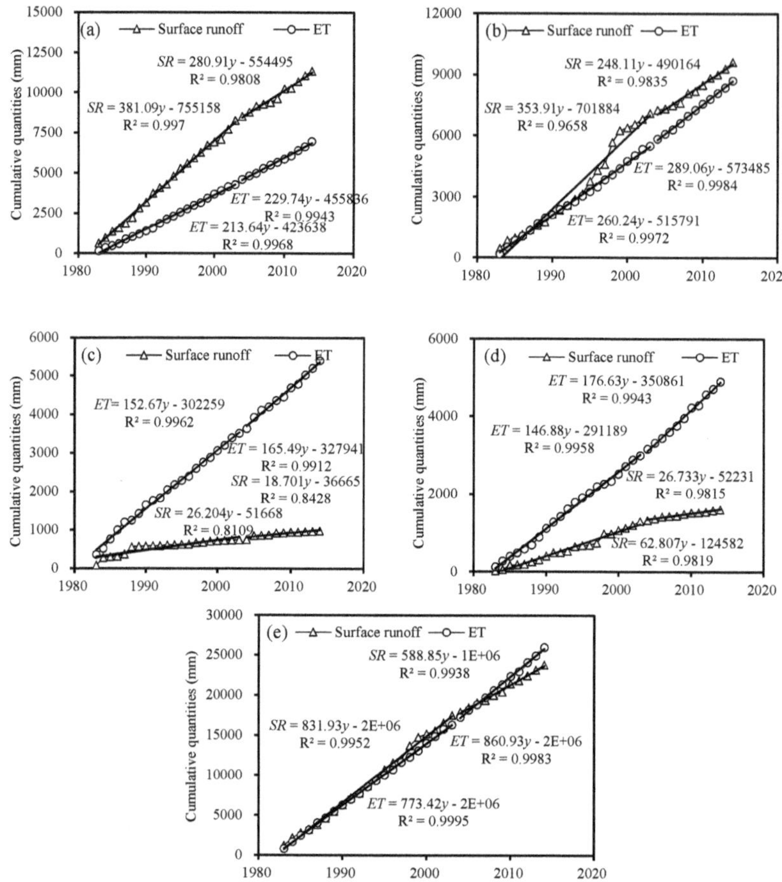

Figure 7. Relationships between year and cumulative surface runoff, ET in spring (**a**), summer (**b**), autumn (**c**), winter (**d**) and whole year (**e**) during two periods in Pengchongjian small watershed.

Table 2. Slopes of the relations between year and cumulative quantities of ET ($K_{1,2}$), surface runoff ($K_{3,4}$), base flow ($K'_{3,4}$) and contribution rates of ET to surface runoff and base flow decrease.

Slope (mm·year^{-1})	Spring	Summer	Autumn	Winter	Whole Year
K_1	213.6	260.2	152.7	146.9	773.4
K_2	229.7	289.1	165.5	176.6	860.9
ΔK_{ET}	0.075	0.111	0.084	0.202	0.113
K_3	381.1	353.9	26.2	62.8	831.9
K_4	280.9	248.1	18.7	26.7	588.9
ΔK_{SR}	−0.263	−0.299	−0.286	−0.575	−0.292
K'_3	166.5	150.1	18.2	49.5	384.3
K'_4	136.7	118	14.2	24.1	293.7
ΔK_B	−0.179	−0.214	−0.220	−0.513	−0.236
C_{ET} to surface runoff (%)	28.5	37.1	29.4	35.1	38.7
C_{ET} to base flow (%)	41.9	51.9	38.2	39.4	47.9

Note: ΔK_{ET}, ΔK_{SR}, ΔK_B is the variation rate of slopes of ET, surface runoff and base flow compared to baseline period, respectively.

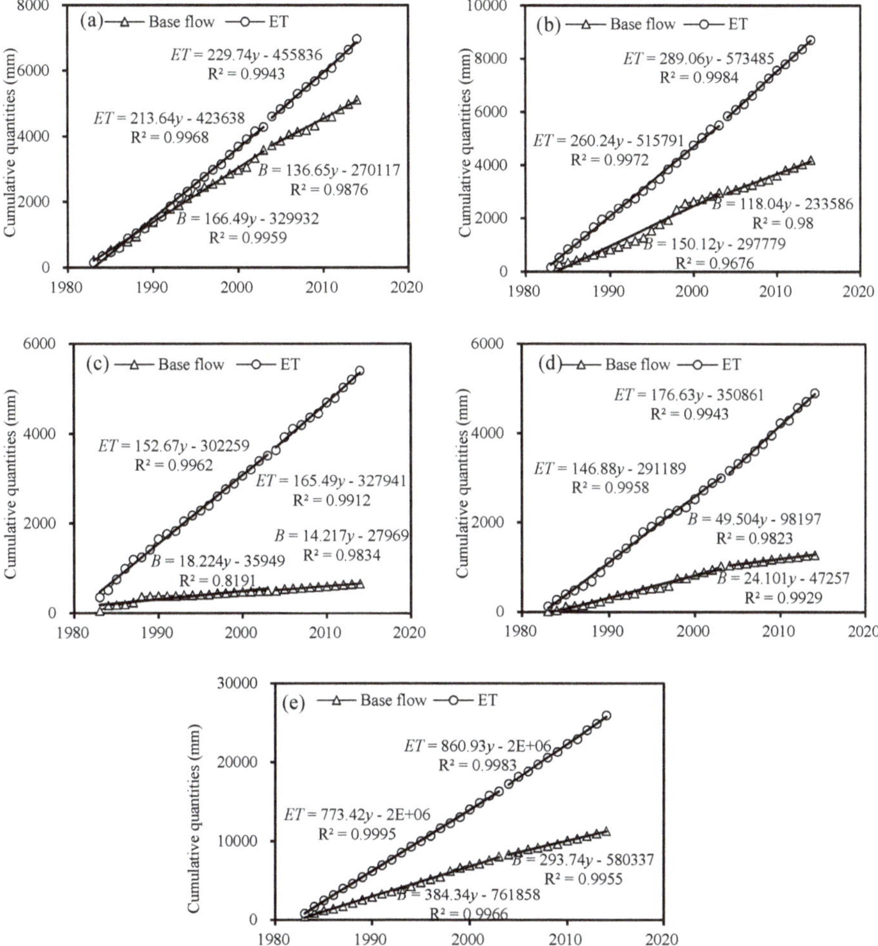

Figure 8. Relationships between year and cumulative base flow, ET in spring (**a**), summer (**b**), autumn (**c**), winter (**d**) and whole year (**e**) during two periods in Pengchongjian small watershed.

3.4. Contribution Rates of Precipitation Variation and Vegetation Restoration to Surface Runoff Decrease

The regression analysis between mean surface runoff and precipitation in seasonal and whole year scale during baseline period were carried out, and the linear equations were shown in Figure 9.

The mean annual and seasonal precipitation during changing period were then substituted into the corresponding equations, and the simulated mean surface runoffs were obtained. The calculation results of contribution rates of precipitation variation and vegetation restoration to surface runoff are shown in Table 3.

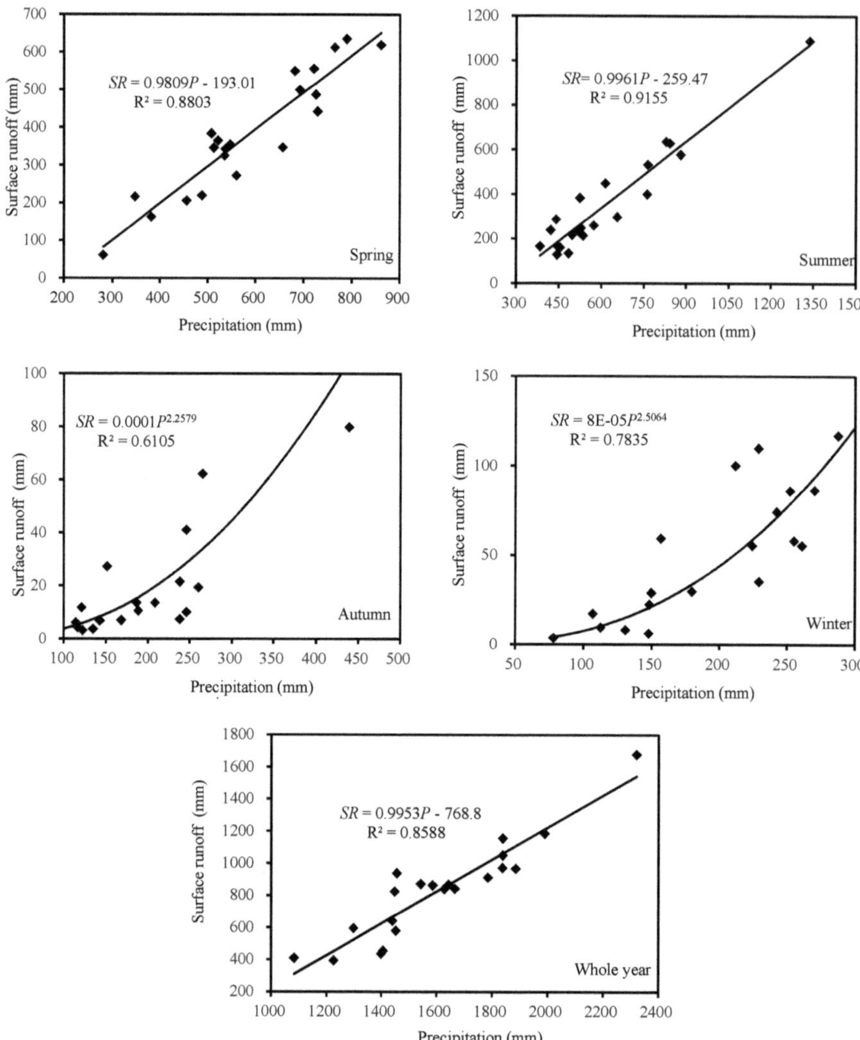

Figure 9. Relationship between observed surface runoff and precipitation in seasonal and annual scale during baseline period.

Table 3. Calculating results of contribution rates of precipitation variation and vegetation restoration to surface runoff decrease.

Time Scale	Simulated Mean Surface Runoff (mm)	Observed Mean Surface Runoff (mm)	Impact of Precipitation Variation		Impact of Vegetation Restoration	
			Variation (mm)	Contribution (%)	Variation (mm)	Contribution (%)
Spring	322.1	280.9	−59.0	42.1	−41.2	29.4
Summer	278.2	248.1	−75.7	45.0	−30.1	17.9
Autumn	23.0	18.7	−3.2	30.1	−4.3	40.5
Winter	53.5	26.7	−9.3	16.7	−26.8	48.2
Year	693.2	588.9	−138.7	35.0	−104.3	26.3

Note: The observed mean surface runoff during baseline period is 381.1, 353.9, 26.2, 62.8 and 831.9 mm in spring, summer, autumn, winter and whole year respectively.

Compared to the baseline period, average surface runoff in spring, summer, autumn, winter and the whole year decreased by 100.2, 105.8, 7.5, 36.1 and 243 mm, respectively. The surface runoff decrease caused by precipitation variation was 59.0, 75.7, 3.2, 9.3 and 138.7 mm, respectively, while by vegetation restoration was 41.2, 30.1, 4.3, 26.8 and 104.3 mm. The contribution rates of precipitation variation and vegetation restoration to surface runoff were 42.1%, 29.4% in spring, 45.0%, 17.9% in summer, 30.1%, 40.5% in autumn, 16.7%, 48.2% in winter, and 35.0%, 26.3% in annual scale, respectively (Table 3). Obviously, in seasonal scale, attribution of surface runoff reduction to precipitation in spring and summer was larger than that in autumn and winter, and vice versus for attribution to vegetation restoration. Herein the contribution rate of precipitation in summer reached the maximum, which may be related to much more precipitation, heavier rainfall intensity, and more rainfall days. For this reason, on the contrary, vegetation restoration counted less (only 44% of that in autumn) than precipitation for surface runoff decease in summer (Table 3). The water diminishing effect of vegetation in autumn and winter was significantly strengthened, which is likely because of less precipitation in autumn and winter, and vegetation enriched the understory and forest litter, and enhanced soil infiltration and water storage [34], hence reducing surface runoff. On an annual scale, in contrast to the conclusion that vegetation change induced by human activity dominated surface runoff generation in arid areas of northern China [25,26,31,35,36], we found that for Pengchongjian small watershed, the contribution of precipitation (35%) to surface runoff was nearly close to that of ET (38.7%), followed by vegetation restoration (26.3%). The reason why the results varied, on the one hand, is probably due to distinct climatic conditions, topography, vegetation structure and type. On the other hand, human activity there is relatively slight with no local residents or water conservancy facilities and soil and water conservation projects. In general, climatic factors (including precipitation variation and elevated ET caused by rising temperature) are the major drivers for annual surface runoff, which is in accordance to similar results in Poyang Lake basin, China [15,16,37].

3.5. Contribution Rates of Precipitation Variation and Vegetation Restoration to Base Flow Decrease

The regression analysis between mean base flow and precipitation in seasonal and whole year scale during the baseline period were carried out, and the linear equations were shown in Figure 10.

The mean annual and seasonal precipitation during the changing period were then substituted into the corresponding equations, and the simulated mean base flow was obtained. The calculation results of contribution rates of precipitation variation and vegetation restoration to base flow are shown in Table 4.

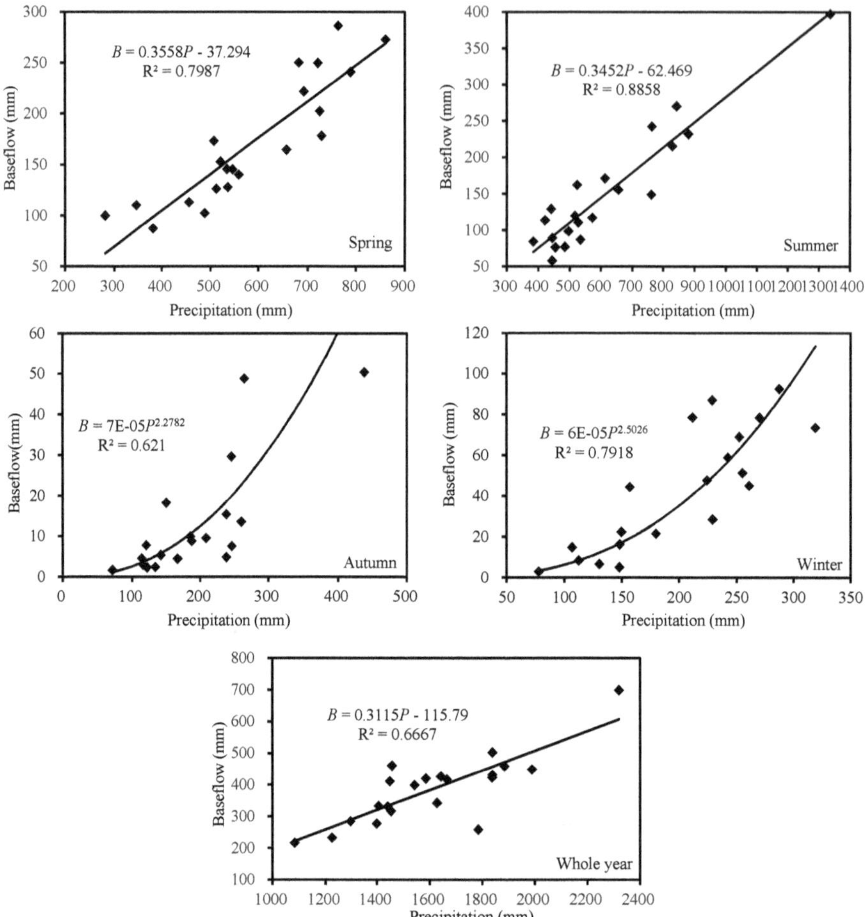

Figure 10. Relationship between observed base flow and precipitation in seasonal and annual scale during baseline period.

Table 4. Calculated results of contribution rates of precipitation variation and vegetation restoration to base flow decrease.

Time Scale	Simulated Mean Base Flow (mm)	Observed Mean Base Flow (mm)	Impact of Precipitation Variation		Impact of Vegetation Restoration	
			Variation (mm)	Contribution (%)	Variation (mm)	Contribution (%)
Spring	149.5	136.7	−17.0	33.1	−12.8	25.0
Summer	123.9	118.0	−26.2	39.3	−5.9	8.8
Autumn	15.6	14.2	−2.6	40.2	−1.4	21.6
Winter	39.3	24.1	−10.2	24.3	−15.2	36.3
Year	341.8	293.7	−42.5	24.4	−48.1	27.7

Note: The observed mean base flow during baseline period is 166.5, 150.1, 18.2, 49.5 and 384.3 mm in spring, summer, autumn, winter and whole year respectively.

Compared to the baseline period, average base flow in spring, summer, autumn, winter and the whole year decreased by 29.8, 32.1, 4.0, 25.4 and 90.6 mm, respectively. Base flow decrease caused by precipitation variation was 17.0, 26.2, 2.6, 10.2 and 42.5 mm, respectively, while by vegetation restoration it was 12.8, 5.9, 1.4, 15.2 and 48.1 mm. The contribution rates of precipitation variation and vegetation restoration to base flow were 33.1%, 25.0% in spring, 39.3%, 8.8% in summer, 40.2%, 21.6% in autumn, 24.3%, 36.3% in winter and 24.4%, 27.7% on an annual scale, respectively (Table 4). It could be seen that, as with surface runoff, the contribution rate of each factor to the reduction of base flow also showed obvious seasonality. The contribution rate of precipitation variation was the lowest in winter while almost the same in other seasons. The contribution rate of vegetation restoration, however, reached the minimum in summer (only 20%~40% of that in other seasons). Moreover, base flow reduction due to vegetation restoration obviously escalated in winter, which was consistent with the conclusion drawn on surface runoff. On an annual scale, the contribution rate of vegetation restoration to base flow reduction was slightly larger than that of precipitation, but ET still dominated. We know that the increase of vegetation coverage and the improvement of stand quality directly led to the increase of ET and thus base flow decrease. So, we assumed that forest may be a main factor with a negative impact on base flow, as mentioned by Huang [38] and Zhang [39]. Anyhow, ET has an important influence on the dynamic change of base flow. Meanwhile, vegetation plays an irreplaceable role in regulating water flow in dry seasons and improving hydrological ecology in watersheds.

4. Conclusions

In this paper, we quantitively calculated the contribution rates of precipitation variation, vegetation restoration and ET to surface runoff and base flow. The results indicated that each factor played a role in reducing surface runoff and base flow, and their contribution rates varied from season to season. For spring and summer, the contribution rate of precipitation was equal to that of ET, which was slightly larger than vegetation restoration; for autumn and winter; however, the effect of vegetation restoration on water yield was obviously stronger than that in spring and summer. On an annual scale, climatic factors including precipitation and ET change dominated surface runoff generation. Meanwhile, the sensitivity of base flow response to ET was higher than surface runoff, so that ET dominated the dynamic change of annual base flow. As presented in this study, ET change is closely associated with vegetation restoration, which fully illustrates that the protection of natural forest ecosystem and the optimization of forest operation and management are important guarantees for ecological water circulation and the sustainable utilization of forest resources.

Author Contributions: S.L., R.W. and Z.L. (Zhihong Lu) designed this study. Z.L. (Zheng Liu), J.Y. and F.S. collected data and performed the data analysis. L.O. wrote this paper.

Funding: This research was funded by the Chinese Natural Science Foundation Program grant number (No. 31460222).

Acknowledgments: The authors are grateful for the support from Jiangxi Hydrological Bureau and the local government in Duchang County, Jiujiang. We also appreciate the reviewers for their helpful comments in polishing this article.

Conflicts of Interest: The authors declare no conflict of interest.

References

1. Smakhtin, V.U. Low flow hydrology: A. review. *J. Hydrol.* **2001**, *240*, 147–186. [CrossRef]
2. Ficklin, D.L.; Robeson, S.M.; Knouft, J.H. Impacts of recent climate change on trends in baseflow and stormflow in United States watersheds. *Geophys. Res. Lett.* **2016**, *43*, 5079–5088. [CrossRef]
3. Guzmán, P.; Batelaan, O.; Huysmans, M.; Wyseure, G. Comparative analysis of baseflow characteristics of two Andean catchments, Ecuador. *Hydrol. Process.* **2015**, *29*, 3051–3064. [CrossRef]
4. Ni, Y.Q.; Zhang, W.H.; Guo, S.L. Discussion on the discharge hydrograph separation methods. *Hydrology* **2005**, *25*, 11–23. (In Chinese)
5. Githui, F.; Gitau, W.; Mutua, F.; Bauwens, W. Climate change impact on SWAT simulated streamflow in western Kenya. *Int. J. Climatol.* **2009**, *29*, 1823–1834. [CrossRef]

6.   Ye, X.C.; Zhang, Q.; Liu, J.; Li, X.H.; Xu, C.Y. Distinguishing the relative impacts of climate change and human activities on variation of streamflow in the Poyang Lake catchment, China. *J. Hydrol.* **2013**, *494*, 83–95. [CrossRef]

7.   Kuk-Hyun Ahn, V.M. Quantifying the relative impact of climate and human activities on streamflow. *J. Hydrol.* **2014**, *515*, 257–266.

8.   Röderstein, M.; Perdomo, L.; Villamil, C.; Hauffe, T.; Schnetter, M. Long-term vegetation changes in a tropical coastal lagoon system after interventions in the hydrological conditions. *Aquat. Bot.* **2014**, *113*, 19–31. [CrossRef]

9.   Chaplot, V.; Saleh, A.; Jaynes, D.B. Effect of the accuracy of spatial rainfall information on the modeling of water, sediment, and NO_3–N loads at the watershed level. *J. Hydrol.* **2005**, *312*, 223–234. [CrossRef]

10. Alvarenga, L.A.; de Mello, C.R.; Colombo, A.; Cuartas, L.A.; Bowling, L.C. Assessment of land cover change on the hydrology of a Brazilian headwater watershed using the Distributed Hydrology-Soil-Vegetation Model. *Catena* **2016**, *143*, 7–17. [CrossRef]

11. Chaplot, V. Water and soil resources response to rising levels of atmospheric CO_2 concentration and to changes in precipitation and air temperature. *J. Hydrol.* **2007**, *337*, 159–171. [CrossRef]

12. Zeng, S.D.; Xia, J.; Du, H. Separating the effects of climate change and human activities on runoff over different time scales in the Zhang River basin. *Stoch. Environ. Res. Risk Assess.* **2014**, *28*, 401–413. [CrossRef]

13. Zhang, X.C.; Nearing, M.A. Impact of climate change on soil erosion, runoff, and wheat productivity in central Oklahoma. *Catena* **2005**, *61*, 185–195. [CrossRef]

14. Zhang, Q.; Liu, J.Y.; Singh, V.P.; Gu, X.H.; Chen, X.H. Evaluation of impacts of climate change and human activities on streamflow in the Poyang Lake basin, China. *Hydrol. Process.* **2016**, *30*, 2562–2576. [CrossRef]

15. Gu, C.J.; Mu, X.M.; Gao, P.; Zhao, G.J.; Sun, W.Y.; Li, P.F. Effects of climate change and human activities on runoff and sediment inputs of the largest freshwater lake in China, Poyang Lake. *Hydrol. Sci. J.* **2017**, *62*, 2313–2330. [CrossRef]

16. Rogger, M.; Agnoletti, M.; Alaoui, A.; Bathurst, J.C.; Bodner, G.; Borga, M.; Chaplot, V.; Gallart, F.; Glatzel, G.; Hall, J.; et al. Land use change impacts on floods at the catchment scale: Challenges and opportunities for future research. *Water Resour. Res.* **2017**, *53*, 5209–5219. [CrossRef] [PubMed]

17. Jiang, S.H.; Ren, L.L.; Yong, B.; Singh, V.P.; Yang, X.L.; Yuan, F. Quantifying the effects of climate variability and human activities on runoff from the Laohahe basin in northern China using three different methods. *Hydrol. Process.* **2011**, *25*, 2492–2505. [CrossRef]

18. Xia, J.; Zeng, S.; Du, H.; Zhan, C.S. Quantifying the effects of climate change and human activities on runoff in the water source area of Beijing, China. *Hydrol. Sci. J.* **2014**, *59*, 1794–1807. [CrossRef]

19. Koskelo, A.I.; Fisher, T.R.; Utz, R.M.; Jordan, T.E. A new precipitation-based method of baseflow separation and event identification for small watersheds (<50 km^2). *J. Hydrol.* **2012**, *450*, 267–278. [CrossRef]

20. Rumsey, C.A.; Miller, M.P.; Susong, D.D.; Tillman, F.D.; Anning, D.W. Regional scale estimates of baseflow and factors influencing baseflow in the Upper Colorado River Basin. *J. Hydrol. Reg. Stud.* **2015**, *4*, 91–107. [CrossRef]

21. Ding, Z.L.; Hu, K.D.; Fang, Y.Y. Analysis and discussion of dividing up ground water by the kalinlin improving Method. *JiangXi Hydraul. Sci. Technol.* **2003**, *29*, 211–215. (In Chinese)

22. Chattopadhyay, G.; Chakraborthy, P.; Chattopadhyay, S. Mann-Kendall trend analysis of tropospheric ozone and its modeling using ARIMA. *Theor. Appl. Climatol.* **2012**, *110*, 321–328. [CrossRef]

23. Rahman, M.A.; Yunsheng, L.; Sultana, N. Analysis and prediction of rainfall trends over Bangladesh using Mann-Kendall, Spearman's rho tests and ARIMA model. *Meteorol. Atmos. Phys.* **2017**, *129*, 409–424. [CrossRef]

24. Nourani, V.; Danandeh Mehr, A.; Azad, N. Trend analysis of hydroclimatological variables in Urmia lake basin using hybrid wavelet Mann–Kendall and Şen tests. *Environ. Earth. Sci.* **2018**, *77*, 207. [CrossRef]

25. Kamruzzaman, M.; Rahman, A.T.M.S.; Ahmed, M.S.; Kabir, M.E.; Mazumder, Q.H.; Rahman, M.S.; Jahan, C.S. Spatio-temporal analysis of climatic variables in the western part of Bangladesh. *Environ. Dev. Sustain.* **2018**, *20*, 89–108. [CrossRef]

26. Wang, S.J.; Yan, Y.J.; Yan, M.; Zhao, X.K. Quantitative estimation of the impact of precipitation and human activities on runoff change of the Huangfuchuan River Basin. *J. Geogr. Sci.* **2012**, *22*, 906–918. [CrossRef]

27. Wang, S.J.; Wang, Y.J.; Ran, L.S.; Su, T. Climatic and anthropogenic impacts on runoff changes in the Songhua River basin over the last 56years (1955–2010), Northeastern China. *Catena* **2015**, *127*, 258–269. [CrossRef]

28. Pang, S.J.; Wang, X.Y. The characteristics and attribution of runoff change in the Miyun reservoir watershed. *J. Arid Land Resour. Environ.* **2016**, *30*, 144–148. (In Chinese)

29. Wu, L.H.; Wang, S.J.; Bai, X.Y.; Luo, W.J.; Tian, Y.C.; Zeng, C.; Luo, G.J.; He, S.Y. Quantitative assessment of the impacts of climate change and human activities on runoff change in a typical karst watershed, SW China. *Sci. Total Environ.* **2017**, *601*, 1449–1465. [CrossRef] [PubMed]

30. Cheng, J.X.; Xu, L.G.; Jiang, J.H.; Tan, Z.Q.; Yu, Q.W.; Fan, H.X. The research of runoff responses to climate change and human activities in the Dongting Lake catchment. *J. Agro-Environ. Sci.* **2016**, *35*, 2146–2153. (In Chinese)

31. Yang, C.X.; Zhang, Z.D.; Zhu, R.X.; Wan, L.W.; Ye, C.; Zhang, J. Response of runoff to climate change and human activities in the river basin of southern humid area-A case study of Ning Jiang. *Res. Soil Water Conserv.* **2017**, *24*, 113–119. (In Chinese)

32. Wang, S.J.; Yan, M.; Yan, Y.X.; Shi, C.X.; He, L. Contributions of climate change and human activities to the changes in runoff increment in different sections of the Yellow River. *Quat. Int.* **2012**, *282*, 66–77. [CrossRef]

33. Lin, P.R.; Rajib, M.A.; Yang, Z.L.; Somos-Valenzuela, M.; Merwade, V.; Maidment, D.R.; Wang, Y.; Chen, L. Spatiotemporal evaluation of simulated evapotranspiration and streamflow over Texas using the WRF-Hydro-RAPID Modeling Framework. *J. Am. Water Resour. Assoc.* **2018**, *54*, 40–54. [CrossRef]

34. Hou, Y.P.; Zhang, M.F.; Meng, Z.Z.; Liu, S.R.; Sun, P.S.; Yang, T.L. Assessing the impact of forest change and climate variability on dry season runoff by an improved single watershed approach: A comparative study in two large watersheds, China. *Forests* **2018**, *9*, 46. [CrossRef]

35. Wu, J.W.; Miao, C.Y.; Zhang, X.M.; Yang, T.T.; Duan, Q.Y. Detecting the quantitative hydrological response to changes in climate and human activities. *Sci. Total Environ.* **2017**, *586*, 328–337. [CrossRef] [PubMed]

36. Xu, X.Y.; Yang, D.W.; Yang, H.B.; Lei, H.M. Attribution analysis based on the Budyko hypothesis for detecting the dominant cause of runoff decline in Haihe basin. *J. Hydrol.* **2014**, *510*, 530–540. [CrossRef]

37. Sun, S.L.; Chen, H.S.; Ju, W.M.; Song, J.; Zhang, H.; Sun, J.; Fang, Y.J. Effects of climate change on annual streamflow using climate elasticity in Poyang Lake Basin, China. *Theor. Appl. Climatol.* **2013**, *112*, 169–183. [CrossRef]

38. Huang, X.; Shi, Z.; Fang, N.; Li, X. Influences of Land Use Change on Baseflow in Mountainous Watersheds. *Forests* **2016**, *7*, 16. [CrossRef]

39. Zhang, Y.K.; Schilling, K.E. Increasing streamflow and baseflow in Mississippi River since the 1940s: Effect of land use change. *J. Hydrol.* **2006**, *324*, 412–422. [CrossRef]

Article

Nursery Production of *Pinus engelmannii* Carr. with Substrates Based on Fresh Sawdust

María Mónica González-Orozco [1], José Ángel Prieto-Ruíz [2,*], Arnulfo Aldrete [3], José Ciro Hernández-Díaz [4], Jorge Armando Chávez-Simental [4] and Rodrigo Rodríguez-Laguna [5]

[1] Programa Institucional de Doctorado en Ciencias Agropecuarias y Forestales, Universidad Juárez del Estado de Durango, Constitución 404 sur Zona centro, Durango 34000, Mexico; m_gonzalez@ujed.mx

[2] Facultad de Ciencias Forestales, Universidad Juárez del Estado de Durango, Río Papaloapan y Boulevard, Durango S/N Col. Valle del Sur, Durango 34120, Mexico

[3] Colegio de Postgraduados, Campus Montecillo, Carretera México-Texcoco Km 36.5, Montecillo, Texcoco, Estado de México 56230, Mexico; aaldrete@colpos.mx

[4] Instituto de Silvicultura e Industria de la Madera, Universidad Juárez del Estado de Durango, Boulevard del Guadiana 501, Ciudad Universitaria, Torre de Investigación, Durango 34120, Mexico; jciroh@ujed.mx (J.C.H.-D.); jorge.chavez@ujed.mx (J.A.C.-S.)

[5] Instituto de Ciencias Agropecuarias, Universidad Autónoma del Estado de Hidalgo, Universidad Km. 1, Ex Hacienda de Aquetzalpa, Tulancingo, Hidalgo 43600, Mexico; rodris71@yahoo.com

* Correspondence: jprieto@ujed.mx; Tel.: +52-618-827-12-20

Received: 18 September 2018; Accepted: 26 October 2018; Published: 29 October 2018

Abstract: Substrate is a factor that significantly influences the quality and production costs of nursery seedlings. The objective of this study was to evaluate combinations of peat moss, composted pine bark, and fresh pine sawdust in order to identify the proportions that favour the quality of *Pinus engelmannii* Carr. seedlings and minimise the production costs in the nursery. Substrates were formed using mixtures of peat moss (15% to 50%), composted pine bark (15% to 50%) and fresh pine sawdust (20% to 70%), with 2, 4 and 6 g L^{-1} of controlled release fertilizer (Multicote®, Haifa, Israel). A completely randomised experimental design with a factorial arrangement of 7×3 was used. The evaluated factors are root collar diameter, biomass, N-P-K content, and production costs of the substrates which were determined based on the container volume and three commercial quotations. Significant differences were found in root collar diameter and biomass, highlighting the treatments using 50% to 70% sawdust with 6 g L^{-1} of fertilizer. Assimilated values of N-P-K were acceptable in all treatments with 4 and 6 g L^{-1} of fertilizer. In the substrates with high percentages of sawdust, seedlings with morphological characteristics and nutritional levels within the values recommended for conifers were produced. In addition, it was possible to reduce the production cost of the substrates by up to 67%.

Keywords: composted pine bark; fresh pine sawdust; seedling quality; peat moss; *Pinus engelmannii* Carr.

---

1. Introduction

Mexico has 42% of the world's total number of *Pinus* species, of which 47% are found naturally in the state of Durango in northwest Mexico [1]. Within this great diversity, *Pinus engelmannii* Carr. stands out due to its wide distribution in semi-dry and temperate forests of the Sierra Madre Occidental, which is a system of mountains in the west of Mexico. This species is found on hillsides of between 1900 and 2700 m in altitude, in areas with deep or poor and rocky soils [2]. During its initial growth, *P. engelmannii* has increases in diameter in relation to height [3]. From an economic point of view, this species is important due to the characteristics of the wood, which makes it widely used in reforestation programmes and commercial forest plantations. Hence, it is the second most produced species in forest nurseries in northern Mexico [4–6].

In the nursery, *P. engelmannii* reproduces by seed, and is generally planted in polystyrene containers or polyethylene tubes, with volumes of 170 to 220 mL, with growth media based on peat moss, pine bark, perlite and vermiculite [5,7]. The seedling quality varies in the different nurseries, hence, viable culture alternatives are sought that allow the production of individuals with suitable characteristics; the characteristics include: a well-developed root system, root collar diameter greater than 5 mm, balance between stem-root, foliage adapted to weather conditions, adequate carbohydrate reserves, and sufficient mineral nutrition [8,9]. These attributes, together with mineral nutrition, influence the growth and adaptation of individuals at the plantation site [10]. Therefore, it is important that fertilization is adequate, as it will promote growth and strengthen tolerance to stress in unfavourable environments [11].

Another important component in the seedling production process in the nursery is the substrate, since the properties of the constituent materials have an effect on seedling growth and future development [12]; and the materials are related to substrate quality [13]. In addition, substrate has a direct relationship with the costs of plant production. Research on the substitution of peat by local substrates, as a culture medium in nurseries, is of great interest [14–17], since this material has a significant influence on the final cost of seedling production [9]. Importantly, excessive extraction of peat in ecosystems with wetlands, where it is obtained, causes great environmental damage [14,18].

Recent studies have shown that fresh sawdust, as a component of the culture medium, serves as an alternative in the production of forest species, such as *Cedrela odorata* L. [19], *Pinus greggii* Engelm. [20], *Pinus montezumae* Lamb. [21] and *Pinus pseudostrobus* Lindl., among others [16,22]. However, while these types of studies have evaluated a few *Pinus* species that are located in the centre of the country, to our knowledge, there have been no such studies of species such as *P. engelmannii* in northern Mexico.

These considerations support the objective of evaluating combinations of peat moss, composted pine bark, and fresh pine sawdust, in combination with three doses of controlled-release fertilizer, in order to identify the proportions that favour seedling quality and minimise costs of production in the nursery. We propose the hypothesis that substrates which include high proportions of fresh sawdust in combination with controlled release fertilizers, allow the production of *P. engelmannii* seedlings with adequate morphological characteristics for optimal development, while reducing production costs.

2. Materials and Methods

2.1. Study Site

This study was carried out in the "General Francisco Villa" Forest Nursery, which is located at coordinates 23°58′20.38″ N and 104°35′55.83″ W, at an altitude of 1875 m, in the City of Durango, State of Durango, Mexico. The experiment was established in a greenhouse covered with 720 gauge plastic and 50% shade mesh, with overhead, lateral and frontal ventilation, and with an automated irrigation system having micro-sprinklers.

2.2. Establishment of the Experiment

The production seedling cycle was from 15 September 2015 to 20 June 2016 (10 months), and in this period the temperature and relative humidity were monitored (Figure 1). The sowing was performed in polystyrene containers of 77 cavities with 170 mL per cavity, using seeds of *P. engelmannii* that had been collected in the Coyote de Calaveras, an area which is located in Durango. Before sowing, the seeds received a pre-germinative treatment (soaking in water for 24 h at room temperature) and then they were disinfected for 5 min in a solution of 90% water and 10% chlorine. Finally, 2.5 g L^{-1} of fungicide (Captán® *N*-trichloromethylthio-4-cyclohexene-1,2-dicarboximide, Industrial engineer, Mexico City, Mexico) dissolved in water was applied.

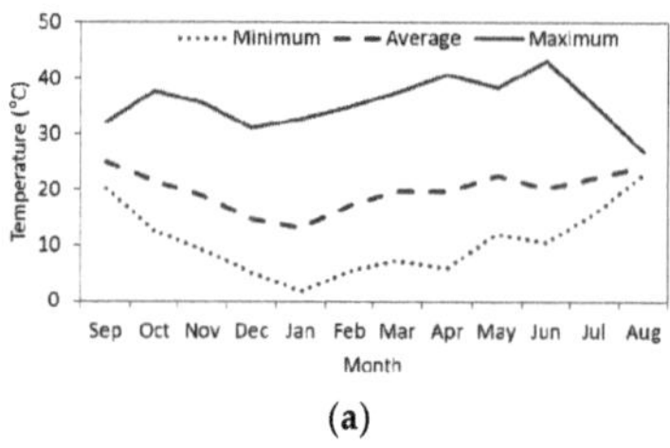 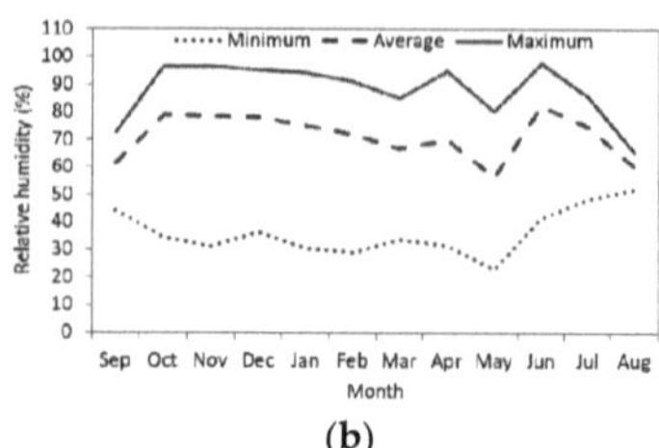

(**a**)   (**b**)

Figure 1. Temperature (**a**) and relative humidity (**b**) recorded during the production of *Pinus engelmannii* from September 2015 to June 2016.

2.3. Substrate and Fertilizer

Seven substrates composed of peat moss, composted bark of *Pinus douglasiana* Martínez, and fresh pine sawdust were evaluated, three weeks after being obtained from the sawing of logs of *Pinus engelmannii* Carr., *Pinus cooperi* Blanco, and *Pinus durangensis* Martínez (Table 1). Multicote® 18 N - 6 P_2O_5 + - 12 K_2O + 2 Mg + ME fertilizer (Haifa Chemicals Ltd., Hifa, Israel) was used, with 8 to 9 months of nutrient release, in three doses: 2 g L^{-1} (M2), 4 g L^{-1} (M4) and 6 g L^{-1} (M6). The said doses of fertilizer were incorporated into the substrate when preparing the mixture. With the interaction of the seven substrates and the three doses of fertilization, 21 treatments were formed, which were identified using the following keys: S1-M2, S1-M4, S1-M6, S2-M2, S2-M4, S2-M6, S3-M2, S3-M4, S3-M6, S4-M2, S4-M4, S4-M6, S5-M2, S5-M4, S5-M6, S6-M2, S6-M4, S6-M6, S7-M2, S7-M4 and S7-M6, where S refers to the substrate and M refers to the Multicote® fertilizer in its different mixtures and doses, respectively.

Table 1. Substrates evaluated in the production of *Pinus engelmannii* seedlings in the nursery, during the crop cycle of September 2015 to June 2016.

Code	Substrate
S1 (Control)	50% Peat moss + 50% Composted pine bark
S2	40% Peat moss + 40% Composted pine bark + 20% Fresh pine sawdust
S3	35% Peat moss + 35% Composted pine bark + 30% Fresh pine sawdust
S4	30% Peat moss + 30% Composted pine bark + 40% Fresh pine sawdust
S5	25% Peat moss + 25% Composted pine bark + 50% Fresh pine sawdust
S6	20% Peat moss + 20% Composted pine bark + 60% Fresh pine sawdust
S7	15% Peat moss + 15% Composted pine bark + 70% Fresh pine sawdust

In addition to the doses of controlled-release fertilizer, the seedlings in all of the treatments were irrigated using water-soluble fertilizers (Poly-Feed® Foliar, Haifa, Israel) during the production cycle, according to the growth phases of the seedlings. In the first or establishment phase (8%-52%-17%, N-P-K), the dose was from 40 to 70 ppm of nitrogen with a duration of 2.5 months. In the rapid growth phase (20%-9%-20%, N-P-K), the dose was 100 to 200 ppm of nitrogen with a duration of 2.5 months and, in the last or pre-conditioning phase (4%-25%-40%, N-P-K), the dose was 20 to 40 ppm of nitrogen, which lasted for two months. The three doses were applied during irrigation by using an automated system. This regimen of foliar fertilization is the one used in the Francisco Villa Nursery, combined with 3 g L^{-1} of Multicote®, when it is used as a substrate of 50% moss peat + 50% composted pine bark. In addition, the presence of pests and diseases was prevented with the application of insecticides (LORSBAN®, Tlaxcala, Mexico, CONFIDOR®, Leverkusen, Germany, ENGEO®, Mexico City, Mexico) and fungicides (CAPTAN®, Mexico City, Mexico, TECTO®, Mexico City, Mexico, DEROSAL®, Leverkusen, Germany) every 4 days.

2.4. Experimental Design

A completely randomised experimental design was used with a factorial arrangement of 7×3, with seven substrates and three doses of fertilization. The experimental unit consisted of 12 seedlings with four replications per treatment, with a total of 1008 seedlings being evaluated in the trial.

2.5. Variables Evaluated

2.5.1. Physical and Chemical Characteristics of Substrate Mixtures

The physical characteristics that were determined for the substrates were: aeration porosity (%), humidity retention porosity (%), and total porosity (%), using the method described by Landis [23], and the chemical properties were: pH and electrical conductivity (dS m^{-1}), which were based on NOM-021-RECNAT-2000. These studies were carried out at the beginning of production, in the Environmental Sciences Laboratory, of the Integral Rural Centre for Interdisciplinary Research for Integral Rural Development (CIIDIR is the Spanish-language acronym), which is a Durango-based unit of the National Polytechnic Institute (IPN is the Spanish-language acronym).

2.5.2. Morphological Variables

When the seedlings were ten months of age, the root collar diameter (mm) was measured with a digital vernier caliper, together with the dry aerial biomass, dry root biomass, and dry total biomass (g). To determine the biomass, the plants were dehydrated in a drying oven at 70 °C for 72 h, until reaching a constant weight. Then, the samples were weighed on an Ohaus® (Ohaus of Mexico, Mexico City, Mexico) analytical balance with an accuracy of 0.0001 g.

2.5.3. Concentrations of Nitrogen (N), Phosphorus (P) and Potassium (K)

Representative samples of dry foliage were integrated to determine the content of N, P and K; these consisted of needles from the middle part of each seedling, up to 5 g per treatment, with three repetitions. N content was determined by using the *Kjeldahl* method [24]. P content was determined by complex determination of yellow colour production of vanadomolybdate reagent in reaction with phosphates, and K content was determined by using atomic emission (the samples were digested using 60% nitric acid and 40% perchloric acid) [25]. With the obtaining of the concentration of each element, vector nomograms were constructed, based on the weight of 100 needles of each treatment [26].

2.5.4. Cost of the Substrate

The cost of the substrate was determined based on the volume of the cavity (170 mL) of the container, and was valued in US dollars based on three commercial quotations (Date: 15 March 2018).

2.6. Statistical Analysis

The data of the morphological variables were evaluated by the non-parametric statistical test of Kruskal-Wallis, due to the non-compliance of the normality assumptions; therefore, the Bonferroni-Dunn means separation test ($p \leq 0.05$) was used. The data were analyzed using the R statistical software version 3.2.3, Vienna, Austria [27].

3. Results

3.1. Physical and Chemical Characteristics of Substrate Mixtures

Aeration porosity ranged from 23.6% in S6 to 29.2% in S1; humidity retention porosity varied from 33.6% in S1 to 49.4% in S7; while total porosity presented a range of 62.8% in S1 to 76.3% in S7; in the latter two variables, the porosity increased as the proportion of sawdust increased. Regarding the chemical characteristics of the substrate mixtures, the pH and electrical conductivity presented slightly higher values in the substrates which included sawdust, in comparison with the control substrate (S1) (Table 2).

Table 2. Physical and chemical properties of the substrates evaluated in the production of *Pinus engelmannii*.

Substrate	Physical (%)			Chemical	
	Aeration Porosity	Humidity Retention Porosity	Total Porosity	pH	Electrical Conductivity (dS m^{-1})
S1	29.2	33.6	62.8	4.07	0.06
S2	28.1	39.2	67.3	4.47	0.08
S3	25.5	41.3	66.8	4.54	0.09
S4	26.4	40.6	67.0	4.65	0.09
S5	25.6	39.8	65.4	4.68	0.09
S6	23.6	46.1	69.7	4.82	0.07
S7	26.9	49.4	76.3	4.74	0.08
RV	25 to 35	25 to 55	60 to 80	5 to 6.5	<1.0

S1: 50% peat moss + 50% composted pine bark, S2: 40% peat moss + 40% composted pine bark + 20% fresh pine sawdust, S3: 35% peat moss + 35% composted pine bark + 30% fresh pine sawdust, S4: 30% peat moss + 30% composted pine bark + 40% fresh pine sawdust, S5: 25% peat moss + 25% composted pine bark + 50% fresh pine sawdust, S6: 20% peat moss + 20% composted pine bark + 60% fresh pine sawdust, S7: 15% peat moss + 15% composted pine bark + 70% fresh pine sawdust. RV: Recommended values [23,28].

3.2. Morphological Variables

3.2.1. Root Collar Diameter

The root collar diameter was not influenced significantly by the substrate or fertilizer doses. The interaction of the factors presented significant differences, with the highest values being related to the seedlings that were produced in the treatments S5-M2 (7.41 mm) and S4-M4 (7.35 mm) (Table 3, Figure 2).

3.2.2. Dry Aerial Biomass

The dry aerial biomass did not present significant differences at the substrate level, while at the fertilizer level it was significantly higher at the 6 g L^{-1} dose with 4.74 g. The interaction of the factors showed significant differences, highlighting the S7-M6 treatment (5.40 g) (Table 3, Figure 2).

3.2.3. Dry Root Biomass

The dry root biomass showed significant differences at the substrate level, in which the substrate with the highest proportion of sawdust S7 (1.35 g) was notable, while the effect of the fertilising factor was non-significant. The interaction of the factors presented highly significant differences, and the treatment with the highest value was S7-M6 (1.56 g) (Table 3, Figure 2).

3.2.4. Dry Total Biomass

The dry total biomass did not show significant differences in its influence over the substrate, while the fertilizer significantly influenced with the dose of 6 g L^{-1} with 5.96 g in the upper level. The substrate and fertilizer interaction was significant, within which the treatment S7-M6 (6.96 g) stood out. In all the morphological variables that were evaluated, the treatments that included sawdust were superior to the control, whereas the control only included peat moss and composted pine bark (Table 3, Figure 2).

Table 3. Significance values (*p* value) based on the response of the substrate, the fertilizer and its interaction, in the variables evaluated in *Pinus engelmannii*, with the non-parametric Kruskal-Wallis test.

Variable	Substrate	Fertilizer	Substrate × Fertilizer
Root collar diameter	0.3276 [ns]	0.3506 [ns]	<0.0001 ***
Dry aerial biomass	0.6821 [ns]	<.0010 **	0.0010 **
Dry root biomass	0.0223 *	0.4116 [ns]	<0.0010 **
Dry total biomass	0.7577 [ns]	0.0057 **	0.0060 **

Significant differences are denoted by: *** $p < 0.001$, ** $p < 0.01$, * $p < 0.05$, and [ns]: non-significant.

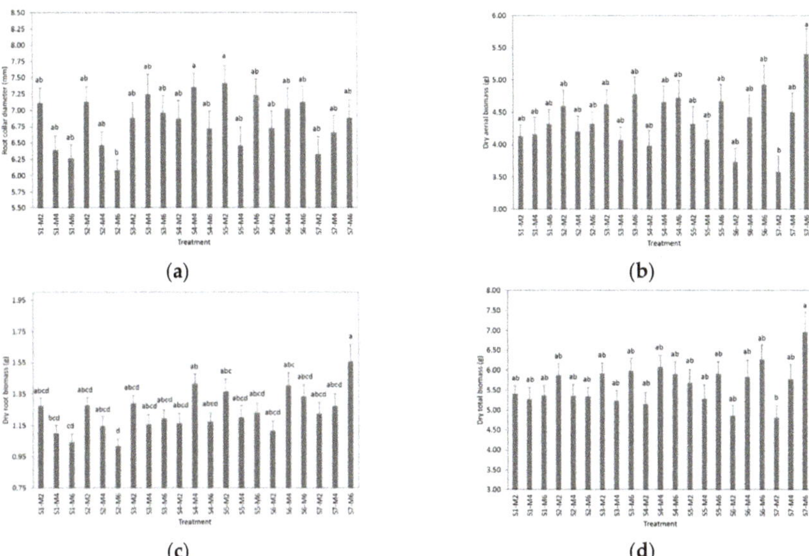

Figure 2. Variables evaluated in *Pinus engelmannii*: Root collar diameter (**a**); dry aerial biomass (**b**); dry root biomass (**c**) and dry total biomass (**d**). Differences between letters indicate significant differences between treatments (Bonferroni, $\alpha = 0.05$).

3.3. Nutrimental Analysis of the Foliage

In the concentrations of N, P and K, no significant differences were found between treatments (Table 4). Despite this, relationship of concentrations and needles weight generated variable responses in the vector nomograms of each element, based on the substrates and doses evaluated.

3.3.1. Nitrogen (N)

The nomogram of N with 2 g L^{-1} Multicote®, indicates that the treatments S2-M2, S3-M2, S4-M2, and S5-M2 generated an increase in the biomass of needles, but had a lower concentration of the element with respect to S1-M2 (substrate without sawdust). Meanwhile, the treatments S6-M2 and S7-M2 presented an amount of biomass of needles that was similar to S1-M2, but with a lower concentration of N (Figure 3a).

Table 4. Means of N, P and K concentrations in foliage of *Pinus engelmannii*.

Treatment	N (%)	P (%)	K (%)
S1-M2	1.6 ± 0.01	0.31 ± 0.01	1.3 ± 0.16
S1-M4	1.7 ± 0.15	0.30 ± 0.02	1.2 ± 0.02
S1-M6	1.7 ± 0.03	0.31 ± 0.01	1.2 ± 0.01
S2-M2	1.3 ± 0.17	0.31 ± 0.03	1.2 ± 0.08
S2-M4	1.7 ± 0.11	0.31 ± 0.00	1.1 ± 0.08
S2-M6	1.7 ± 0.02	0.30 ± 0.01	1.2 ± 0.11
S3-M2	1.4 ± 0.04	0.27 ± 0.00	1.2 ± 0.12
S3-M4	1.6 ± 0.10	0.29 ± 0.03	1.1 ± 0.10
S3-M6	1.5 ± 0.20	0.31 ± 0.01	1.2 ± 0.14
S4-M2	1.3 ± 0.12	0.30 ± 0.03	1.5 ± 0.14
S4-M4	1.5 ± 0.17	0.31 ± 0.00	1.2 ± 0.14
S4-M6	1.6 ± 0.08	0.29 ± 0.01	1.4 ± 0.16
S5-M2	1.4 ± 0.05	0.26 ± 0.02	1.1 ± 0.01
S5-M4	1.6 ± 0.04	0.29 ± 0.01	1.1 ± 0.10
S5-M6	1.7 ± 0.16	0.29 ± 0.03	1.2 ± 0.18
S6-M2	1.3 ± 0.01	0.27 ± 0.02	1.3 ± 0.05
S6-M4	1.4 ± 0.07	0.29 ± 0.01	1.1 ± 0.08
S6-M6	1.5 ± 0.14	0.28 ± 0.02	1.3 ± 0.06
S7-M2	1.3 ± 0.05	0.28 ± 0.01	1.2 ± 0.04
S7-M4	1.4 ± 0.07	0.28 ± 0.01	1.2 ± 0.06
S7-M6	1.5 ± 0.04	0.27 ± 0.00	1.2 ± 0.03
p value	0.7543	0.9037	0.9976

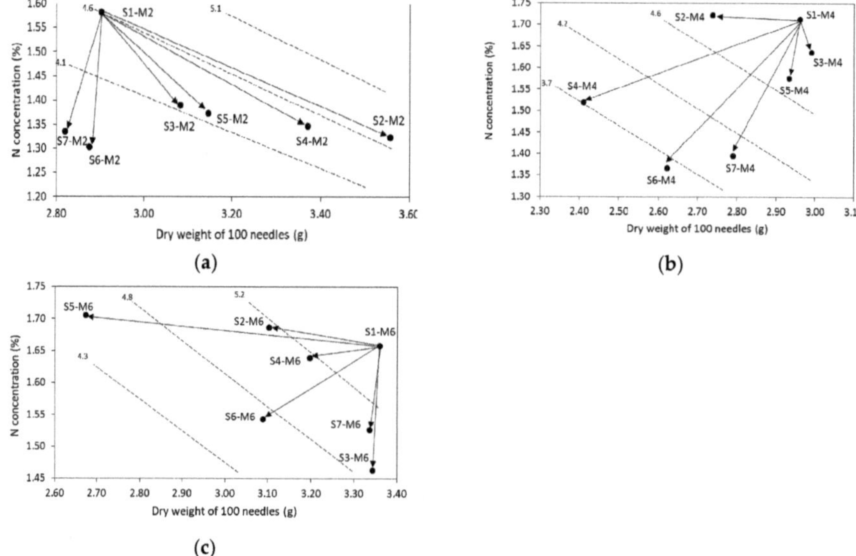

Figure 3. Nomograms of nitrogen (N) concentration vectors in the foliage of *Pinus engelmannii*, produced on seven substrates with the addition of three fertilization doses: (**a**) 2 g L^{-1} Multicote®; (**b**) 4 g L^{-1} Multicote® and (**c**) 6 g L^{-1} Multicote®.

On the other hand, according to the nomogram where 4 g L^{-1} of Multicote® was added, the S3-M4 treatment showed a greater increase in the biomass of needles in relation to S1-M4, but had a slight decrease in N; treatments S4-M4, S5-M4, S6-M4 and S7-M4 showed a smaller increase in the biomass of needles and lower concentration of N than S1-M4; and finally, the S2-M4 treatment showed a lower increase in biomass of needles, and the same concentration of N (Figure 3b).

In the nomogram with 6 g L^{-1} of Multicote®, it can be observed that the treatments S3-M6 and S7-M6 present a biomass of needles similar to S1-M6, but with a lower concentration of N. On the other hand, the treatments S4-M6 and S6-M6 show a lower biomass of needles and a lower concentration of N. Finally, treatments S2-M6 and S5-M6 present a lower biomass of needles and a similar concentration of N, compared to S1-M6 (Figure 3c).

3.3.2. Phosphorus (P)

The nomogram with the dose of 2 g L^{-1} Multicote®, shows that the treatments S2-M2 and S4-M2 generated a greater increase in the biomass of needles and similar concentration, with respect to the S1-M2. On the other hand, treatments S3-M2 and S5-M2 show a higher biomass of needles, but with a lower concentration of P, while treatments S6-M2 and S7-M2 present a lower biomass of needles and a lower concentration of P (Figure 4a).

On the other hand, in the nomogram with the addition of 4 g L^{-1} of Multicote®, only S3-M4 presents a biomass in needles that is similar to S1-M4, but with a lower concentration of P; treatments S5-M4, S6-M4 and S7-M4 show a lower increase in the biomass of needles and a lower concentration than S1-M4; finally, in the treatments S2-M4 and S4-M4 there was a smaller increase in biomass of needles, and a slight increase in the concentration of P (Figure 4b).

In the nomogram with 6 g L^{-1} of Multicote®, treatment S3-M6 has generated the same amount of biomass of needles and concentration of P as SI-M6, while S7-M6 shows the same biomass of needles but lower concentration of P; finally, the treatments S2-M6, S4-M6, S5-M6 and S6-M6 have a lower biomass of needles and a lower concentration of P, all in comparison to S1-M6 (Figure 4c).

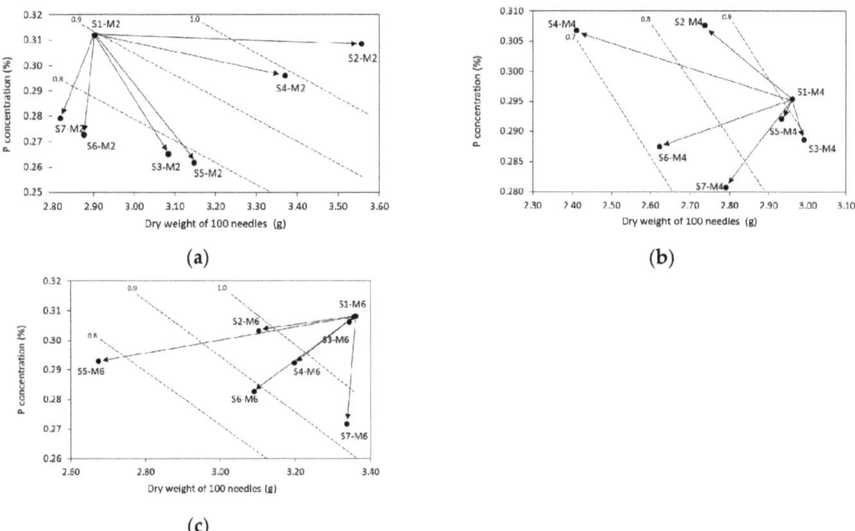

Figure 4. Vector nomograms of the concentration of phosphorus (P) in the foliage of *Pinus engelmannii*, produced in seven substrates with the addition of three doses of fertilization: (**a**) 2 g L^{-1} Multicote®; (**b**) 4 g L^{-1} Multicote® and (**c**) 6 g L^{-1} Multicote®.

3.3.3. Potassium (K)

The nomogram with 2 g L^{-1} of Multicote®, shows that the treatment S4-M2 generated a greater increase in the biomass of needles and a higher concentration of K with respect to the S1-M2. On the other hand, the treatments S2-M2, S3-M2 and S5-M2 achieved a greater biomass of needles than S1-M2, but with a lower concentration of K; while the treatments S6-M2 and S7-M2 had a lower biomass of needles and lower concentration of K (Figure 5a).

On the other hand, in the nomogram with the addition of 4 g L^{-1} of Multicote®, only the S3-M4 attained a greater increase of biomass in needles in relation to S1-M4, but with a lower concentration of K; treatments S2-M4, S5-M4 and S6-M4 show a lower increase in the biomass of needles and a lower concentration of K than S1-M4. Finally, treatments S4-M4 and S7-M4 show a lower increase in biomass of needles, but a higher concentration of K (Figure 5b).

In the nomogram with 6 g L^{-1} Multicote®, the treatments S3-M6 and S7-M6 have practically the same amount of biomass and concentration as the S1-M6 treatment, while the treatments S4-M6 and S6-M6 express a lower biomass of needles but with higher concentrations of K; the S5-M6 treatment has a lower biomass of needles and an equal concentration of K. Finally, the S2-M6 treatment presents a lower biomass of needles and a lower concentration of K (Figure 5c).

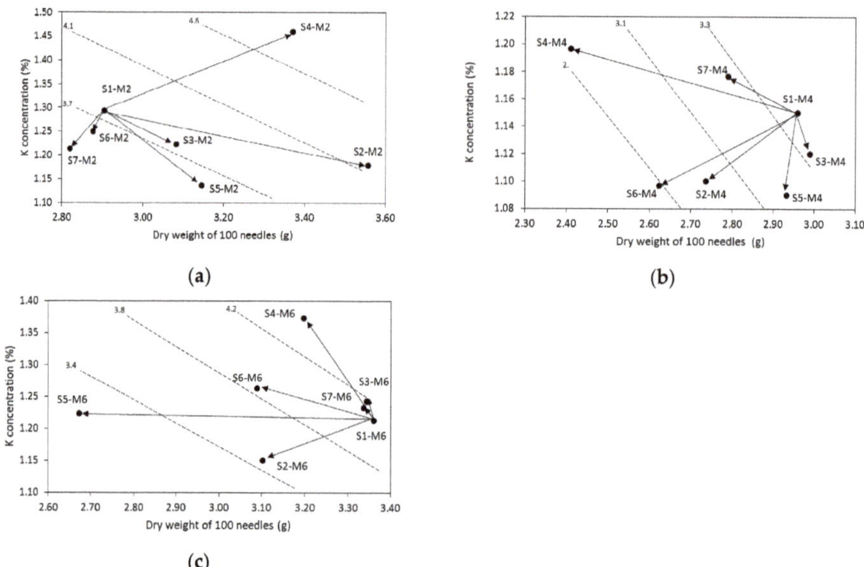

Figure 5. Vector nomograms of the potassium concentration (K) in the foliage of *Pinus engelmannii*, produced in seven substrates with the addition of three doses of fertilization; (a) 2 g L^{-1} Multicote®; (b) 4 g L^{-1} Multicote® and (c) 6 g L^{-1} Multicote®.

3.4. Substrate Cost

The cost per litre of substrate was valued in dollars as follows: S1 = 3.49, S2 = 2.82, S3 = 2.49, S4 = 2.15, S5 = 1.82, S6 = 1.48 and S7 = 1.15 ($). Based on the previous quotations, the substrates that contain sawdust are between 23% and 67% more economical, in relation to the substrate that contains only peat moss and composted pine bark. With the costs obtained from the substrate, an extrapolation was carried out in a production of 100,000 plants (Figure 6), where the investment in the seedling production by substrates appreciates when including the fresh sawdust.

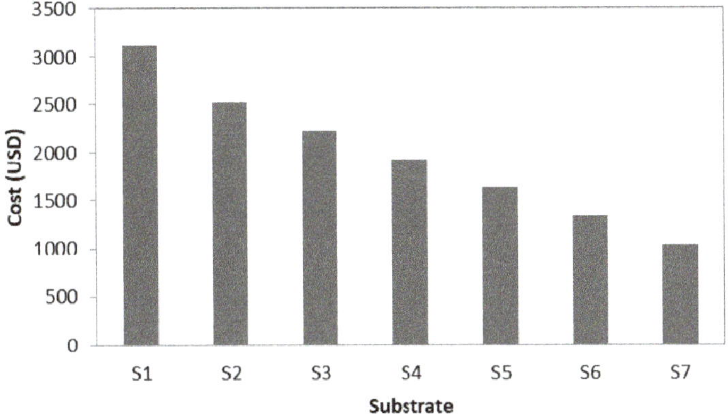

Figure 6. Projected substrate cost in a production of 100,000 seedlings of *Pinus engelmannii*, produced in polystyrene containers with 77 cavities of 170 mL.

4. Discussion

4.1. Physical and Chemical Characteristics of Substrate Mixtures

In general, the substrates that were evaluated in the present study had percentages of moisture retention within the recommended values, which were increasing in the mixtures when increasing the percentage of sawdust, due to the size of the fine particles of that sawdust. As a consequence, aeration porosities were lower in the mixtures with higher percentage of sawdust, without presenting root rot problems due to excess moisture, since the size and composition of the pine bark particles provide a greater flow of air, and greater water filtration [29], although it is inevitable that the porosity decreases with time due to the compaction and settlement of the substrate by way of constant humidity changes [30].

The three porosities that were evaluated are within the recommended ranges for producing coniferous seedlings (Table 1) [23,28]. This indicates that the evaluated substrates retain sufficient moisture, due to the degree of porosity that was determined in this study, which generated a broad root system that favoured the growth of the seedlings, and that root system is expected to help withstand severe growing conditions in the field [15]. Other works in the body of literature agree on the feasibility of including fresh sawdust in substrates when producing *Pinus spp.*, and on which substrates can be formulated regarding their appropriate physical and chemical characteristics [16,17,21].

Because these properties can change according to time and source, they must always be tested for each condition [9], since the physical and chemical properties of organic waste can change according to type and origin [31]. In this case, the materials used in this study did not cause problems of diseases and pests, nor did they cause toxicity in the seedlings when using sawdust without composting.

The pH and electrical conductivity are also important properties, as they influence the availability of nutrients in growth media [32]. On the one hand, the pH presented values with a strong acidity in all of the treatments, but the S6 substrate (pH of 4.82) with sawdust presented a value that was closer to the recommended value (pH of 5). However, pH values increase during the production process, due to watering. Substrates including composted pine bark and sawdust have been found with pH values of 4.3 to 4.7 [21], that is, pH values that are very close to those reported in the present study. In general, pH values within the range of 4.0 to 6.0 are typical in substrates with pine bark and moss in the production of various species under nursery conditions [33].

On the other hand, the values that were reached in the electrical conductivity are also low in comparison with the recommended values; however, these values must not exceed 3.5 dS m^{-1}, since they could cause salinity problems [28].

4.2. Morphological Variables

4.2.1. Root Collar Diameter

The most widely used morphological attribute with which to estimate the quality of the seedlings is root collar diameter [8,11,34]. Because the survival and growth of plants in the field is associated with the root system of the seedling, the diameter has been defined as an important indicator [35,36]. In this sense, treatments with S4-M4 (30% peat + 30% bark + 40% sawdust + 4 g L^{-1} of Multicote$^{®}$) and S5-M2 (25% peat + 25% bark + 50% sawdust + 2 g L^{-1} of Multicote$^{®}$) stood out from the rest with respect to this variable; however, the criteria established by the NMX-AA-170-SCFI-2016 standard to produce *P. engelmannii*, establishes a root collar diameter of $\geq$ 5 mm [37]; that is, all treatments exceed this parameter.

For this variable, in *Pinus montezumae* Lamb. (species of similar growth), average values of 7.26 mm have been reported, and they were produced in substrate with composted pine sawdust (70%) + composted pine bark (15%) + vermiculite (15%) in combination with 4, 6 and 8 g L^{-1} of Multicote$^{®}$ [17]. Also, in the same species, values of 9.7 to 11.4 mm are reported in substrates with different proportions of fresh pine sawdust (20% to 80%) in combination with peat moss (10% to 80%), composted pine bark (10%), perlite (10%), and vermiculite (10%) [21]. The information in the above indicates that the inclusion of sawdust in high proportions does promote suitable conditions for the seedlings to reach the appropriate diameter.

4.2.2. Dry Biomass (Aerial, Root, and Total)

In general, dry biomass was favoured in the treatments with the highest proportion of sawdust (40% to 70%), with addition of 4 and 6 g L^{-1} of Multicote$^{®}$; that is, with values higher than those found in the treatments that include the control substrate (S1-M2, S1-M4, and S1-M6). This can be attributed to the fact that the substrates with sawdust retained more moisture, a condition that could increase the availability of nutrients in these seedlings. This indicates that the fertilization scheme that was used (controlled-release fertilization plus foliar fertilization) was sufficient to prevent competition between the microorganisms and the seedlings, based on their dry biomass and diameter. Aguilera-Rodríguez et al. [17] reported similar values in aerial biomass (3.28–4.41), root biomass (0.91–1.11) and total biomass (4.19–5.51), when evaluating substrates with composted pine sawdust (70%) + composted pine bark (15%) + vermiculite (15%) in combination with 4, 6 and 8 g L^{-1} of Multicote$^{®}$ in the production of *P. montezumae*.

In general, the root system of the seedlings produced in the different treatments, presented good physiognomy based on the points of growth observed in the root, possibly due to the ratio of the dry root biomass and the growth potential [8]. These results suggest a good performance of these seedlings in the field, since at the beginning of their establishment they will depend on the root system that was created in nursery [38], because the growth of the root is a determining factor in ensuring the development and establishment of the plantation [8].

4.3. Nutrimental Analysis of the Foliage

When observing the nomograms of N, P and K, it is evident that with the dose of 2 g L^{-1} of Multicote$^{®}$ an increase of biomass of needles is generated in most of the treatments that include fresh pine sawdust, but with effects of dilution of the elements, with reference to the treatment that only includes peat moss and composted bark (S1-M2). On the other hand, in the doses of 4 and 6 g L^{-1} of Multicote$^{®}$, it can be seen that the treatments containing sawdust are below their referents (S1-M4 and S1-M6) for the three elements. This decrease in the concentration of the elements may be due to the greater growth in aerial biomass, root biomass, and total biomass in plants grown on substrates that include sawdust, derived from their use in the production of photosynthates.

On the other hand, these concentrations of N, P and K are within the recommended levels of sufficiency for conifers (N 1.40%–2.20%, P 0.20%–0.40%, and K 0.40%–1.50%, respectively) [39], in 17 treatments, and only in the case of N did the treatments S2-M2, S4-M2, S6-M2 and S7-M2 not reach the minimum value. However, in the ranges recommended for the species (N 1.1%–3.5%, P 0.1%–0.6% and K 0.2%–2.5%) [7], all treatments presented an acceptable value.

In *P. montezumae* produced in composite substrate with 70% compost pine sawdust + 15% compost pine bark + 15% vermiculite, with added high fertilization doses (8 g L^{-1} of Multicote$^{®}$ and 8 g L^{-1} of Osmocote Plus$^{®}$), Aguilera-Rodríguez et al. [17] reported slightly lower values (0.79% to 1.33% in N, 0.06% to 1.16% in P, and 0.12% to 0.31% in K) compared to the values in the present study, despite using composted sawdust. It is possible that this was due to the fact that the controlled release fertilizers were not accompanied by foliar fertilization, or it is possible that this was because they had already been assimilated by the seedlings.

4.4. Costs

In this study, sawdust substrates were up to 67% cheaper than those substrates that include peat moss and bark, due to the low cost of sawdust, being that it is a locally-produced material. Otherwise, peat increases its cost due to increases in fuel consumption related to logistics, since the peat has to be imported from other countries, mainly from the U.S.A., hence it is important to replace or gradually reduce its use [40,41].

The main objective of the seedlings producers is to minimise the cost and to improve the quality of the seedlings [42]. Therefore, it is recommended that the managers of the forest nurseries use local materials in their culture media [41,43,44]. In addition, special attention should be paid to the use of simple and environmentally acceptable materials that are easily accessible to nursery producers [40,45]. Composted materials have disadvantages, because developing countries are not necessarily equipped with the appropriate infrastructure to treat and compost solid waste, which generates pollutants that negatively affect human health and that contaminate the water table [40]. For this reason, the inclusion of high proportions of fresh pine sawdust turns out to be a viable option and with favourable results in the morphology of *P. engelmannii*, likewise presenting no toxic waste or health risks during its production in the nursery.

5. Conclusions

The species *P. engelmannii* can be produced in a nursery in substrates with high proportions of fresh sawdust (up to 70%), with nutrient reserves (N, P and K) within the ranges recommended for the species. Based on the price of peat moss, composted bark and fresh sawdust, substrates with these materials can be 23% to 67% cheaper than the control substrate, depending on the proportion of sawdust in the mix. Finally, it is recommended to carry out tests to corroborate whether, after planting in the field, the same tendency of the best substrates and fertilization doses found in the nursery is maintained.

Author Contributions: M.M.G.O. was involved in the establishment of the experiment, collection, capture and statistical analysis of the data and in the writing of the manuscript; J.Á.P.R. was involved in the design and establishment of the experiment and a review of the manuscript; A.A. was involved in the counselling regarding the treatments to be evaluated and also in a review of the manuscript; J.C.H.D. was involved in the counselling on the results of the analysis of costs and also in a review of the manuscript; J.A.C.S. was involved in a review of the manuscript; R.R.L. was involved in a review of the manuscript.

Acknowledgments: The authors wish to express their gratitude to Roberto Trujillo and Roberto Trujillo Ayala, who are managers of the "General Francisco Villa" Forest Nursery, for the facilities provided there for carrying out the experiments related to this study. Appreciation goes to Manuel Aguilera Rodríguez for the advice in the establishment of the experiment. Finally, thanks go to CONACyT, for the scholarship granted to the first author.

Conflicts of Interest: The authors declare no conflict of interest.

References

1. Sánchez-González, A. Diversity and distribution of Mexican pines, an overview. *Madera Bosques* **2008**, *14*, 107–120. [CrossRef]
2. González-Elizondo, M.S.; González-Elizondo, M.; Tena-Flores, J.A.; Ruacho-González, L.; López-Enríquez, I.L. Vegetación de la Sierra Madre Occidental, México: Una síntesis. *Acta Bot. Mex.* **2012**, 351–405. [CrossRef]
3. García, A.A.; González, E.M.S. *Pináceas de Durango*; Instituto de Ecología, Comisión Nacional Forestal, Ed.; Instituto de Ecología A.C. y Comisión Nacional Forestal: León, Guanajuato, Mexico, 2003; ISBN 968-6021-18-3.
4. Ávila-Flores, I.J.; Hernández-Díaz, J.C.; González-Elizondo, M.S.; Prieto-Ruíz, J.Á.; Wehenkel, C. Degree of hybridization in seed stands of *Pinus engelmannii* Carr. in the Sierra Madre Occidental, Durango, Mexico. *PLoS ONE* **2016**, *11*, e0152651. [CrossRef] [PubMed]
5. Ávila-Flores, I.J.; Prieto-Ruíz, J.A.; Hernández-Díaz, J.C.; Wehenkel, C.A.; Corral-Rivas, J.J. Preconditioning *Pinus engelmannii* Carr. seedlings by irrigation deficit in nursery. *Rev. Chapingo Ser. Cienc. For. y del Ambient.* **2014**, *20*, 237–245. [CrossRef]
6. Prieto, R.J.A.; Cornejo, O.E.H.; Domínguez, C.P.A.; de J. Návar, C.J.; Marmolejo, M.J.G.; Jiménez, P.J. Water stress in *Pinus engelmannii* Carr., under nursery conditions. *Investig. Agrar. Sist. Recur.* **2004**, *13*, 443–451.
7. Prieto, R.J.Á; Sáenz, R.T.J. *Indicadores de calidad de planta en viveros forestales de la Sierra Madre Occidental*, Primera ed.; Centro Nacional de Investigaciones Forestales Agrícolas y Pecuarias, Instituto Nacional de Investigaciones Forestales, Agrícolas y Pecuarias, Campo Experimental Uruapan. del G.: Durango, Mexico, 2011; ISBN 978-607-425-716-8.
8. Grossnickle, S.C.; MacDonald, J.E. Why seedlings grow: Influence of plant attributes. *New For.* **2017**. [CrossRef]
9. Mañas, P.; Castro, E.; de Las Heras, J. Quality of maritime pine (*Pinus pinaster* Ait.) seedlings using waste materials as nursery growing media. *New For.* **2009**, *37*, 295–311. [CrossRef]
10. Viera, M.; Ruíz Fernández, F.; Rodríguez-Soalleiro, R. Nutritional prescriptions for *Eucalyptus* plantations: Lessons learned from Spain. *Forests* **2016**, *7*, 84. [CrossRef]
11. Merine, A.K.; Rodríguez-García, E.; Alía, R.; Pando, V.; Bravo, F. Effects of water stress and substrate fertility on the early growth of *Acacia senegal* and *Acacia seyal* from Ethiopian Savanna woodlands. *Trees* **2015**, *29*, 593–604. [CrossRef]
12. Mondragón-Valero, A.; Lopéz-Cortés, I.; Salazar, D.M.; de Córdova, P.F. Physical mechanisms produced in the development of nursery almond trees (*Prunus dulcis* Miller) as a response to the plant adaptation to different substrates. *Rhizosphere* **2017**, *3*, 44–49. [CrossRef]
13. Dumroese, R.K.; Sung, S.J.S.; Pinto, J.R.; Ross-Davis, A.; Scott, D.A. Morphology, gas exchange, and chlorophyll content of longleaf pine seedlings in response to rooting volume, copper root pruning, and nitrogen supply in a container nursery. *New For.* **2013**, *44*, 881–897. [CrossRef]
14. Aleandri, M.P.; Chilosi, G.; Muganu, M.; Vettraino, A.; Marinari, S.; Paolocci, M.; Luccioli, E.; Vannini, A. On farm production of compost from nursery green residues and its use to reduce peat for the production of olive pot plants. *Sci. Hortic.* **2015**, *193*, 301–307. [CrossRef]
15. Wisdom, B.; Nyembezi, M.; Agathar, K. Effect of different vermiculite and pine bark media substrates mixtures on physical properties and spiral rooting of radish (*Raphanus sativus* L.) in float tray system. *Rhizosphere* **2017**, *3*, 67–74. [CrossRef]
16. Aguilera, R.M.; Aldrete, A.; Martínez, T.T.; Ordaz, C.V.M. Production of *Pinus pseudostrobus* Lindl. with sawdust substrates and controlled release fertilizers. *Rev. Mex. Cienc. For.* **2016**, *7*, 7–19.
17. Aguilera-Rodríguez, M.; Aldrete, A.; Martínez-Trinidad, T.; Ordáz-Chaparro, V.M. Production of *Pinus montezumae* Lamb. with different substrates and controlled release fertilizers. *Agrociencia* **2016**, *50*, 107–118.
18. Mañas, P.; Castro, E.; Vila, P.; de las Heras, J. Use of waste materials as nursery growing media for *Pinus halepensis* production. *Eur. J. For. Res.* **2010**, *129*, 521–530. [CrossRef]
19. Mateo-Sánchez, J.J.; Bonifacio-Vázquez, R.; Pérez-Ríos, S.R.; Mohedano-Caballero, L.; Capulín-Grande, J. *Cedrela odorata* L. production in a raw sawdust substrate in technician system at Tecpan de Galeana, Guerrero, México. *Ra Ximhai* **2011**, *7*, 123–132.

20. Maldonado-Benitez, K.R.; Aldrete, A.; López-Upton, J.; Vaquera-Huerta, H.; Cetina-Alcalá, V.M. Nursery production of *Pinus greggii* Engelm. in mixtures of substrate with hydrogel and irrigation levels. *Agrociencia* **2011**, *45*, 389–398.

21. Hernandez-Zarate, L.; Aldrete, A.; Ordaz-Chaparro, V.M.; López-Upton, J.; López-López, M.Á. Nursery growth of *Pinus montezumae* Lamb. influenced by different substrate mixtures. *Agrociencia* **2014**, *48*, 627–637.

22. Reyes-Reyes, J.; Aldrete, A.; Cetina-Alcalá, V.M.; López-Upton, J. Seedlings production of *Pinus pseudostrobus* var. *Apulcensis in substrates based on sawdust. Rev. Chapingo Ser. Cienc. For. Ambient.* **2005**, *11*, 105–110.

23. Landis, T.D. Containers: Types and Functions. In *The Container Tree Nursery Manual, Volume 2*; Landis, T.D., Tinus, R.W., McDonald, S.E., Barnett, J.P., Eds.; U.S. Department of Agriculture, Forest Service: Washington, DC, USA, 1990; Volume 2, pp. 41–89.

24. Kjeldahl, J. Neue methode zur bestimmung des stickstoffs in organischen körpern. [New method for the determination of nitrogen in organic substances.]. *Z. Anal. Chem.* **1883**, *22*, 366–382. [CrossRef]

25. Mckean, S.J. *Manual de Análisis de Suelos Y Tejido Vegetal (Soils and Vegetal Tissue Analysis Manual)*; Laboratorio de Servicios Analíticos. Centro Internacional de Agricultura Tropical (CIAT): Cali, Colombia, 1993; p. 103.

26. Timmer, V. Interpretation of seedling analysis and visual symptoms. In *Mineral Nutrition of Conifer Seedlings*; van den Driessche, R., Ed.; CRC Press: Boca Raton, FL, USA, 1991; pp. 113–134.

27. R Core Team R: A Language and Environment For Statistical Computing. R Foundation for Statistical Computing. Available online: https://www.r-project.org/ (accessed on 2 July 2018).

28. Mathers, H.M.; Lowe, S.B.; Scagel, C.; Struve, D.K.; Case, L.T. Abiotic factors influencing root growth of woody nursery plants in containers. *Hort. Technol.* **2007**, *17*, 151–162.

29. Cruz-Crespo, E.; Can-Chulim, A.; Sandoval-Villa, M.; Bugarín-Montoya, R.; Robles-Bermúdez, A.; Juárez-López, P. Substrates in horticulture. *Rev. Biol. Cienc.* **2013**, *2*, 17–26.

30. Bakry, M.; Lamhamedi, M.S.; Caron, J.; Bernier, P.Y.; El Abidine, A.Z.; Stowe, D.C.; Margolis, H.A. Changes in the physical properties of two Acacia compost-based growing media and their effects on carob (*Ceratonia siliqua* L.) seedling development. *New For.* **2013**, *44*, 827–847. [CrossRef]

31. Ostos, J.C.; López-Garrido, R.; Murillo, J.M.; López, R. Substitution of peat for municipal solid waste- and sewage sludge-based composts in nursery growing media: Effects on growth and nutrition of the native shrub *Pistacia lentiscus* L. *Bioresour. Technol.* **2008**, *99*, 1793–1800. [CrossRef] [PubMed]

32. Manh, V.H.; Wang, C.H. Vermicompost as an important component in substrate: Effects on seedling quality and growth of muskmelon (*Cucumis Melo* L.). *Procedia APCBEE* **2014**, *8*, 32–40. [CrossRef]

33. Altland, J.E.; Locke, J.C.; Krause, C.R. Influence of pine bark particle size and pH on cation exchange capacity. *Hort. Technol.* **2014**, *24*, 554–559.

34. Luis, V.C.; Puértolas, J.; Climent, J.; Peters, J.; González-Rodríguez, Á.M.; Morales, D.; Jiménez, M.S. Nursery fertilization enhances survival and physiological status in Canary Island pine (*Pinus canariensis*) seedlings planted in a semiarid environment. *Eur. J. For. Res.* **2009**, *128*, 221–229. [CrossRef]

35. Thompson, B.E. Seedling morphological evaluation—What you can tell by looking. In *Proceedings: Evaluating Seedling Quality: Principles, Procedures, and Predictive Abilities of Major Tests*; Durges, M.L., Ed.; Forest Research Laboratory, Oregon State University: Corvallis, OR, USA, 1985; pp. 59–71. ISBN 0874370000.

36. Batziou, M.; Milios, E.; Kitikidou, K. Is diameter at the base of the root collar a key characteristic of seedling sprouts in a *Quercus pubescens–Quercus frainetto* grazed forest in north-eastern Greece? A morphological analysis. *New For.* **2017**, *48*, 1–16. [CrossRef]

37. Secretaría de Economía NMX-AA-170-SCFI-2016. Certificación de la Operación de Viveros Forestales. Diario Oficial de la Federación. Available online: http://www.dof.gob.mx/nota_detalle.php?codigo=5464460&fecha=07/12/2016 (accessed on 7 December 2016).

38. Grossnickle, S.C. Importance of root growth in overcoming planting stress. *New For.* **2005**, *30*, 273–294. [CrossRef]

39. Landis, T.D. Mineral nutrients and fertilization. In *The Container Tree Nursery Manual, Volume 4*; Landis, T.D., Tinus, R.W., McDonald, S.E., Barnett, J.P., Eds.; U.S. Department of Agriculture, Forest Service: Washington, DC, USA, 1989; pp. 1–67.

40. Bakry, M.; Lamhamedi, M.S.; Caron, J.; Margolis, H.; El Abidine, A.Z.; Bellaka, M.; Stowe, D.C. Are composts from shredded leafy branches of fast-growing forest species suitable as nursery growing media in arid regions? *New For.* **2012**, *43*, 267–286. [CrossRef]

41. Tian, N.; Fang, S.; Yang, W.; Shang, X.; Fu, X. Influence of container type and growth medium on seedling growth and root morphology of *Cyclocarya paliurus* during nursery culture. *Forests* **2017**, *8*, 387. [CrossRef]
42. Bashir, A.; Qaisar, K.N.; Khan, M.A.; Majeed, M. Benefit-cost analysis of raising *Pinus wallichiana* seedlings in different capacities/sizes of root trainers in the nursery. *For. Stud. China* **2009**, *11*, 118–121. [CrossRef]
43. Akpo, E.; Stomph, T.J.; Kossou, D.K.; Omore, A.O.; Struik, P.C. Effects of nursery management practices on morphological quality attributes of tree seedlings at planting: The case of oil palm (*Elaeis guineensis* Jacq.). *For. Ecol. Manag.* **2014**, *324*, 28–36. [CrossRef]
44. Heiskanen, J. Effects of compost additive in *sphagnum* peat growing medium on Norway spruce container seedlings. *New For.* **2013**, *44*, 101–118. [CrossRef]
45. Reyes, R.J.; Pimienta de la Torre, D.d.J.; Morales, R.J.A.; Pérez, F.M.A.; Figueroa, P.E. Plant quality of *Gmelina arborea* Roxb. produced with different substrate mixtures at the nursery. *Rev. Mex. Cienc. For.* **2018**, *9*, 111–130.

MDPI
St. Alban-Anlage 66
4052 Basel
Switzerland
Tel. +41 61 683 77 34
Fax +41 61 302 89 18
www.mdpi.com

Forests Editorial Office
E-mail: forests@mdpi.com
www.mdpi.com/journal/forests

www.ingramcontent.com/pod-product-compliance
Lightning Source LLC
LaVergne TN
LVHW071518180726
843512LV00014B/1111